DNA
TECHNOLOGY
SECOND EDITION

SECOND EDITION

DNA TECHNOLOGY

The Awesome Skill

I. Edward Alcamo

State University of New York at Farmingdale
Farmingdale, New York

San Diego　　San Francisco　　New York　　Boston　　London　　Sydney　　Tokyo

Cover image: John Labbe/The Image Bank © 1999.

This book is printed on acid-free paper. ∞

Copyright © 2001, 1996 by ACADEMIC PRESS

Academic Press
A Harcourt Science and Technology Company
525 B Street, Suite 1900, San Diego, California 92101-4495, USA
http://www.academicpress.com

Academic Press
24-28 Oval Road, London NW1 7DX, UK
http://www.hbuk.co.uk/ap/

Harcourt/Academic Press
200 Wheeler Road, Burlington, MA 01803
http://www.harcourt-ap.com

Library of Congress Catalog Card Number: 99-68792

International Standard Book Number: 0-12-048920-1

PRINTED IN THE UNITED STATES OF AMERICA
01 02 03 04 05 06 CO 9 8 7 6 5 4 3 2 1

Dedication

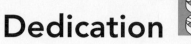

I am pleased to dedicate this book to the bacterium *Escherichia coli* (*E. coli*, as it is commonly known). For over a half-century *E. coli* has been the hammer-and-nail of DNA technology. Scientists used it as a model organism in the 1950s to prove that DNA is the hereditary material; they employed it during the 1960s to learn how genes work and to decipher the genetic code; in the 1970s, *E. coli* was the first organism to have its genes altered biochemically; in the 1980s, molecular biologists put it to work as a living factory to produce an array of genetically-engineered drugs and medicines; during the 1990s, scientists used it to develop transgenic animals and plants; and now, in the twenty-first century, *E. coli* continues to illustrate how bacteria and other microorganisms play key roles in the interest of science and for the betterment of humanity.

Contents

Preface

Most Americans would recognize the following words without too much difficulty.

"*We hold these truths to be self-evident, that all men are created equal…*"

Few, however, would identify the source of these words correctly:

"*We wish to suggest a structure for the salt of deoxyribonucleic acid (DNA)…*"

The first line is taken from the Declaration of Independence, a document that sparked a revolution in history and politics and gave birth to a country. The second is the opening line from a document that stimulated a different revolution—a revolution in science and medicine. This is from a scientific article published on April 23, 1953, in *Nature* magazine. The authors were James D. Watson, an American graduate student, and Francis H. C. Crick, a British biochemist and Watson's mentor. In the article, Watson and Crick suggested a structure for DNA, and in so doing, they spurred the age of DNA technology. Like the American revolution, the DNA revolution changed the scheme of things forever.

In the fifty years since publication of the Watson-Crick article, scientists have attained an understanding of genes and their activity that has stretched their imaginations to the limit. Where the gene was once an incomprehensible bead on a string, it has since become a segment of DNA that can be identified, removed, altered, and treated almost like a plaything. The gene research conducted since 1953 has encouraged even the most conservative scientists to speak in glowing terms of the practical applications of DNA technology. To some, the implications of gene knowledge are more important to the future of the human species than any scientific knowledge uncovered to date. Indeed, in 1983, the editors of Time magazine referred to DNA technology as "…the most awesome and powerful skill acquired by man since the splitting of the atom."

AUDIENCE AND ORGANIZATION

DNA Technology: The Awesome Skill explores the scientific revolution fostered by the myriad uses of DNA research. Written for the mature reader with a limited science background, the book presents the applications and implications of DNA technology. Laboratory techniques are deemphasized in favor of discussions with practical significance. It is hoped that the reader will come away thinking: "So that's what DNA technology is all about!"

DNA Technology: The Awesome Skill is divided into three major sections. In the first section (three chapters), the reading flows through several themes: the discovery of the laws of genetics by Gregor Mendel, the realization that DNA is the hereditary material, and the Watson-Crick work on the structure and replication of DNA. The next section (two chapters) summarizes the development of thought leading to DNA technology and recounts some of the technical problems requiring attention before DNA technology could bear fruit.

In the third section, seven chapters survey contemporary DNA technology and its uses in various fields. For example, one chapter describes the pharmaceutical products of DNA technology, and another explains genetic diagnoses and therapies; still other chapters survey medical forensics, gene detectives, and genetically engineered plants and animals; and a final chapter details the effort to decipher all 100,000 human genes. It is hoped that on completing this book you will better understand and appreciate the breathtaking developments in DNA technology going on in today's world.

SPECIAL FEATURES

To increase the effectiveness of your learning experience, several features have been incorporated into the chapters. Perhaps they will reduce the anxiety of approaching a new topic and enhance your ability to absorb the concepts.

1. Each chapter opens with a **Chapter Outline** that lets you know how the main topics interrelate with one another and with the subtopics.

2. The section entitled **Looking Ahead** presents several broad objectives that you should attain by reading the chapter.

3. **Pronunciations** of difficult terms are presented in the margins of the text so you can have more confidence in using them.

4. A **Marginal Glossary** provides review definitions of terms from the nearby text or from an earlier chapter.

5. Major terms and the names of investigators have been **boldfaced** to point up the key elements and researchers in DNA technology.

6. The **Questionline** series presents questions about DNA technology as they might be asked by your contemporaries. The answers contain brief summaries of information in the text.

7. The **Summary** brings together the key points in the chapter and relates them to one another in a few paragraphs. It is always a good idea to read the summary before beginning the chapter.

8. The **Review** gives you an opportunity to test your mastery of the chapter by presenting ten broad questions that cover its contents.

9. **For Additional Reading** presents a list of current, broad-based articles related to DNA technology. They should be available in most libraries.

10. The **Illustration Program** includes a number of electron micrographs, flow diagrams, and display features to help you put the text information in perspective.

ABOUT THE SECOND EDITION

I vividly recall the summer of 1995. All the ducks seemed to be in line as I completed the first edition of this book. Then I got word that, for the first time, the base sequence of a chromosome had been worked out. Because this was such an important breakthrough in DNA technology, I had to call and hold the presses while I warmed up my trusty Mac.

But that's the way it is with DNA technology. Rarely does a day go by when I'm not clipping an article from a magazine and stuffing it into the file. Keeping up with DNA technology is an exhilarating but unending battle. As soon as I finish marveling at one momentous discovery, it's time to marvel at something new. The "oohs and ahs" never seem to stop. Indeed, it has been said that the developments in DNA technology will outrace the ability of writers to report them.

And so it's been with this book. When the first edition appeared on bookshelves, gene therapy was in its infancy, Neanderthals were still on the succession line to modern humans, and a complex organism had never been cloned from a single cell. Today, all that has changed: thousands of patients are receiving gene therapy; DNA analyses show that Neanderthals evolved separately from modern humans; and Dolly, the first mammal cloned, is a historical landmark (and the mother of her own offspring).

For those reasons and so many more, the second edition of *DNA Technology: The Awesome Skill* is more a necessity than a luxury. I've tried to keep you abreast of what's going on in the field so you can be a more educated citizen, and I've enhanced the features of the first edition with more of what readers found useful. Broad strokes are used to identify the principles of DNA technology, and they are combined with some salient terminology and insights; the research leaders have been noted prominently so you can watch for their names in the media; and enough fields have been surveyed so you can be confident in discussing all of them.

Scientists are dreamers, and DNA technology has given them much to dream about. We hope that, as you tune into the world of DNA technology, your mind will also begin to stir. Perhaps you will be at the forefront of the next generation of discovery.

ACKNOWLEDGMENTS

Though authors generally receive credit (and sometimes vilification) for books as this, the truth of the matter is that many gifted and talented individuals lend their expertise to the final product. I am pleased to acknowledge the contributions of Fran Agliata, Tom Gagliano, Connie Mueller, Judi Wolken, and Barbara Elliot. Fran designed the book and supervised its production, while lending her creativity to each page. Tom is the artistic guru who translated my ideas to much of the finished artwork in this edition. Connie hunted down the beautiful photographs reproduced in this book. Judi edited the work, and Barbara typed the manuscript with care and precision.

I was particularly fortunate to work with Jeremy Hayhurst, senior editor at Harcourt/Academic Press. Jeremy brought a wealth of experience to the project and provided a boost when it was needed most. He is a knowledgeable, attentive, and well-versed editor; I respect him greatly, and I hold him in high esteem.

I also benefited from the substantial talents of numerous colleagues who shared their insights and served as reviewers of the manuscript. They include Professors John Lammert of Gustavus Adolphus College, Victor Fet of Marshall University, Dennis Bogyo of Valdosta State University, Pattle Pun of Wheaton College, and Robert J. Sullivan of Marist College.

My wife and children continue to be my inspiration. Michael Christopher is a corporate attorney in New York; Patricia Joy is a business consultant, also in New York; Tracey Lynn is a computer whiz; and Elizabeth Ann has just completed her doctorate at MIT in molecular biology (they say the apple doesn't fall far from the tree).

Charlene Alice, my wife of six months, is my future. Her warm heart, gentle smile, and firm resolve encourage me to keep writing when I'd prefer to vege out. She's there to help me tolerate the bad times and celebrate the successes. My life has taken on new meaning since she became my copilot.

To each of the above, I extend a warm and gracious word of thanks.

And I am pleased that you, the reader, have picked up this book. I hope it gives you a glimpse of the world of DNA technology and spurs you to become part of the biological revolution that is now taking place.

AN INVITATION TO READERS

I would love to say I'm taking a vacation after this edition, but the truth of the matter is that I'm already thinking about the third edition. I have no idea what it will contain, but that goes with the territory. Science is an ever-changing buffet table with something for everyone. And keeping up with the directions and applications of DNA is a never-ending job.

And so I would like to enlist your help. You could help me immensely if you would send along copies of any articles you spot in your community newspapers or magazines and help me keep this book as up-to-date as possible. I'd also like to know how well the book fills your needs and how I can improve the next edition. I can be reached at the Department of Biology, State University of New York, Farmingdale, New York 11735. If you care to give me a buzz, I'm at 631-420-2423. And if you have an e-mail connection, you can try me at alcamoie@farmingdale.edu.

Best wishes for a successful learning experience in today's most exciting field of science, and welcome to the wild and wonderful world of DNA technology.

E. Alcamo
Fall, 1999

The Roots of DNA Research  1

LOOKING AHEAD

DNA technology has its foundations in genetics, the science of heredity. It is appropriate, therefore, to open this book by exploring the insights and experiments that led scientists to recognize DNA as the hereditary substance. When you have completed the chapter, you should be able to:

■ recognize how the experiments of Gregor Mendel focused attention on cellular factors as the basis for inheritance.

■ understand the circumstances under which Mendel's experiments were verified and how Sutton related Mendel's "factors" to cellular units called chromosomes.

■ show how Morgan related eye color in fruit flies to chromosomes.

■ appreciate the origin of the term "gene" and describe how the gene concept emerged.

■ recount Miescher's work on nuclei and conceptualize how Feulgen and Mirsky contributed to the insight that genes are composed of DNA.

■ understand the significance of Griffith's experiments in bacterial transformation and conceptualize how the transforming principle was identified as DNA.

■ explain the seminal experiments of Hershey and Chase and describe why their results pointed to DNA as the substance controlling protein and nucleic acid synthesis.

■ increase your vocabulary of terms relating to DNA technology.

INTRODUCTION

In past centuries, it was customary to explain inheritance by saying, "it's all in the blood." People believed that children received blood from their parents and that a union of bloods led to the blending they saw in one's characteristics. Such expressions as "blood relations," "blood will tell," and "bloodlines" reflect this belief.

However, by the end of the 1800s, the blood basis of heredity was challenged and eventually discarded. In its place, scientists developed an interest in nucleic acid molecules organized into functional units called **genes**. Scientists guessed that genes control heredity by specifying the production of proteins. But even the gene basis of heredity was hard to believe because the amount of nucleic acid in the cell seemed insignificant.

The gene basis for heredity has been strengthened in the past fifty years, and it has become one of the foundation principles of biology. In the pages ahead, we will explore the development of the gene theory and note how interest grew in DNA as the substance of the gene. Long before scientists could apply the fruits of DNA research to modern technology, they had to learn what DNA was all about. "What purpose," they asked, "did DNA serve in a living cell?"

INHERITANCE FACTORS

By the 1850s, scientists were questioning the blood theory of inheritance: They could see quite clearly that semen contained no blood, and it was apparent that blood was not being transferred to the offspring. But if blood was not the hereditary substance, then what was?

At that time, a relatively obscure Austrian monk named **Gregor Mendel** (pictured in Figure 1.1) was conducting experiments to reveal the statistical pattern of inheritance. Mendel's great contribution to science was the discovery of a predictable mechanism by which inherited characteristics move from parents to offspring. His work with plants laid the groundwork for intensive studies in genetics, a science that would blossom in the early part of the twentieth century.

The region in which Mendel lived relied heavily on agriculture, so it was not uncommon for educated individuals to have an interest in animal and plant breeding. Mendel had studied plant science at the University of Vienna, and he continued his interest in plants at the monastery at Brno (now a part of the Czech Republic). He began a series of experiments to learn more about the breeding patterns of **pea plants**. Peas were well suited for his work because they were easy to cultivate. Moreover, they had a short growing season, they could be fertilized artificially, and they resisted interference by foreign pollen.

Other important features of pea plants were their easily distinguished traits. Mendel observed, for example, that in his garden were pea plants with wrinkled seeds and other plants with smooth seeds; some had green pods, and others had yellow pods; some had white flowers, and others had red flowers. Figure 1.2 shows this diversity. The more Mendel pondered the source of variations, the more his curiosity was aroused. He set out to determine how the variations originated and how the traits were passed to the next generation.

Mendel studied pea plants by crossing plants having a certain characteristic with others having a contrasting characteristic. He then studied how traits were expressed in the offspring plants. Mendel found, for example, that by breeding selected tall plants to

gene

a segment of a DNA molecule that, among other functions, provides chemical information for the synthesis of protein in a cell.

DNA

an ancronym for deoxyribonucleic acid, the organic substance of heredity and the material of which genes are composed.

(a)

FIGURE 1.1

Mendel and his pea plants. (a) Gregor Mendel (1822–1884), the Austrian monk who established the principles of genetics through meticulous experiments with pea plants. (b) Anatomy of the pea plant showing the growth cycle and the reproductive features that make artificial pollination feasible.

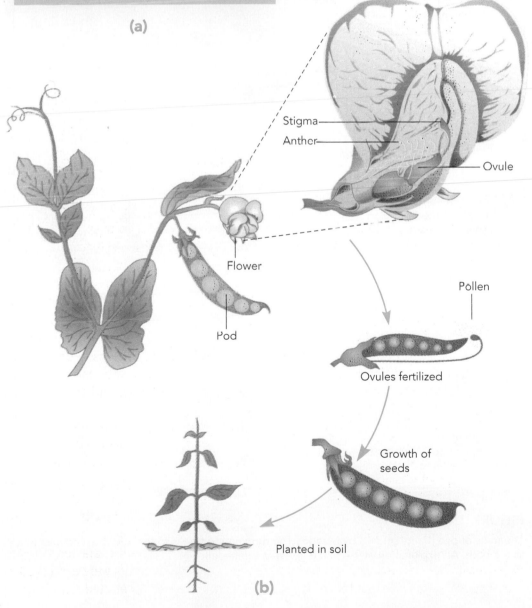

Stigma

Anther

Ovule

Flower

Pod

Pollen

Ovules fertilized

Growth of seeds

Planted in soil

(b)

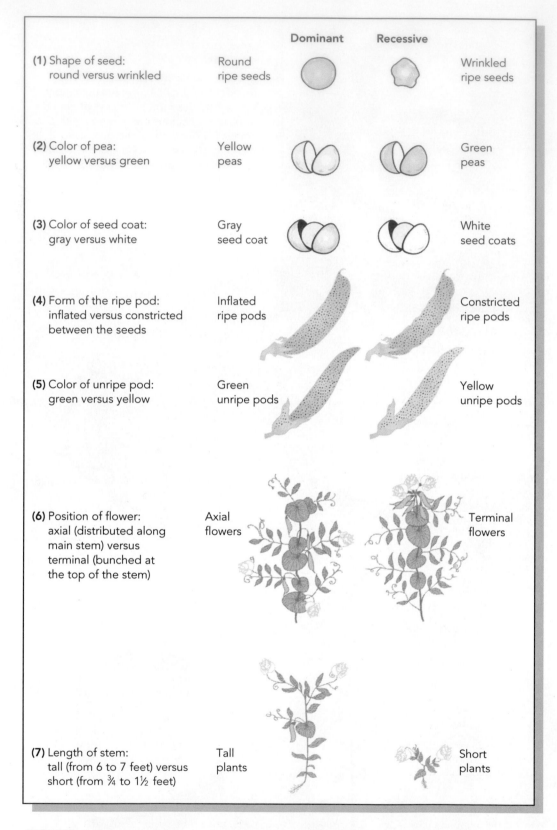

FIGURE 1.2

The traits of pea plants studied by Mendel. The dominating trait is to the left, the recessive trait to the right. An explanation of the trait is given at the far left.

selected short plants he could obtain plants that were exclusively tall. The trait for shortness had apparently disappeared. But when he bred the tall plants from this first generation among themselves, some short plants reappeared in the next generation among the tall plants. These results were unexpected and perplexing.

Mendel's forte was mathematics. He carefully counted the plants displaying a particular characteristic and the plants having the contrasting characteristic (for example, tall plants and short plants); using his mathematical skills, he discovered **similar ratios of traits** among the offspring. He noted, for example, that crossing the first generation's tall plants among themselves always seemed to yield three tall plants for every short plant, as Figure 1.3 shows. (By that time, the monks in the monastery noted that peas had become a fairly regular item in the dinner menu.)

Many scientists of the 1850s believed that a single factor controlled a trait, but Mendel began with the assumption that each trait was controlled by two factors (although the nature of the factor was unknown). He reasoned that one of the factors was obtained from the male and one from the female. He guessed that the factors express themselves in the offspring, but one dominates over the other. For example, the factor for tall plants dominates over the factor for short plants (it suppresses the short-plant factor). The factors are then passed on to the next generation. Today we know Mendel's factors as genes.

From his work, Mendel developed a **theory of inheritance** completely at odds with the blood basis of heredity. Mendel's results implied that sperm and egg cells, not blood cells, carry the factors of inheritance. Moreover, Mendel surmised that the factors are discrete units, not some vague, mysterious elements of the blood. Aware of the

factor

Mendel's term for transmissible hereditary units now know as genes.

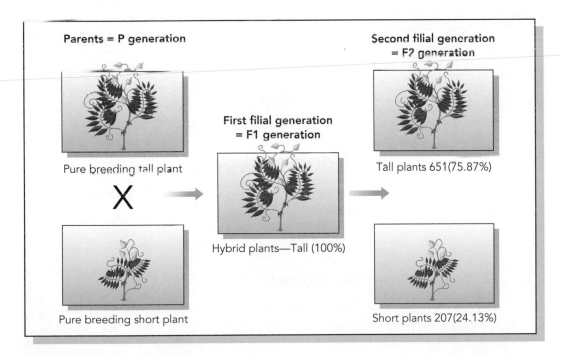

FIGURE 1.3

Mendel's experiments with tall and short pea plants. Mendel bred pure-bred tall plants (of the P generation) to pure-bred short plants. He discovered that all the offspring plants were tall in the first filial (F1) generation. He then bred the tall plants of the F1 generation among themselves and found that short plants grew among the tall plants in the F2 generation. His meticulous calculations revealed that about 75% of the plants in the F2 generation were tall and 25% were short. This 75% to 25% ratio was equivalent to 3:1. Partly on this basis, Mendel guessed that two "factors" for height exist in pea plants, and one factor dominates over the other.

unconventional nature of his suppositions, Mendel avoided controversy by keeping his suppositions largely to himself.

Mendel's theory came to be known as the **theory of transmissible factors**. Although it was revolutionary for the times, Mendel did not stop here. For many years he studied how one factor in the pair dominates the other factor and how a pair of factors separates during transmission to the next generation. He experimented up to the early 1860s and published his results in 1866 in the *Proceedings of the Society of Natural Sciences* in Brno. Mendel included a detailed analysis of his theories in the publication, and he communicated his findings to other scientists of the times through a series of letters. In retrospect, Mendel's observations are regarded as one of the great insights in science and the beginning of the discipline of genetics (Questionline 1.1).

Unfortunately, scientists of his time paid little attention to Mendel's work or its implications. One probable reason is that they had little understanding of biological chemistry. Another is that they failed to appreciate the significance of the cellular nucleus, the chromosomes, or the process of fertilization. Also, during the late 1800s, biologists were largely immersed in studying the theory of evolution, first promulgated in 1859 in Charles Darwin's epic work *On the Origin of Species*. Research on inheritance and breeding was placed on the proverbial "back burner" as the biological, social, and economic implications of the theory of evolution continued to capture the attention and imagination of scientists and lay people. Not until the year 1900 would interest in genetics once again come to the forefront of science.

In the spring of 1900, three European botanists, working independently of each other, repeated and verified Mendel's work. Each botanist cited Mendel's article in his research, and each awakened the scientific community to the work of the pioneering monk. It was not so unusual that all three should be aware of Mendel's work, but it was remarkable that the rediscovery of his theories was made almost simultaneously by three investigators; indeed, the happenstance remains one of the unusual coincidences of scientific history. Within weeks, a wave of enthusiasm for

theory of transmissible factors

Mendel's theory that inheritance is controlled by cellular factors passed from parents to offspring.

evolution

the biological principle that all living things have descended from a common ancestor over the eons of time and are continuing to evolve.

QUESTIONLINE 1.1

1. **Q.** When did interest in DNA as the material of heredity begin to surface?

    A. The study of heredity was rather primitive before Gregor Mendel's work in the 1860s. Mendel proposed that heredity is based on the transfer of several "factors" from parents to their offspring. His theories, however, were not studied further until the early 1890s, and real interest in DNA did not develop until the 1940s.

2. **Q.** Why was Mendel's theory novel?

    A. Until Mendel's time, scientists were unsure how heredity worked; some believed that blood transfers inherited characteristics. Mendel's work focused attention on identifiable factors in the cells and provided a viable alternative to the blood theory.

3. **Q.** When did scientists relate chromosomes to Mendel's factors?

    A. In the early 1900s, T. H. Morgan and his colleagues proved that white eye color, an inheritable trait in fruit flies, depends on the transfer of a single chromosome from the insect parent to its offspring. Morgan's work indicated that Mendel's "factors" and chromosomes are one and the same.

inheritance research sprang up. The discoveries made by Mendel had been forgotten for almost 40 years. Now they would change scientific thinking forever.

■ CHROMOSOMES AND FACTORS

During the first years of the twentieth century, Mendel's experiments were carefully studied, and the belief emerged that Mendel's factors were related to parts of the cell called **chromosomes**. Chromosomes (literally "colored bodies") are threadlike strands of chemical material located in the cell nucleus. They are clearly visible under the microscope when a cell is dividing, as Figure 1.4 shows. With few exceptions, all human body cells have 46 chromosomes, and the 46 chromosomes are organized in 23 pairs. (Red blood cells have no chromosomes, and sperm and egg cells have only 23 chromosomes). It is now known that chromosomes contain the DNA that carries the cell's genetic message.

Among the leaders in chromosome research at the turn of the century was the American biologist **W. H. Sutton**. In 1902, Sutton wrote that certain of Mendel's rules of inheritance could be explained if Mendel's factors were located on or in the chromosomes. Mendel had written, for instance, that inheritance factors occur in pairs, one member of the pair received from each parent. By 1900, cell biologists had established that chromosomes also occur in pairs, one chromosome derived from each parent. Moreover, Mendel theorized that during the production of sperm and egg cells, the paired factors separate and move as units to each cell. Studies in cell biology showed that chromosomes behave similarly during reproductive cell formation. Sutton pointed out that chromosomes could be the hypothetical inheritance factors Mendel thought responsible

chromosome

a structural element of protein and DNA that serves as the repository of hereditary information in the cell.

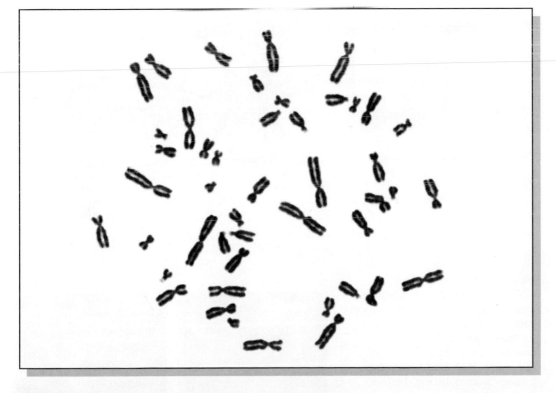

FIGURE 1.4

Human chromosomes seen under a microscope after separation from the nucleus. The cell was in a reproductive process called mitosis. Note that during this portion of mitosis the chromosomes occur in pairs joined at one point by an area called the centromere. With the notable exceptions of reproductive cells and red blood cells, 46 chromosomes are present in each human cell.

for heredity. Perhaps, he suggested, chromosomes and inheritance factors were identical.

To demonstrate the validity of the chromosomal theory of inheritance, scientists had to relate at least one trait to a cell's chromosome. But chromosomes come in pairs, and in the early 1900s, the members of the pair could not be distinguished visually. Thus it was impossible to relate a single trait to a single chromosome by sight alone.

The problem was resolved in 1910 by **Thomas Hunt Morgan** of Columbia University (pictured in Figure 1.5). Morgan used the fruit fly *Drosophila melanogaster* in his work. By careful observations, he determined that one of the four pairs of chromosomes in the fruit fly determines its sex. This chromosome pair, he discovered, also determines colorless white eyes. Through an exhaustive series of genetic crosses and statistical analyses, Morgan determined that the male **fruit fly** inherits only one chromosome for sex determination. Thus it must also inherit only one chromosome for white eye color. Therefore, white eye color must depend on a single chromosome. By providing statistical evidence for the relationship between sex and eye color in *Drosophila*, Morgan placed the chromosomal theory of inheritance on a firm footing and enhanced the role of the chromosome as the possible vehicle of inheritance.

The next question was whether the whole chromosome or a part of a chromosome is responsible for an inherited trait. Writing in 1903, Sutton proposed that merely a part of a chromosome is the basis for a trait because not enough chromosomes are possible to account for all an individual's traits. Sutton suggested that "...the chromosome may be divisible into smaller entities." Most other scientists agreed, and before long, the concept of the gene as the "smaller entity" gained prominence.

Drosophila
dro-sof'il-ah

melanogaster
mel"an-o-gas'ter

fruit fly

a tiny insect commonly used in genetics studies because of its relatively large chromosomes.

(a)

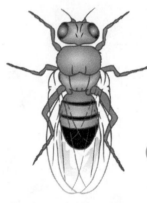

(b)

adult male

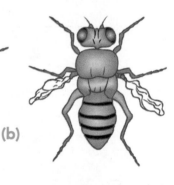

adult female

FIGURE 1.5

Morgan and his fruit flies.
(a) Thomas Hunt Morgan, whose experiments revealed that colorless white eyes in fruit flies are based on the presence of a single chromo-some. His experiments related an inherited characteristic to a chromosome. (b) The fruit flies used by Morgan. (c) Close-up light micrographs of (l) the eye of a wild-type fruit fly dark from pigmentation and (2) the eye of a white-eyed mutant fruit fly.

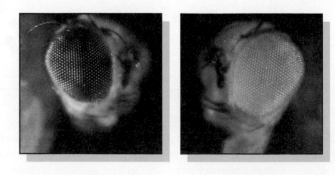

(1)   (2)

(c)

■ GENES AND DNA

In the early 1900s, geneticists began using the terms "inheritance unit" and "genetic particle" to describe the factors occurring on the chromosomes of Mendel's pea plants. By the 1920s, however, these terms had been discarded, and at the suggestion of the Scandinavian scientist **Willard Johannsen**, geneticists agreed to use the word **gene** instead. ("Gene" is derived from the Greek *gennan* meaning "to produce".) The term was originally used as part of Darwin's word "pangenesis" to describe the theory that the whole body (including every atom and unit) "produces" itself over and over. In a 1910 article, Johannsen suggested using only the last syllable "gene" because it was completely free of any hypothesis. The word was and continues to be a less cumbersome term than "inheritance unit" (Questionline 1.2).

Scientists of the 1920s viewed the gene as a specific and separate entity located on the cell's chromosomes. (Figure 1.6 shows the relationship of the gene to other aspects of the body.) One could not help visualizing genes as beads on a string, the string being the chromosome. It stood to reason that if genes are associated with chromosomes, a first step in learning the chemical nature of the gene would be to learn the chemical composition of the chromosomes; this is precisely what researchers attempted to do in the early 1900s.

One possible chemical component of chromosomes was a seemingly unique organic compound of the cell nucleus called **nucleic acid**. Nucleic acid was first described in 1869 by the Swiss researcher **Johann Friedrich Miescher**. With great difficulty, Miescher separated nuclei from human white blood cells, and he searched for evidence of protein within these nuclei. Instead of protein, however, he found a substance chemically unlike any class of chemicals then known. Miescher named the new substance **nuclein** (relating to its source). When he identified phosphorus in

Miescher

Mi'sher

QUESTIONLINE 1.2

1. **Q.** What is the difference between a chromosome and a gene?

   A. A chromosome is primarily a strand of deoxyribonucleic acid (DNA) that carries the genetic messages for synthesizing many cellular proteins. A gene is a selected region of the chromosome where the message for a single protein is located.

2. **Q.** How did scientists conclude that chromosomes are made of DNA?

   A. Scientists first had to pinpoint chromosomes in the nucleus of the cell. Then they extracted the nuclei from cells and tested the chemical composition of the nucleus. They narrowed their search down to a class of organic compounds called nucleic acids, and they discovered that chromosomes contain one type of nucleic acid called deoxyribonucleic acid, or DNA.

3. **Q.** Did a single scientist supply the proof that chromosomes are composed of DNA?

   A. No. Various scientists contributed to the theory that chromosomes are composed of DNA. For example, certain scientists discovered that sperm cells, which carry the hereditary traits, have half the amount of DNA as normal cells. Other scientists used dyes to ascertain that DNA is concentrated in chromosomes.

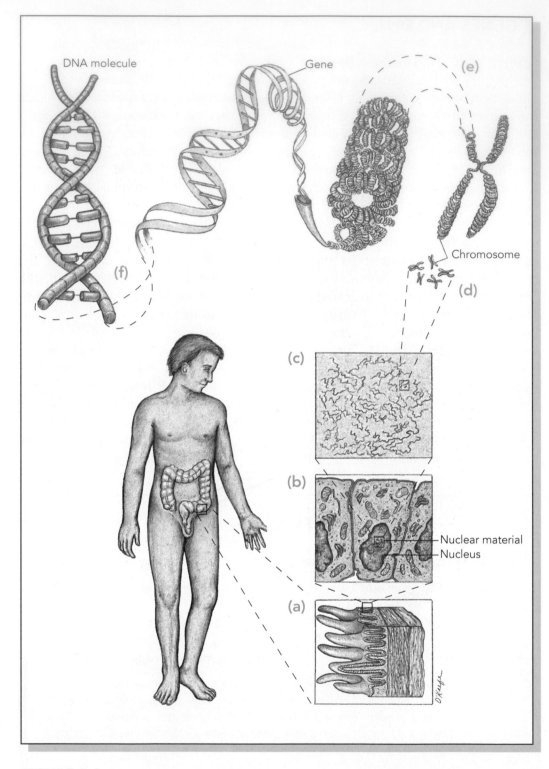

FIGURE 1.6

The structural relationships among genetic elements in cells. **(a)** A section of intestinal tract yields tissue, which **(b)** reveals cells, each having a nucleus. **(c)** When the nuclear material is chemically treated and examined, chromosomes are seen. **(d)** The chromosomes are enlarged and just before cell division, each is revealed as a pair of chromosomes joined at one point. **(e)** A section of one chromosome is enlarged, and an area is denoted as a gene. **(f)** The gene is composed of DNA.

nuclein, he postulated that the substance was a storehouse for phosphorus in the cell.

As Miescher continued his study of nuclein, he located it in yeast cells and kidney, liver, and testicular cells. He also made the notable observation that nuclein was abundant in sperm cells obtained from a salmon. Some years later, chemists led by Phoebus Levene (Chapter 2) used this information in their studies and determined the components of nuclein. They gave it the more descriptive and technical name **deoxyribonucleic acid (DNA)**. Coincidentally, Levene was born in 1869, the year of Miescher's first report of nuclein.

By the 1920s, it was clear that chromosomes had a role in heredity, and the isolation of DNA from cell nuclei made this organic substance a good candidate for the hereditary substance. Interest in DNA was further strengthened by a 1924 discovery attributed to the German biochemist **Robert Feulgen**. Feulgen observed that a dye (now called Feulgen stain) turns bright purple when it reacts with DNA. (Figure 1.7 shows an example of this staining characteristic.) The dye could be used to locate cellular DNA and to observe DNA's concentration at various times in the cell's life cycle. Isolating DNA was a difficult chore at that time, but Feulgen's dye technique allowed DNA research to leap ahead without the burden of complex chemical isolations.

Another observation in that period helped forge the link between DNA and heredity. In the late 1920s, **Alfred Mirsky** and his coworkers at New York's Rockefeller Institute reported that, with only two exceptions, all cells of an organism have virtually the same amount of DNA in their nuclei. The two exceptions are reproductive **sperm** and **egg** cells. These cells contain precisely half the amount of DNA found in nonreproductive cells such as muscle cells. Table 1.1 presents these data. Mirsky's observation correlated with the theory that sex cells are the vehicle for bringing half

deoxyribonucleic

de-ox′e-ri″bo-new-kla′ik

Feulgen

Full′gen

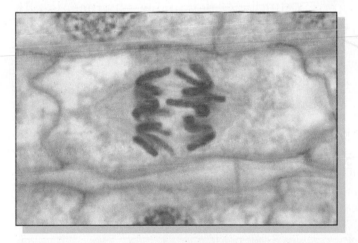

FIGURE 1.7

Stained chromosomes. A photomicrograph of a plant cell stained with Feulgen stain to highlight the chromosomes of the cell nucleus. In this view, the chromosomes have replicated and are in the process of separating into two newly forming cells.

TABLE 1.1

A Comparison of the DNA Content in the Tissue Cells and Sperm Cells of Various Animals.

ORGANISM	TISSUE CELLS	SPERM CELLS
Cow	6.6	3.3
Human	6.4	3.2
Chicken	2.6	1.3
Frog	15.0	7.5

DNA amount in picograms (trillionths of a gram).

the genetic information from each parent to the offspring.

zygote
a fertilized egg cell.

The Rockefeller group led by Mirsky also experimented with the **zygote**, the cell resulting from the union of sperm and egg cells. The researchers found that the zygote contains the same amount of DNA as other cells of the body. This observation reinforced the notion that the DNA from two parents comes together during fertilization. In effect, it also verified the work of Mendel performed more than sixty years before because Mendel proposed that genetic factors separate when reproductive cells form, then come together in the offspring. The visual observations and chemical analyses of DNA led to the conclusion that the DNA was reduced by half during formation of the reproductive cells, then reconstituted in the offspring zygote.

RELATING DNA TO HEREDITY

By the late 1920s, two concepts were evolving: genes are involved in heredity, and genes are composed of DNA. However, most evidence continued to be indirect and based on guesswork from laboratory deductions. Scientists needed direct evidence linking DNA to the development of observable traits. Two series of studies would provide that evidence. The first series was performed by Griffith, Alloway, Avery, and their colleagues. It drew a connection between DNA and the appearance of traits in bacteria. The second, performed by Hershey and Chase, linked DNA to the synthesis of protein. Both were landmark achievements in DNA technology; both connected enough dots to show that DNA is the substance of heredity.

BACTERIAL TRANSFORMATIONS

Streptococcus
strep-to-kok'us

pneumoniae
new-mo'ne-a

medium
the substance on which or in which bacteria are cultivated.

pneumococci
new-mo-kok'si
spherical bacteria that can cause pneumonia.

In 1928, a British medical officer named **Frederick Griffith** reported some puzzling results in his work with bacteria. Griffith was performing experiments with *Streptococcus pneumoniae*, a cause of bacterial pneumonia. The bacterium, commonly known as the pneumococcus, occurs in two strains. One strain is designated **S** because it forms **smooth colonies** when growing on bacteriological medium; the other strain is designated **R** because it forms **rough colonies**. (Both forms are shown in Figure 1.8). It is well known that S strain pneumococci are lethal to mice (as well as to humans), whereas R strain pneumococci are harmless.

Griffith's initial experiments produced the expected results. He confirmed that S strain pneumococci were deadly to mice and that R strain bacteria were harmless. Then he performed some variations of these experiments. Griffith found that he could inject debris from dead S strain bacteria into mice without harming the animals. As before, these results were not surprising because the S strain bacteria were dead.

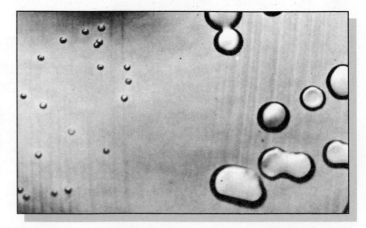

FIGURE 1.8

A macroscopic view of colonies of the two types of bacteria used by Griffith in his experiments in transformation. The bacteria are growing on the surface of a nutrient medium in a Petri dish. On the left are the R (rough) strain colonies of *Streptococcus pneumoniae*; on the right are S (smooth) strain colonies.

However, what happened next was unusual.

Griffith mixed a sample of live R strain pneumococci (harmless) together with debris from dead S strain pneumococci (also harmless because no living cells were present) Then he injected the mixture into the mice. By all expectations, the mice should have lived (both the S strain debris and the R strain bacteria were harmless). But pneumonia developed in the mice and they died, as Figure 1.9 displays. Griffith wondered why the animals died. His answer came when he performed an autopsy on the animals: their lungs were full of live S strain pneumococci, the deadly strain. Apparently, the harmless R strain bacteria had changed into deadly S strain bacteria, which killed the animal.

Griffith's "biochemical magic" appeared to work each time he performed the

S strain pneumococci

the deadly strain of pneumonia bacteria that form smooth colonies when grown in the laboratory.

R strain pneumococci

the harmless strain of pneumonia bacteria that form rough colonies when grown in the laboratory.

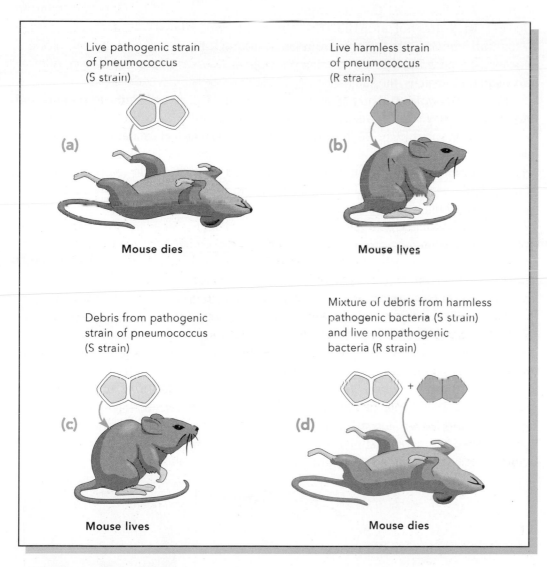

FIGURE 1.9

The transformation experiments performed by Griffith. (a) When live pathogenic pneumococci (S strain) were injected into mice, the animals died. (b) When live, harmless pneumococci (R strain) were injected, the animals remained healthy. (c) When debris from heat-killed S strain bacteria were injected into animals, they lived and were healthy. All these results were as anticipated. (d) However, when live, harmless bacteria (R strain) were mixed with cell debris of heat-killed S strain bacteria and the mixture was injected into animals, the animals died. On autopsy, Griffith isolated live, pathogenic bacteria (S strain). Apparently some of the live R strain bacteria had been transformed to live S strain bacteria.

experiment. Moreover, other researchers were able to confirm his findings shortly thereafter. Griffith postulated that something in the S strain chemical debris (a protein, he believed) was entering the R strain bacteria and "transforming" them by changing their biochemistry. Unfortunately, he was unable to identify the transforming substance. Nor would he live to see the significance of his work. Griffith died during the German air attacks on London in 1941.

Griffith's work did not go unnoticed by microbial geneticists. In 1933, **James Lionel Alloway** and his group at Rockefeller Institute successfully purified cell debris from S strain pneumococci and produced a **cell-free extract.** Then they used the extract to transform R strain bacteria to S strain bacteria. Moreover, they performed the transformation in test tubes, a procedure freeing them from the rigors of working with animals. Alloway noted that the transforming principle could be extracted from the mixture with alcohol and that the chemical appeared as "...a thick, stringy precipitate...which slowly settled out on standing." From the description, modern biochemists recognize the transforming principle as DNA. At the time, however, Alloway was inclined to believe the substance was made of protein.

deoxyribonucleic

de-ox'e-ri"bo-new-kla'ik

In 1935, **Oswald T. Avery** (Figure 1.10) and his associates, **Colin MacLeod** and **Maclyn McCarty**, began a series of exhaustive chemical analyses to identify the transforming substance. Beginning with crude extracts from bacteria, they used a process of elimination to determine the nature of the chemical substance by finding out what it was not. As the months and years unfolded, they dismissed proteins, fats, and carbohydrates as possible candidates for the transforming substance. It appeared that the only possible thing left was nucleic acid. Finally, the search narrowed down to one nucleic acid: **deoxyribonucleic acid (DNA)**. In a seminal article published in 1944, Avery and his colleagues presented evidence that DNA is the transforming principle (and opened the way to modern DNA technology). The researchers were not bold enough to claim that DNA was the hereditary material, but the implication was clear: DNA was apparently able to transform bacteria so dramatically that a harmless strain changed to a deadly strain.

Most scientists were blind to Avery's discovery and were reluctant to accept DNA as the hereditary substance. The majority of geneticists of the 1940s were not trained in biochemistry, and Avery's experiments were difficult for some to repeat. In addition, many scientists were hard pressed to believe that results obtained from bacteria could be applied to more complex organisms such as humans. Moreover, the results were published in the *Journal of Experimental Medicine*, a publication not usually read by geneticists and bacteriologists of that period; and the preoccupation with World War II had restricted the dissemination and flow of scientific knowledge, while limiting funds for scientific research. Therefore, the impact of Avery's finding was lost.

FIGURE 1.10

Oswald Avery, the Rockefeller Institute researcher who led the effort to identify DNA as the transforming principle. Avery's success focused attention on DNA as the chemical material of heredity.

Nevertheless, Avery's group continued their efforts to prove that DNA was more than an inert chemical of the cell nucleus. With great difficulty they isolated a quantity of DNase from beef pancreas cells. DNase is a biological catalyst, an enzyme that destroys DNA but has no effect on other molecules. The investigators mixed their transforming substance with DNase and noted that the mixture lost its ability to transform bacteria. Still, biochemists were not convinced

Opposition was also voiced from the "**protein supporters**." These scientists chose to believe that protein was the genetic material because protein appeared to have the necessary complexity to encode all the biochemical information in the cell's nucleus. Proteins are chains of twenty different and relatively simple organic molecules called amino acids (Chapter 2). A protein can be made of ten amino acids or 1000 amino acids or 10,000 amino acids. The crucial element of any protein is the sequence of amino acids, that is, which amino acid follows which in the chain. This sequence of amino acids is as important to proteins as is the sequence of letters in a word ("ton" has a much different meaning than "not"). Protein supporters were of the opinion that the amino acid sequence in one protein serves as a model for constructing a new protein. Their outlook would be dealt a severe blow by the 1952 experiments of Hershey and Chase.

DNase

an enzyme that breaks down DNA.

■ THE HERSHEY-CHASE EXPERIMENT

Alfred Hershey (Figure 1.11) worked with **Martha Chase** at the Cold Spring Harbor Laboratory in New York. The pair studied bacteria and the **viruses** that multiply within those bacteria. In 1952, scientists knew that certain viruses use bacteria as chemical factories for producing new viruses, as Figure 1.12 displays; however, the actual mechanism was uncertain. Biochemists were also aware that bacterial viruses are composed of a core of DNA enshrouded in a protein coat. What they did not know was whether the nucleic acid or the protein (or both) directs replication of the virus. Hershey and Chase would answer that question and in so doing, they would establish the essential role played by DNA in cellular biochemistry and inheritance.

virus

a particle of nucleic acid enclosed in protein; replicates within a living cell and usually destroys the cell.

Hershey and Chase made use of the observation that viral DNA contains phosphorus (P) but no sulfur (S). By contrast, the outer protein coat of the virus has sulfur (S) but no phosphorus (P). In their first experiments, Hershey and Chase cultivated viruses with the radioactive forms of phosphorus (^{32}P) and sulfur (^{35}S). They successfully prepared viruses whose nucleic acid was radioactive with ^{32}P and whose protein was radioactive with ^{35}S.

radioactive

emitting detectable radiations.

Now came the seminal experiments. Hershey and Chase mixed the radioactive viruses with a population of bacteria. Then they waited just long enough for viral

FIGURE 1.11

Alfred Hershey, who worked with Martha Chase and performed the key experiments demonstrating that DNA is responsible for directing the synthesis of viral protein in a host cell.

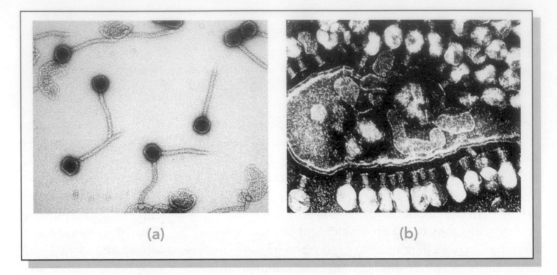

(a)   (b)

FIGURE 1.12

Bacterial viruses. (a) An electron microscopic view of viruses that bind to bacterial cells and replicate within them, thereby producing new viral particles. Viruses such as these are called bacteriophages ("bacteria-eaters"). Note the somewhat hexagonal heads and the elongated tails. Viruses consist of little more that a fragment of nucleic acid (such as DNA) enclosed in a protein coat. (b) A bacterium infected with viruses. The viruses are attached by the "tails" to the surface of the bacterium. Scientists agree that the protein coat remains outside the bacterium, while the DNA enters the bacterium to direct the replication of viruses.

replication to begin. At this point they used an ordinary household blender (now a museum piece) to shear away any viruses and debris clinging to the bacterial surface.

Then the analysis began. Hershey and Chase tested the bacteria and surrounding fluid to find out where the radioactivity was. This would enable them to develop a biochemical glimpse of viral replication. After experimentally bursting the bacteria, the researchers found most of the ^{32}P within the contents of the bacterial cytoplasm. This finding indicated that viral DNA was entering the bacteria. Then they discovered that the ^{35}S was largely in the sheared-away remains of the viruses and in the surrounding fluid. This observation indicated that the protein part of the viruses was remaining outside the bacteria. The results led Hershey and Chase to the inescapable conclusion that viral DNA enters the bacterium, whereas the viral protein remains outside. (Figure 1.13 shows the process.) Thus, DNA was the sole element responsible for viral replication. Protein had no place in the process.

Certain experiments stand out as turning points in scientific history, and the experiments performed by Hershey and Chase are one such turning point. In retrospect, we can see how their results had substantial impact on the thinking of that era. Hershey and Chase clarified the important aspect of viral replication that nucleic acid goes inside the cell, whereas the protein coat remains outside. But in broader terms, their results strengthened the place of DNA in cellular biochemistry. Bacterial viruses, it should be remembered, are composed solely of nucleic acid and protein, and the Hershey-Chase experiments reinforced the concept that DNA, and only DNA, is involved in the synthesis of both nucleic acid and protein (Questionline 1.3).

It was becoming clear that all the biochemical information for the synthesis of both nucleic acid and protein is stored in the DNA. Avery's experiments performed eight years earlier had linked DNA to the genetic material. The work of Hershey and Chase considerably strengthened that link.

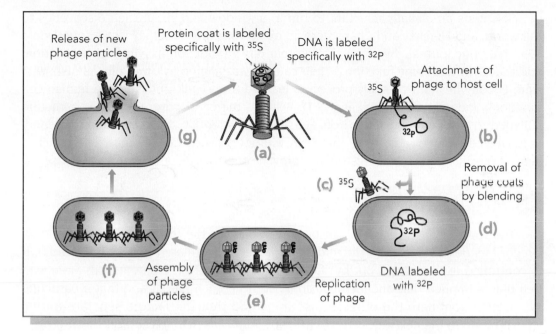

FIGURE 1.13

The Hershey-Chase experiment with viruses (bacteriophages) and bacteria. (a) Viruses are prepared with two radioactive labels, one in the coat and one in the DNA. (b) The viruses are combined with host bacterial cells and given an opportunity to interact. (c) Then the mixture is agitated in a blender, and the empty viral protein coats are removed. They are later found to have the radioactive sulfur. (d) The DNA carrying the radioactive phosphorus is found within the host cell cytoplasm. (e) Here it directs the synthesis of new viral components, (f) which are assembled to new viruses. (g) The viruses are then released from the cell cytoplasm. The results indicate that DNA directs the synthesis of both the DNA and the protein of the new viruses.

QUESTIONLINE 1.3

1. **Q.** Why was Frederick Griffith's work important?

    A. Griffith's work indicated that it was possible to change the inherited characteristics of a living organism. He showed that a transforming factor in the debris of harmful bacteria could transform harmless bacteria to pathogenic (disease-causing) bacteria.

2. **Q.** What was the "transforming factor" in the debris that Griffith worked with?

    A. In 1944, Avery and his coworkers reported that the "transforming factor" in Griffith's experiments was DNA. Avery's work showed that DNA could change the inherited characteristics of a bacterium.

3. **Q.** Why was the work of Hershey and Chase significant?

    A. Hershey and Chase performed the seminal experiments indicating that DNA can direct the synthesis of protein in bacterial cells. By directing protein synthesis, DNA profoundly influences the appearance of an organism's characteristics. After the work of Hershey and Chase, evidence mounted that DNA controls inherited traits by controlling the synthesis of proteins.

Scientists are usually reluctant to run to a window and shout their discoveries to the world, and Hershey and Chase were no exceptions. Instead, they wrote soberly in a 1952 scientific journal: "We infer that protein has no function in [viral] multiplication, and that DNA has some function." Their caution was shared by other biochemists who were equally hesitant to apply processes learned from viral replication to human cell functions. Indeed, late in 1952 James D. Watson, the co-discoverer of DNA's structure (Chapter 2), read a telling letter from Hershey to a scientific group at Oxford, England. Said Hershey: "my own guess is that DNA will not prove to be a unique determiner of genetic specificity." Scientific history has proven otherwise—considerably otherwise.

SUMMARY

The roots of DNA research can be traced back to the innovative experiments of Gregor Mendel conducted in the mid-1800s. Mendel postulated that an inherited trait is controlled not by blood, but by "factors" obtained from the parents. His work with pea plants implied that one factor is obtained from each parent and that a particular factor may dominate the other. It also appeared that the factors separate during transmission to the next generation. Little attention was paid to the work, however.

At the beginning of the 1900s, Mendel's experiments were repeated and verified, and scientists postulated that his "factors" were really chromosomes. Morgan's work of 1910 showed that white eye color in fruit flies is determined by a single chromosome, and he postulated that a single chromosome determines a trait. But because all traits could not be explained by individual chromosomes, molecular geneticists came to believe that entities on the chromosome called "genes" were the hereditary factors. Evidence presented by Miescher, Feulgen, and Mirsky indicated that chromosomes are composed of deoxyribonucleic acid, or DNA.

Experiments performed by other biologists and chemists strengthened the link between genes and DNA. In 1928, for example, Griffith showed that the characteristics of certain bacteria could be changed (i.e., the bacteria could be transformed) if they were mixed with debris from another type of bacterium. Alloway and his group found they could purify the debris to increase its potential to transform bacteria, and Avery and his colleagues discovered that the transforming material in the debris was DNA.

The experiments of Hershey and Chase provided the final proof for DNA's involvement in heredity. Hershey and Chase performed experiments with bacteria and the viruses that replicate within them. Bacterial viruses are composed primarily of protein and DNA. The experimental results showed that DNA alone directs the replication of new viruses; that is, DNA contains the biochemical information for the synthesis of both the viral protein and the viral DNA. The results obtained by Hershey and Chase confirmed that DNA is the hereditary material and stimulated additional interest in the study of molecular genetics.

REVIEW

This chapter has recounted the process by which DNA was identified as the substance of heredity. To test your comprehension of the chapter's major ideas, answer the following review questions:

1. What was Mendel's great contribution to science, and how did his work lay the groundwork for studies in genetics?

2. State several important facts that are true of chromosomes.

3. How did Morgan go about demonstrating that a trait can be associated with a chromosome?

4. What is the origin of the term "gene," and how was a gene viewed in the 1920s?

5. Describe the contributions of Johann Miescher to the study of molecular genetics.

6. What observations made by Feulgen and Mirsky helped forge a link between DNA and heredity?

7. Describe the experiments performed by Frederick Griffith, and explain why they were significant to the development of genetics.

8. Identify the accomplishment of Avery and his collaborators, and indicate why their work did not receive acclaim.

9. Explain the experiments performed by Hershey and Chase, indicating what the "tools" of the experiment were and what the results were.

10. Summarize how the Hershey-Chase experiment was a deciding factor in linking DNA to protein and DNA synthesis.

FOR ADDITIONAL READING

Cummings, M. C. *Human Heredity*, 4th ed. Belmont CA: Wadsworth Publishing Co., 1997.

Dubos, R. *The Professor, The Institute, and DNA*. New York: Rockefeller University Press, 1976.

Flannery, M. "Who Could Have Guessed It?" *The American Biology Teacher* 49 (1987): 57–63.

Griffith, F. "The Significance of Pneumococcal Types." *Journal of Hygiene* 27 (1928): 113–115.

Lewis, R. *Human Genetics*, 2d ed. Dubuque, IA: WC Brown, Mc Graw-Hill Publishing Co., 1997

Olby, R. *The Path to the Double Helix*. Seattle, WA: University of Washington Press, 1974.

Portugal, F., and Cohen, J. *A Century of DNA*. Cambridge, MA: MIT University Press, 1977.

The Double Helix

 CHAPTER OUTLINE

LOOKING AHEAD

Determining the structure of DNA was one of the major scientific achievements of the twentieth century. Knowing DNA's structure gave scientists insight to how heredity works and stimulated the development of molecular biology and DNA technology. Moreover, it helped them understand how DNA replicates during cellular reproduction. On completing this chapter, you should be able to:

- recognize the three elements of DNA and how they combine with one another to form nucleotides.

- describe how sugars and phosphate groups link together to form the "backbone" of the DNA molecule.

- summarize Erwin Chargaff's findings and indicate why they were important in determining DNA's structure.

- understand the contributions of Franklin, Wilkins, Watson, and Crick in determining the structure of DNA.

- explain how DNA replicates by the semiconservative method and summarize the experimental process by which Meselson and Stahl determined this method.

- use your newly acquired vocabulary to speak confidently about DNA structure.

INTRODUCTION

By the 1950s, it was becoming increasingly clear that deoxyribonucleic acid (DNA) is the molecule of heredity. As the evidence mounted, scientists realized they had to know the structure of DNA for two reasons: first, knowing the structure of DNA might explain how the molecule functions in the hereditary process; and second, understanding DNA's structure might shed light on how the molecule duplicates during cellular reproduction. Both heredity and cellular reproduction are among the fundamental processes of biology; if DNA was intimately involved in each, it stood to reason that exhaustive studies into the nature of DNA should be ongoing.

But in the 1950s, exhaustive studies on DNA were not ongoing. Part of the reason was that scientists were not absolutely certain about the involvement of DNA in heredity and cellular reproduction. Their studies were tentative rather than sure and directed. Data were accumulating, but very slowly, and a united front was lacking as various research centers tended to move in their own directions.

Amid the uncertainty, two researchers used the available data together with their own intuition and guesswork and put forth a suggested structure for DNA. The two researchers, James D. Watson and Francis H. C. Crick, performed no laboratory experiments, but they managed to guess the sum by analyzing the parts. In so doing, they gained international fame and put biochemists on the track to unlocking the secrets of heredity. As we will see in this chapter, the work of Watson and Crick was the jumping-off point for the science of DNA technology.

THE STRUCTURE OF DNA

Establishing the structure of DNA has been one of the major achievements of the twentieth century. Not only did it yield myriad practical benefits, but it also gave scientists the philosophical pride of understanding how heredity works. Biology has many bedrock principles—the cellular basis of living things, the germ theory of disease, and the process of evolution are three—and the chemical basis of heredity is another of those principles. Unlocking the secret of DNA was the key to understanding this principle.

THE COMPONENTS OF DNA

Although the structure of DNA was unknown in the early 1950s, its chemical components had been known for thirty years. In the 1920s, the basic chemistry of nucleic acids was determined by **Phoebus A. T. Levene.** Working with his colleagues at Rockefeller Institute in New York City, Levene isolated from yeast cells and thymus tissue two types of nucleic acid: ribonucleic acid (RNA) and deoxyribonucleic acid (DNA).

Levene's analyses revealed that both nucleic acids contain three basic components illustrated in Figure 2.1: (1) a five-carbon sugar, which could be either **ribose** (in RNA) or **deoxyribose** (in DNA); (2) a series of **phosphate groups**, that is, chemical groups derived from phosphoric acid molecules; and (3) four different compounds containing nitrogen and having the chemical properties of bases. Because of their nitrogen content and basic qualities, the four compounds are known as **nitrogenous bases**. In DNA the four bases include adenine, thymine, guanine, and cytosine in DNA; and in RNA, they are adenine, uracil, guanine, and cytosine. Adenine and guanine are double-ring molecules known as **purines**; cytosine, thymine, and uracil are single-ring molecules called **pyrimidines** (Questionline 2.1).

Research performed by Levene's group indicated that DNA contains roughly equal proportions of phosphate groups, deoxyribose molecules, and nitrogenous bases. Levene

thymus

an organ at the base of the neck of vertebrate animals that functions in the development and activities of the immune system.

nitrogenous

ni-troj'en-us

guanine

gwan'in

cytosine

si'to-sin

uracil

u'ra-cil

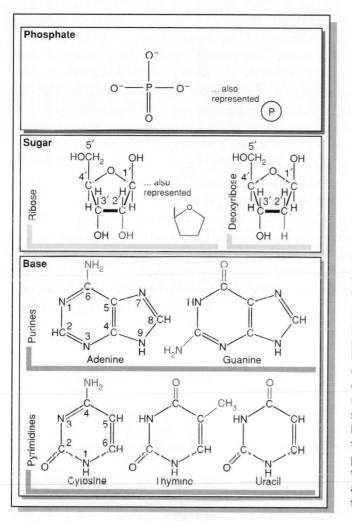

FIGURE 2.1

The components of nucleic acids. The first component is a phosphate group, a derivative of phosphoric acid composed of phosphorus, oxygen, and hydrogen atoms. The second component is a five-carbon sugar, either deoxyribose (in DNA) or ribose (in RNA). The third is a series of the five nitrogenous bases: adenine, guanine, cytosine, thymine, and uracil. Note the presence of nitrogen in each of the molecules. The first two bases are known as purines; the last three are pyrimidines.

QUESTIONLINE 2.1

1. **Q. Why is it important for scientists and lay people to know the structure of deoxyribonucleic acid (DNA)?**

   A. Knowing the structure of DNA is important because in biology, structure is always related to function. In the case of DNA, knowing the structure gives insight as to how the molecule functions in hereditary processes and helps us understand how DNA replicates.

2. **Q. What are the components of DNA?**

   A. DNA is composed of three basic components: a five-carbon sugar called deoxyribose; phosphate groups (containing phosphorous, oxygen, and hydrogen); and four different chemical compounds called nitrogenous bases ("nitrogenous" because they contain nitrogen; "base" because they act as bases act when dissolved in solution).

3. **Q. How are the components organized in DNA?**

   A. To form a strand of DNA, the deoxyribose molecules are linked to one another by phosphate groups. This structure yields a chain of alternating deoxyribose molecules and phosphate groups that form a "backbone" of the DNA strand. Nitrogenous bases are connected to the "backbone" by linking to the deoxyribose molecules and extending toward the side. The combination of a deoxyribose molecule, phosphate group, and nitrogenous base is called a nucleotide.

therefore concluded (correctly, in retrospect) that DNA is composed of three essential components and that they are used to form units. He surmised that the units are strung together to form a long chain. In contemporary biochemistry, the units are called **nucleotides**. (Figure 2.2 shows how a nucleotide is constructed.) A nucleotide of DNA consists of a deoxyribose molecule attached to a phosphate group and to a nitrogenous base. The identity of the base is the only thing distinguishing one nucleotide from another.

To identify the various chemical components of DNA it is customary to number the carbon atoms of the nitrogenous base and deoxyribose molecule, then refer to the number of the carbon atom when specifying where a chemical group is attached. As illustrated in Figure 2.3, the carbon atoms in deoxyribose are numbered 1′ to 5′ (pronounced one-prime to five-prime). The numbering begins to the right of the oxygen atom and proceeds clockwise. From Figure 2.3 it can be seen that the phosphate group is attached to the 5′ carbon atom of deoxyribose. Furthermore, a nitrogenous base is connected to the 1′ carbon atom. In addition, a free —OH group (referred to in chemistry as a hydroxyl group) exists at the 3′ carbon atom of the deoxyribose molecule.

When a cell constructs a molecule of **nucleic acid** from nucleotides, it forges linkages between the 5′–carbon atom of one deoxyribose and a phosphate group. It then links the

nucleotide

a building block unit of a nucleic acid consisting of a five-carbon carbohydrate molecule, a phosphate group, and one of four available nitrogenous bases.

hydroxyl

hi-drox′il

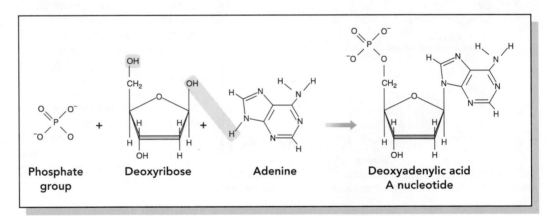

FIGURE 2.2

Construction of a nucleotide using a phosphate group, a deoxyribose molecule, and an adenine molecule. The shaded –OH groups and –H are lost during the synthesis of the nucleotide. This nucleotide is called deoxyadenylic acid, or adenosine monophosphate.

FIGURE 2.3

The numbering system in a nucleotide. In the ribose molecule, the numbering begins to the right of the oxygen atom and proceeds clockwise. A prime (′) is placed next to each number. Note that no prime occurs in the numbering system for the base. Also note the attachment points of the phosphate group and the adenine molecule similar to those in Figure 2.2. When constructing a strand of DNA, another nucleotide will attach at the 3′ carbon atom (as shown in Figure 2.4).

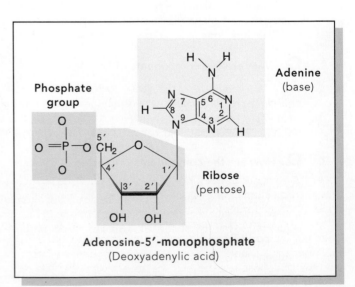

phosphate group and the 3′–carbon atom of a second deoxyribose, as Figure 2.4 illustrates. The phosphate group thereby bridges two deoxyribose molecules. The binding is accompanied by the loss of a hydroxyl (—OH) group from the 3′–carbon atom of the second deoxyribose molecule.

To continue building the DNA molecule, linkages to other nucleotides are forged in the same way; that is, the phosphate group links to the 5′–carbon atom of one deoxyribose and the hydroxyl group at the 3′–carbon of another. In this way, tens or thousands (or millions, or any number) of nucleotides bond to one another in a long chain to form a DNA molecule. By agreement among scientists, the sequence of DNA's bases is usually expressed in the 5″ to 3′ direction as one proceeds along the molecule.

Levene's studies in the 1920s indicated that nucleotides with all four nitrogenous bases were present in DNA. His studies also showed that the bases were present in virtually the

phosphate group

a molecule of phosphorus, oxygen, and hydrogen atoms derived from phosphoric acid and present in both DNA and RNA.

deoxyribose

a five-carbon carbohydrate molecule and one of the components of DNA; similar to ribose except with one less oxygen atom.

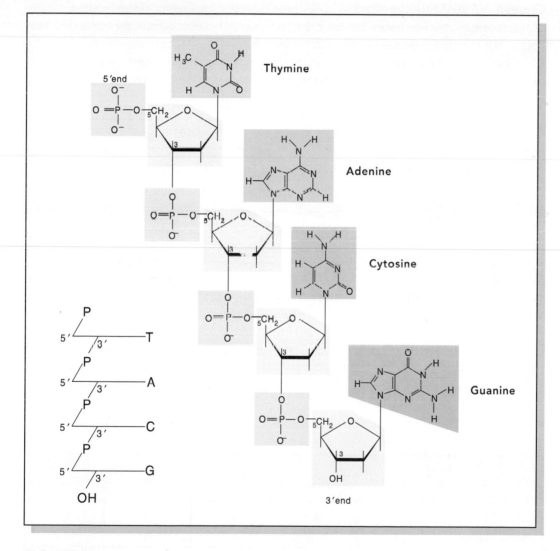

FIGURE 2.4

The binding of nucleotides to form a nucleic acid. The phosphate group forms a bridge between the 5′carbon atom of one nucleotide and the 3′carbon atom of the next nucleotide. A water molecule (H₂O) results from union of the hydroxyl group (–OH) formerly at the 3′-carbon atom and a hydrogen atom (–H) formerly in the phosphate group. The linkage between nucleotides is a "3′-5′ linkage"; the bond is called a phosphodiester bond. Note that the 3′carbon of the lowest nucleotide is available for linking to another nucleotide (this is called the 3′ end of the molecule) and that the phosphate group of the uppermost nucleotide can link to still another nucleotide (this is the 5′ end).

polymer

a chemical molecule consisting of repeating units of a particular substance.

same amounts. This conclusion was later proven untrue, but at the time it encouraged the belief that DNA was simply a polymer of repeating nucleotide units (e.g., TGACTGACTGAC, for thymine-guanine-adenine-cytosine-thymine-guanine-adenine-etc.). Without sequence variation in a repeating chain, it was difficult to see how DNA could provide any biochemical or hereditary information. This was one reason why Avery's identification of DNA as the transforming substance in bacteria, as noted in Chapter 1, did not receive immediate acclaim. (Ironically, Avery and Levene both had laboratories at Rockefeller Institute, but on different floors). It appeared that DNA was nothing more than a structural element of chromosomes.

After World War II, Levene's chemical analyses were repeated with more sophisticated equipment and with quite a different set of results. Tests indicated that the DNA's four nitrogenous bases were present in unequal amounts. In addition, biochemists led by **Erwin Chargaff** reported that chromosomal DNA from different organisms has different amounts of the four nitrogenous bases, as Table 2.1 indicates. These observations suggested that DNA is not simply a repeating polymer. Moreover, if the base amounts vary in an organism's chromosomes, perhaps DNA might have an information-coding property. And if different organisms have different amounts of bases in their DNA, maybe the bases have something to do with difference in the organisms.

Chargaff's experiments, reported in 1949, resulted in other data that would in later years weigh heavily on determining the structure of DNA. His research indicated that in DNA the amount of adenine is always equal to the amount of thymine regardless of the source of the DNA (as Table 2.1 shows). Moreover, the amount of cytosine is consistently equal to the amount of guanine. It appeared that for every adenine molecule there was a thymine molecule (and vice versa), and for every cytosine molecule there was a guanine molecule (and vice versa). Four years would pass before the significance of this observation was understood.

DNA IN THREE DIMENSIONS

Students of the twenty-first century are taught the structure of DNA as if it has always been known. They learn about DNA's components and they study its double-stranded spiral form known as the double helix. But in the early 1950s, biochemists were unaware of either the strandedness or helical arrangement of DNA, nor did they have a clear understanding of its functions. Although the components of DNA and their relative amounts were known, the spatial arrangements of the components was still a mystery.

helix

a spiral coil.

Against this backdrop, a young American graduate student named **James D. Watson** arrived at Cambridge University in London to study with **Francis H. C. Crick**, a prominent

TABLE 2.1
The Base Compositions of DNA from Various Species, as Determined by Erwin Chargaff.

SPECIES	A	T	G	C
Homo sapiens	31.0	31.5	19.1	18.4
Drosophila melanogaster	27.3	27.6	22.5	22.5
Zea mays	25.6	25.3	24.5	24.6
Neurospora crassa	23.0	23.3	27.1	26.6
Escherichia coli	24.6	24.3	25.5	25.6
Bacillus subtilis	28.4	29.0	21.0	21.6

Note that the percents of adenine and thymine are consistently similar, as are the percents of cytosine and guanine.

biochemist, (Both researchers are pictured in Figure 2.5.) Watson and Crick would do no laboratory bench work—their great contribution to science was interpreting the available data and putting them together to postulate a structure for DNA.

In the early 1950s, a new technique called **X-ray diffraction** was being used by chemists and other scientists in their molecular analyses. In X-ray diffraction, crystals of a chemical substance are bombarded with X rays as the crystals rotate within the X-ray field. Electrons in the chemical substance scatter (or "diffract") the X rays, and a diffraction pattern develops on a photographic plate. The pattern gives a strong clue to the three-dimensional structure of the chemical substance. The effect is somewhat similar to creating ripples in a lake by tossing a rock into the water. (The ripples give an idea of the size and shape of the rock.) Figure 2.6 shows how X ray diffraction technology works.

Among the leading experts on the diffraction patterns of DNA were British

X-ray diffraction

a laboratory technique in which X rays pass through a substance and are deflected, thereby creating a pattern that reflects the substance's structure.

(a)   (b)

FIGURE 2.5

James D. Watson and Francis H. C. Crick, the American graduate student and the British biochemist, who correctly explained the structure of DNA. (a) The scientists as they appeared in 1952, when the structure of DNA was formulated. (b) Photographs of more recent vintage.

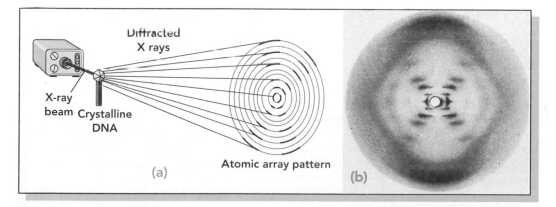

(a)   (b)

FIGURE 2.6

X-ray diffraction technology. (a) How X-ray diffraction works: DNA fibers are placed in an ultrathin tube so that most fibers are oriented in the same direction. X rays are directed through the fibers, and electrons in the DNA scatter the X rays to form a diffraction pattern on the photographic plate. (b) The X-ray diffraction photograph taken of DNA by Rosalind Franklin. Two important clues to DNA's structure were obtained from the pattern: The large smudges at the top and bottom of the pattern indicate that there is a regular spacing of 34 nm between bases along the fiber's length. The "X" pattern formed by the oblique smudges that cross at the center indicates a zigzag pattern that conforms to a helix.

FIGURE 2.7

(a) Rosalind Franklin and (b) Maurice H. F. Wilkins. The two British scientists prepared the diffraction photographs used by Watson and Crick to explain the structure of DNA.

(a)   (b)

biochemists **Maurice H. F. Wilkins** and **Rosalind Franklin** (Figure 2.7). Wilkins had found a way to prepare more uniformly oriented fibers of DNA than available previously, and Franklin was using these fibers to obtain relatively good diffraction patterns of the molecule. The patterns were suggesting that the DNA molecule was a **helix**. (A helix is a spiral or coil similar to a wound telephone cord.) The patterns also indicated that the helix had a diameter of about 2.0 nanometers (nm) (a nanometer is a billionth of a meter). Moreover, the helix appeared to make a complete turn at every 3.4-nm distance, and a repeating pattern at intervals of 0.34 nm seemed to occur.

Franklin's data were scheduled to be published in 1953, but Watson and Crick obtained the data before publication and used them to construct a model of DNA. In constructing their jigsaw puzzle, Watson and Crick theorized that phosphate groups (P) bind with adjacent deoxyribose molecules (D) in alternating fashion (P-D-P-D-P-D-P-etc.). Then they suggested that a nitrogenous base is connected to each deoxyribose molecule, flaring out as a side group from the main backbone chain. Now, using the mathematical data, they concluded that the 0.34 nm distance was the space between successive nucleotides on the chain. Franklin's measurements showed 3.4 nm per turn of the spiral, so Watson and Crick (noting that 3.4 is exactly ten times 0.34) postulated that ten nucleotides exist per turn of the spiral, as Figure 2.8 illustrates.

nanometer

a unit of measurement equivalent to a billionth of a meter.

FIGURE 2.8

What the X-ray diffraction photographs revealed about DNA. Watson and Crick postulated that DNA is composed of two ribbonlike "backbones" composed of alternating deoxyribose and phosphate molecules. They surmised that nucleotides extend out from the backbone chains and that the 0.34 nm distance represents the space between successive nucleotides. The data showed a distance of 34 nm between turns, so they guessed that ten nucleotides exist per turn. One strand of DNA would only encompass 1 nm width, so they postulated that DNA is composed of two stands to conform to the 2 nm diameter observed in the X-ray diffraction photographs. The figure shown is similar to the one presented in Watson and Crick's 1953 article.

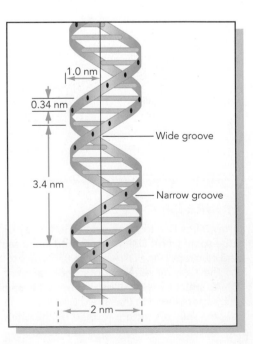

1.0 nm
0.34 nm
Wide groove
3.4 nm
Narrow groove
2 nm

When Watson and Crick attempted to apply the 2.0-nm diameter to the spiral, they encountered a problem: their calculations showed that a single helix with a diameter of 2 nm would have a density only half as great as the known density of DNA. After trying various combinations of molecules and densities, they hit on the idea that DNA was not a single-stranded molecule (a single helix) but rather a double-stranded molecule, a **double helix**. And if the bases were arranged to point inward (not outward, as some biochemists were suggesting), the density of DNA would come close to fitting the 2.0-nm diameter observed in the X-ray studies. The data were saying that a DNA molecule is composed of two nucleotide chains wound like a spiral staircase around a hypothetical cylinder. The deoxyribose-phosphate combinations form the backbones of both chains, and the nitrogenous bases point inward. The model was evolving (Questionline 2.2).

Now the observations made years before by Erwin Chargaff came into the picture. Chargaff had reported that in any DNA molecule, the amounts of adenine and thymine are identical. Watson and Crick envisioned that for every adenine molecule on one DNA strand there must be a thymine molecule on the other DNA strand (and vice versa). Figure 2.9 illustrates this pattern. Similarly, because DNA has equal amounts of guanine and cytosine, there must be a guanine molecule for every cytosine molecule. Moreover, the available space in the 2.0-nm diameter of DNA would accommodate an adenine-thymine pair or a cytosine-guanine pair perfectly. And the weak chemical bonds between the opposing base molecules would make chemical sense and hold the bases together. The model was complete.

So it was that Watson and Crick formulated the accepted structure of a DNA molecule. Scientists now agree that DNA is arranged as a double helix of two

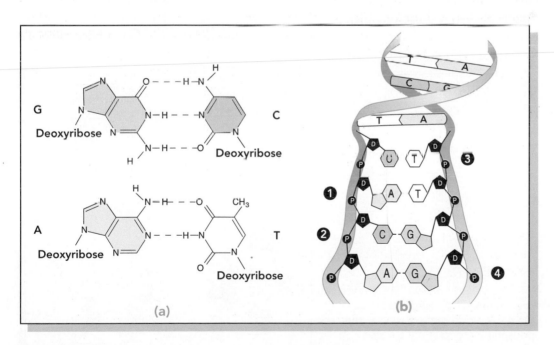

(a)   (b)

FIGURE 2.9

Base pairing in DNA. (a) The guanine-cytosine pair (G-C) has almost the same shape as the adenine-thymine pair (A-T). Weak chemical bonds called hydrogen bonds bind the bases together (dotted lines point out the three hydrogen bonds in G-C and two hydrogen bonds in A-T). The "comfortable" fit of one purine base with one pyrimidine base fulfills Chargaff's principle that for every purine molecule there is a pyrimidine molecule. (b) If cytosine were to pair with thymine (number 3) or adenine with guanine (number 4), the molecule would squeeze together or flare out. The squeezing and flaring would form a "lumpy" molecule not seen in the X-ray diffraction photographs. The A-T and C-G combinations fit the X-ray data.

complementary

being the opposite member of a pair.

reverse polarity

the condition in which the two strands of DNA run opposite to one another in a double-stranded DNA molecule; antiparallel.

intertwined chains, with complementary bases (A-T and G-C) opposing each other. Moreover, the strands run opposite to one another, that is, the strands display the **reverse polarity** illustrated in Figure 2.10. They are said to be "antiparallel." This means that one strand of DNA will have a free phosphate attached to a 5′–carbon atom at the top of the strand and the other strand will have a free 3′–carbon atom at its top. Given the base sequence of one chain of DNA, the base sequence of its partner

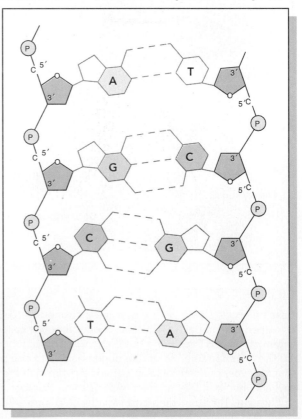

FIGURE 2.10

Reverse polarity in DNA. In DNA the strands run in opposite directions. On the left side of the molecule the phosphate group at the 5′-carbon atom is free to bond with another nucleotide at the top of the strand; whereas on the right side, the strand has a free 3′-carbon atom available for bonding at the bottom. The two strands are said to be antiparallel, with the left strand running 5′ to 3′ top to bottom and the right strand running 5′ to 3′ bottom to top.

chain is automatically determined by simply noting which bases are complementary (adenine-thymine or cytosine-guanine). Furthermore, the structure provides a mechanism by which one chain can serve as a template (a model or pattern) for the synthesis of the other chain (Figure 2.11). We will see how this works presently.

The announcement of the structure of DNA was greeted enthusiastically by the scientific community for its philosophical sake, as well as for another practical reason: knowing the structure of DNA made it easy to see how DNA could provide hereditary information. Biochemists saw that the nitrogenous bases, occurring in highly variable sequences, could provide a **code of heredity**. The sequence was not repeating (TGAC...TGAC...TGAC) as Levene's work had suggested. Rather it was variable (TGGACTTGCCTAAGCGATA...), and such a variable sequence could encode a variable sequence of amino acids in a protein chain. Knowing the coding ability was the real beauty of knowing the structure of DNA. Scientists could now let their minds ponder how a base sequence in DNA could be translated to an amino acid sequence in protein. Was this the key to heredity?

variable sequence
referring to the occurrence of nitrogenous bases in random order in the DNA molecule.

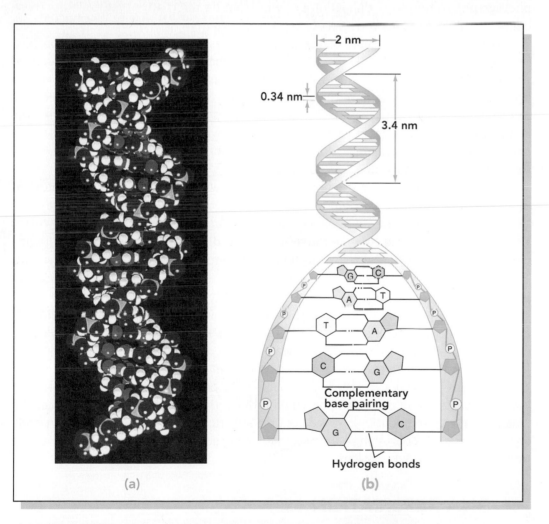

(a) (b)

FIGURE 2.11

The accepted view of the DNA molecule. (a) A space-filling model of the DNA molecule as determined by the X-ray diffraction data. (b) A stylized model of DNA. The molecule is arranged as a double helix of two intertwined strands of deoxyribose molecules and phosphate groups, with complementary nitrogenous bases extending out from the deoxyribose molecules toward one another. In the early 1950s, scientists observed that one strand of the molecule could serve as a template for synthesis of the opposite strand. This was later proven correct.

It should be noted that neither Watson nor Crick were experts in any of the scientific areas they used to construct their DNA model. Franklin was a better crystallographer, Chargaff understood base relationships better, and numerous scientists helped out with the chemistry of the discovery process. Watson and Crick achieved their goal because they were able to see the big picture. They took what they needed from several disciplines and used it to compose something greater than its parts. In effect, they saw the proverbial "forest for the trees."

On April 25, 1953, three short articles appeared in *Nature*, the prominent British publication of science. The first article, by Watson and Crick, opened with the lines quoted in the Preface of this book ("We wish to suggest a structure for the salt of deoxyribose nucleic acid..."). Figure 2.12 shows how the article was seen by readers. The article described the structure of DNA accepted in contemporary biochemistry. The second article, by Wilkins, A. R. Stokes, and H. R. Wilson, presented general evidence from X-ray data to support the helical structure of "deoxypentose nucleic acid," another nucleic acid. The third, by Rosalind Franklin and Raymond Gosling, included photographs of the X-ray diffraction patterns from an alternative type of DNA. These data also supported the proposed helical structure of DNA.

More than fifty years later, controversy continues to swirl about the relationships between Watson, Crick, Franklin, and Wilkins. Questions remain about who influenced whom, whether one individual had the idea about DNA's structure before another, and how the thought process developed. Although the questions will probably never be resolved, the controversy gives us a glimpse of how scientists go about uncovering the truths waiting to be discovered in nature. Readers who wish to learn more about the process and controversy are invited to read the books by Watson, Sayre, Olby, and Gribbon as listed at this Chapter's conclusion.

The contributions to science made by Watson, Crick, Franklin, and Wilkins were original, important, and sufficient to merit a place in the annals of scientific fame. DNA became the charter molecule of molecular biology. In 1962, Watson, Crick, and Wilkins were awarded the **Nobel Prize in Physiology or Medicine**. Unfortunately, Franklin had died of cancer in 1958 and because the Nobel committee does not cite individuals posthumously, she did not share in the award. However, Franklin's contributions to unraveling the structure of DNA have been universally acknowledged.

By the spring of 1953, as the articles in *Nature* were being printed, the evidence favoring the double helix structure of DNA was accumulating rapidly. Few biologists doubted the molecular model, and the ideas expressed in the *Nature* articles encouraged many investigators to think about a genetic code in terms of DNA (Questionline 2.3). But how, they asked, could a chromosomal double helix pass on its hereditary messages to the next generation of cells; and how could it direct the construction of needed cellular materials? It was apparent that working out the structure of DNA was not an end in itself. Rather it was only a beginning—the beginning of the science of molecular biology and in time, its fruits in DNA technology.

DNA REPLICATION

To scientists of 1953, the double helix was a molecule whose architecture of paired bases accommodated the biochemical requirements for genetic replication. Indeed, toward the end of a second 1953 article in *Nature*, Watson and Crick penned one of the great understatements of scientific history when they wrote that knowing the structure of DNA might provide insight on how DNA replicates:

---

deoxypentose

a five-carbon carbohydrate missing one oxygen atom.

Nobel Prize

an annual prize offered in six categories by the countries of Scandinavia; honors achievements in Physiology or Medicine, Chemistry, Physics, Literature, Peace, and Economics.

NATURE

No. 4356 April 25, 1953

MOLECULAR STRUCTURE OF NUCLEIC ACIDS

A structure for Deoxyribose Nucleic Acid

We wish to suggest a structure for the salt of deoxyribose nucleic acid (D.N.A.). This structure has novel features which are of considerable biological interest.

A structure for nucleic acid has already been proposed by Pauling and Corey[1]. They kindly made their manuscript available to us in advance of publication. Their model consists of three intertwined chains, with the phosphates near the fibre axis, and the bases on the outside. In our opinion, this structure is unsatisfactory for two reasons: (1) We believe that the material which gives the X-ray diagrams is the salt, not the free acid. Without the acidic hydrogen atoms it is not clear what forces would hold the structure together, especially as the negatively charged phosphates near the axis will repel each other. (2) Some of the van der Waals distances appear to be too small.

Another three-chain structure has also been suggested by Fraser (in the press). In his model the phosphates are on the outside and the bases on the inside, linked together by hydrogen bonds. This structure as described is rather ill-defined, and for this reason we shall not comment on it.

We wish to put forward a radically different structure for the salt of deoxyribose nucleic acid. This structure has two helical chains each coiled round the same axis (see diagram). We have made the usual chemical assumptions, namely, that each chain consists of phosphate diester groups joining β-D-deoxyribofuranose residues with 3′, 5′ linkages. The two chains (but not their bases) are related by a dyad perpendicular to the fibre axis. Both chains follow right-handed helices, but owing to the dyad the sequences of the atoms in the two chains run in opposite directions. Each chain loosely resembles Furberg's[2] model No. 1; that is the bases are on the inside of the helix and the phosphates on the outside. The configuration of the sugar and the atoms near it is close to Furberg's 'standard configuration', the sugar being roughly perpendicular to the attached base.

This figure is purely diagrammatic. The two ribbons symbolize the two phosphate—sugar chains, and the horizontal rods the pairs of bases holding the chains together. The vertical line marks the fibre axis

There is a residue on each chain every 3·4 A. in the z-direction. We have assumed an angle of 36° between adjacent residues in the same chain, so that the structure repeats after 10 residues on each chain, that is, after 34 A. The distance of a phosphorus atom from the fibre axis is 10 A. As the phosphates are on the outside, cations have easy access to them.

The structure is an open one, and its water content is rather high. At lower water contents we would expect the bases to tilt so that the structure could become more compact.

The novel feature of the structure is the manner in which the two chains are held together by the purine and pyrimidine bases. The planes of the bases are perpendicular to the fibre axis. They are joined together in pairs, a single base from one chain being hydrogen-bonded to a single base from the other chain, so that the two lie side by side with identical z-co-ordinates. One of the

pair must be a purine and the other a pyrimidine for bonding to occur. The hydrogen bonds are made as follows: purine position 1 to pyrimidine position 1; purine position 6 to pyrimidine position 6.

If it is assumed that the bases only occur in the structure in the most plausible tautomeric forms (that is, with the keto rather than the enol configurations) it is found that only specific pairs of bases can bond together. These pairs are: adenine (purine) with thymine (pyrimidine), and guanine (purine) with cytosine (pyrimidine).

In other words, if an adenine forms one member of a pair, on either chain, then on these assumptions the other member must be thymine; similarly for guanine and cytosine. The sequence of bases on a single chain does not appear to be restricted in any way. However, if only specific pairs of bases can be formed, it follows that if the sequence of bases on one chain is given, then the sequence on the other chain is automatically determined.

It has been found experimentally[3,4] that the ratio of the amounts of adenine to thymine, and the ratio of guanine to cytosine, are always very close to unity for deoxyribose nucleic acid.

It is probably impossible to build this structure with a ribose sugar in place of the deoxyribose, as the extra oxygen atom would make too close a van der Waals contact.

The previously published X-ray data[5,6] on deoxyribose nucleic acid are insufficient for a rigorous test of our structure. So far as we can tell, it is roughly compatible with the experimental data, but it must be regarded as unproved until it has been checked against more exact results. Some of these are given in the following communications. We were not aware of the details of the results presented there when we devised our structure, which rests mainly though not entirely on published experimental data and stereochemical arguments.

It has not escaped our notice that the specific pairing we have postulated immediately suggests a possible copying mechanism for the genetic material.

Full details of the structure, including the conditions assumed in building it, together with a set of co-ordinates for the atoms, will be published elsewhere.

We are much indebted to Dr. Jerry Donohue for constant advice and criticism, especially on inter-atomic distances. We have also been stimulated by a knowledge of the general nature of the unpublished experimental results and ideas of Dr. M. H. F. Wilkins, Dr. R. E. Franklin and their co-workers at King's College, London. One of us (J. D. W.) has been aided by a fellowship from the National Foundation for Infantile Paralysis.

J. D. Watson
F. H. C. Crick

Medical Research Council Unit for the
Study of the Molecular Structure of
Biological Systems,
Cavendish Laboratory, Cambridge.
April 2.

[1] Pauling, L., and Corey, R. B., *Nature*, 171, 346 (1953); *Proc. U.S. Nat. Acad. Sci.*, 39, 84 (1953).

[2] Furberg, S., *Acta Chem. Scand.*, 6, 634 (1952).

[3] Chargaff, E., for references see Zamenhof, S., Brawerman, G., and Chargaff, E., *Biochim. et Biophys. Acta*, 9, 402 (1952).

[4] Wyatt, G. R., *J. Gen. Physiol.*, 36, 201 (1952).

[5] Astbury, W. T., Symp. Soc. Exp. Biol. 1, Nucleic Acid, 66 (Camb. Univ. Press, 1947).

[6] Wilkins, M. H. F., and Randall, J. T., *Biochim. et Biophys. Acta*, 10, 192 (1953).

FIGURE 2.12

A copy of the 1953 article in *Nature,* in which Watson and Crick described the double helix structure of DNA. Publication of this article spurred the growth of molecular biology and DNA technology.

> It has not escaped our notice that the specific base pairing we have postulated immediately suggests a possible copying mechanism for the genetic material.

A major feature of the Watson-Crick model of DNA is that it provides a vision of how the molecule can replicate. It is easy to see, for instance, that one strand of the double helix has a base sequence that exactly determines the base sequence of the partner strand. Figure 2.13 pictures how this replication pattern works. Any conceivable sequence of bases can be present in one strand, but a complementary sequence

replicate

produce an exact copy.

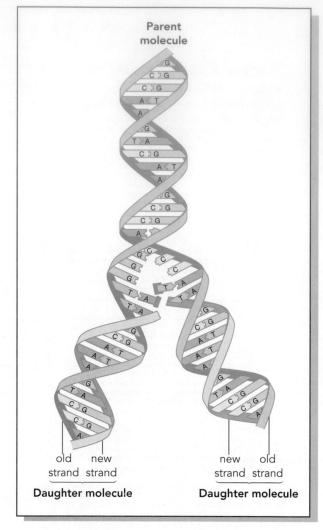

FIGURE 2.13

The general plan of DNA replication. The double helix unwinds, and the two "old" strands serve as templates for the synthesis of "new" strands having complementary bases. In the 1950s biochemists were uncertain whether a new and an old strand combine to form a new double helix (as shown here) or whether the two old strands recombine with one another, leaving the two new strands to join with one another. The experiment of Meselson and Stahl answered this question.

template

model.

isotope

a variant of an atom of the same element; may be radioactive.

of bases must occur in the second strand. For example, if the base sequence in one strand is G-T-A-C-C-A-T..., the base sequence of the partner strand must be C-A-T-G-G-T-A... Each strand in the double helix is a complementary mirror image of the other strand. Thus, for replication purposes, the DNA need only "unzip" and each strand will serve as a template for a new, complementary strand.

But there was still an important question to answer: Does each parent strand combine with the complementary new strand to reform the double helix; or do the parent strands go back and reunite with each other, leaving the new strands to form a second double helix? The answer to this question was determined in 1957 by **Matthew Meselson** and **Franklin W. Stahl** of the California Institute of Technology. Following the pattern set by other researchers, the researchers looked to bacteria to provide answers to a tough biochemical question. Their research is shown in Figure 2.14.

Meselson and Stahl cultivated bacteria for a prescribed period of time in a growth medium containing a heavy isotope of nitrogen called ^{15}N. This made the bacterial DNA denser than normal because atoms of ^{15}N replaced the normal, lighter ^{14}N atoms in all nitrogen-containing molecules (such as the nitrogenous bases of DNA). They then transferred the multiplying cells to a new growth medium containing the normal ^{14}N. Now they harvested bacteria at regular intervals for analysis. For each round of bacterial reproduction, they assumed the DNA would be duplicated.

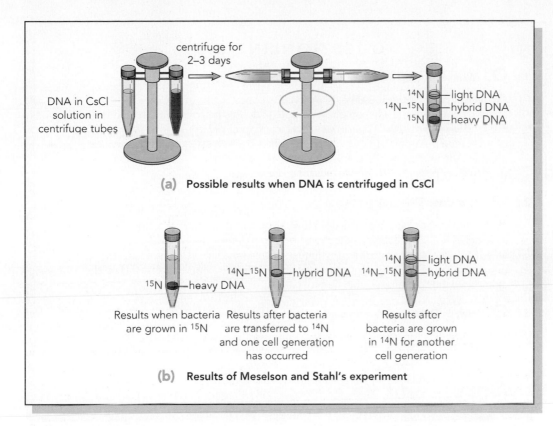

(a)   Possible results when DNA is centrifuged in CsCl

(b)   Results of Meselson and Stahl's experiment

FIGURE 2.14

The experiment of Meselson and Stahl. (a) DNA is placed in a tube containing the chemical compound cesium chloride (CsCl). The dense solution is centrifuged for two to three days. During this time, different forms of DNA containing different nitrogen isotopes separate according to their weights. (b) Meselson and Stahl grew bacteria in heavy nitrogen ^{15}N and found that the DNA occurred at a certain level in the CsCl. Then they permitted the bacteria to replicate one time and synthesize DNA in a medium with ordinary ^{14}N. They discovered that a hybrid DNA containing both ^{14}N and ^{15}N was forming. Thus, the DNA contained one old strand (with ^{15}N) and one new strand (with ^{14}N). After another generation of replication, light DNA containing only ^{14}N was isolated. This was also a product of semiconservative DNA replication.

Meselson and Stahl observed that the DNA first manufactured by the bacteria during their reproduction was all "heavy" because it contained the heavy isotope ^{15}N. When bacterial reproduction continued in the new growth medium, a newer form of DNA was synthesized. The investigators noted that it had a lighter density. This was because the lighter ^{14}N isotope was now being incorporated to the DNA.

But, the investigators noted, the new DNA did not have the density of all-light DNA (with only ^{14}N). Rather it had a density intermediate between that of the all-heavy DNA (with ^{15}N) and that of the all-light DNA (with ^{14}N). The intermediate density indicated that the newest DNA had some ^{14}N and some ^{15}N. This observation implied that each new DNA molecule had one strand of parent DNA (with ^{15}N) and one strand of newly synthesized DNA (with ^{14}N). It was clear that each parent DNA strand was serving as a template for a new, complementary DNA strand then combining with the parent DNA strand to form a double helix.

Nor did the experiments end there. After another round of bacterial reproduction took place, the intermediate-density DNA was present as anticipated (^{14}N^{15}N), but the bacteria were also forming a light-density DNA containing only ^{14}N (^{14}N^{14}N). Once again the double helix was unraveling, and each parent strand was modeling a

---

QUESTIONLINE 2.3

1. **Q. What is the structure of DNA?**

   A. As it exists in a chromosome, DNA consists of two chains of nucleotides wound like a spiral staircase around a hypothetical cylinder (a double helix). Alternating units of deoxyribose and phosphate groups form the backbones of both strands. Nitrogenous bases extend out from the deoxyribose molecules on both strands and oppose one another. Adenine molecules oppose thymine molecules (and vice versa) and guanine molecules oppose cytosine molecules (and vice versa).

2. **Q. How does DNA replicate?**

   A. For replication purposes, the double helix of DNA unwinds. Each strand then serves as a model for constructing a new strand with complementary bases. An enzyme places an adenine molecule opposite a thymine molecule (and vice versa); and an enzyme places a cytosine molecule opposite a guanine molecule (and vice versa). Then the strands rewind to form a double helix.

3. **Q. Why is DNA replication said to be semiconservative?**

   A. DNA replication is described as semiconservative because one old strand combines with the newly synthesized strand to form a double helix. Meanwhile the other old strand has been serving as a model for a second new strand. It combines with the newly synthesized strand to form a second double helix. Where there was one double helix of DNA, there are now two.

---

semiconservative

the process of DNA replication in which each of the original DNA strands occurs in one of the two newly formed double helices.

polymerase

po-lim´er-ase

enzyme

a protein that catalyzes a biological reaction while itself remaining unchanged.

complementary strand then combining with it. The $^{14}N^{14}N$ light-density DNA represented the molecule formed from a parent strand of the intermediate DNA (the ^{14}N strand) and newly synthesized DNA from ^{14}N in the culture medium.

The method of DNA replication suggested by the Watson-Crick model and determined by Meselson and Stahl has been termed **semiconservative replication** because one strand of the original DNA ("semi-") is brought over to each new double helix ("-conserved"), as Figure 2.15 illustrates. Contemporary biologists accept that semiconservative DNA replication occurs in animal and plant cells just before the time that cells undergo division. Each "old" strand of DNA acts as a template for synthesis of a "new" strand. In effect, the old strand dictates that a nucleotide containing adenine will be placed opposite one containing thymine (and vice versa); it dictates that a nucleotide containing guanine will be positioned opposite one containing cytosine (and vice versa).

For replication to occur, the nucleotides must be joined one by one by an enzyme called **DNA polymerase**, first isolated by Arthur Kornberg, Nobel laureate of 1959. Molecules of DNA polymerase move along the template strand of the open helix reading the nucleotide in the template and joining the complementary molecule onto the end of the new strand. The new strand thereby grows in length.

At first, scientists believed that the molecules of DNA polymerase would move along both template strands toward the site where the helix was unzipping, a place called the **replication fork**. Further research showed, however, that the polymerase enzymes for the two strands of DNA moved in opposite directions—toward the replication fork on one side and away from the replication fork on the other side (Figure 2.16). As a result, the new strand growing toward the replication fork is constructed by continuous assembly, whereas the new strand growing away from the replication fork is assembled by a discontinuous method involving pieces (called **Okazaki fragments**) that are formed then linked together. Thus, a number of different enzymes are needed for unwinding the helix, keeping the

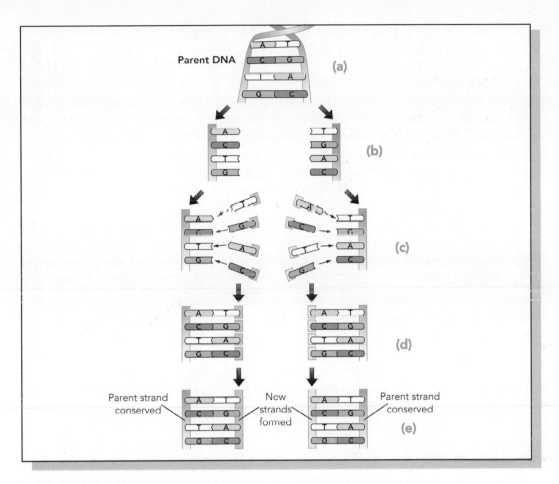

FIGURE 2.15

Semiconservative replication and complementary base pairing in DNA. (a) The parent double helix unwinds, and (b) the strands separate as the hydrogen bonds between the bases break. (c) Free nucleotides approach the strands, and (d) the bases are added one by one, with adenine complementing thymine and cytosine complementing guanine. This complementary base pairing causes the new strand to be a mirror image of the original strand. (e) The enzyme DNA polymerase joins the backbones of the new strands together by forming phosphodiester bonds between the phosphate groups and the deoxyribose molecules. An old parent strand binds to a new strand, whereas the second old parent strain binds to the second new strand. The strands wind to form two double helices. The synthesis of new DNA molecules is now complete.

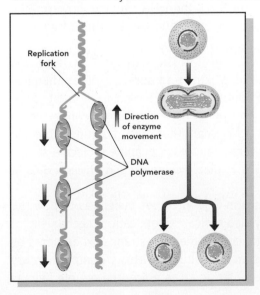

FIGURE 2.16

Details of DNA polymerase activity and DNA replication in a bacterium. The DNA unzips at the replication fork, and DNA polymerase synthesizes a new DNA strand while moving toward this point. Other DNA polymerase molecules synthesize fragments of DNA in pieces, while moving away from the origin of replication. The pieces are then joined to form the complete DNA molecule, and separation occurs as new cells form from the original cell.

origin of replication

the area or areas of a DNA molecule where the replication process begins.

strands separate, and joining DNA segments. DNA polymerase is only one member of this complex enzyme system.

In bacteria, the replication begins at a segment of DNA called the **origin of replication** and proceeds outward as Figure 2.16 demonstrates. At ideal growing temperature, the enzyme complex of a bacterium moves like a locomotive along the DNA at an astonishing rate of 850 nucleotides per second, reading the template and incorporating the complementary nucleotides. For human cells, about 3 billion nucleotides must be positioned. Days would be consumed if there were a single origin of replication. Instead, there are multiple origins, and replication occurs with high efficiency. Once the new strand is completed, the old and new stands combine with one another to reform the double helix.

The semiconservative method for DNA replication has an interesting implication for those who are fond of establishing links to the past. You will note that each helix unwinds, and the two strands each serve as a model for a new double helix. However, it is possible that neither of the original strands is ever lost. Indeed, in all the cell divisions taking place from conception, the original strands of DNA from one's mother (by way of the egg cell) and from one's father (by way of the sperm cell) may still be in existence. (Of course the cell containing the strands may have died and the strands may have been lost.) Each double strand has merely unwound, separated, and rewound with new strands during an inconceivable number of replications. Somewhere in the body are the original strands of DNA that we inherited from our parents—not copies of the strands, but the strands themselves. It can be said that we can never escape our heritage.

SUMMARY

Once scientists agreed that DNA was the hereditary substance, they sought to learn its structure because knowing the structure would give them a clue as to how it functions in heredity and how it replicates.

In the 1920s, Phoebus Levene and his colleagues identified three basic components in DNA: a five-carbon sugar named deoxyribose (ribose in RNA); a series of phosphate groups; and four nitrogenous bases named adenine, thymine, guanine, and cytosine. The researchers suggested that the three components form units called nucleotides and that DNA is a chain of nucleotides. After World War II, studies by Chargaff indicated that the base amounts in DNA vary in different organisms, suggesting that organisms may be different because their DNAs are different. Regardless of the DNA's source, however, the amount of adenine always equals the amount of thymine and the amount of guanine always equals the amount of cytosine.

The structure of DNA was deduced by Watson and Crick in 1953. Using X-ray diffraction photographs obtained by Franklin and Wilkins and applying other chemical data, Watson and Crick proposed that DNA is a double-stranded molecule wound to form a double helix. In the molecule, deoxyribose and phosphate groups are connected to each another and the bases point inward toward one another. Each adenine molecule stands opposite a thymine molecule (and vice versa), and each guanine molecule stands opposite a cytosine molecule (and vice versa). The available chemical and physical data fit this structure.

The variable sequences of nitrogenous bases in DNA provides insight as to how the molecule could provide a code of heredity. They also indicate a replication mechanism for DNA, one in which each DNA strand encodes its complementary strand. Experiments by Meselson and Stahl showed that the DNA strand combines with its complementary strand after serving as a model for its synthesis. As this takes place, the double helix is reformed. The replication method is said to be semiconservative because one of the old strands is conserved in each of the newly formed double helices.

▨ REVIEW

The pages of this chapter have followed the development of thought relating to the structure and replication of DNA. To assess your comprehension of these topic areas, answer the following review questions.

1. According to Levene's research, what are the three basic components of DNA?

2. Explain what is meant by a nucleotide and describe the method by which nucleotides are linked to one another.

3. What two significant observations were made by Erwin Chargaff's group and why were they important to establishing the structure of DNA?

4. Explain how X-ray diffraction studies helped determine DNA's structure.

5. Summarize the chemical structure of DNA, beginning with the deoxyribose-phosphate "backbone," continuing with the complementary bases, and concluding with the helical construction.

6. Given the base sequence of one chain of a DNA molecule, how can the base sequence of the opposite chain be determined?

7. What part of the DNA molecule constitutes a code, and why can that part be considered as having hereditary information?

8. Why is knowing the structure of DNA important to the development of DNA technology?

9. Outline the experiments of Meselson and Stahl indicating a semiconservative method for DNA replication.

10. Explain the method by which DNA replicates itself in the cell.

▨ FOR ADDITIONAL READING

Check, W. "DNA Helix Turns 40." *ASM News*. Washington, DC: ASM Press, 1994

Flannery, M. "The Many Sides of DNA." *American Biology Teacher* 59 (1997): 54–60.

Hall, S. "James Watson and the Search for the Holy Grail." *Smithsonian* (February, 1990)

Hoffman, R. "Unnatural Acts." *Discover* (August, 1993).

Jaroff, L. "Happy Birthday, Double Helix." *Time* (March 15, 1993).

Johnson, S., and Mertens T. R. "An Interview with Nobel Laureate Maurice Wilkens." *The American Biology Teacher* 51 (1989): 151–157.

Olby, R. *The Path To The Double Helix*. Seattle: University of Washinton Press.

Sayre, A. *Rosalind Franklin and DNA*. New York: W. W. Norton and Co., 1975

Watson, J. *The Double Helix*. New York: Athenum Press, 1968.

DNA In Action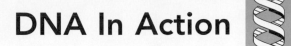

 CHAPTER OUTLINE

LOOKING AHEAD

This chapter explores the succession of observations and experiments that led scientists to relate deoxyribonucleic acid (DNA) to protein synthesis. It also describes the mechanisms by which the biochemical information in DNA is converted to biochemical information in protein. On completing this chapter, you should be able to:

- understand how scientists made the connection between chromosomes and biochemical activities taking place in the cell and between gene activity and the chemistry of enzymes.

- follow the reasoning that led scientists to a genetic code in DNA and helped them relate the code to an amino acid sequence in protein.

- recognize the role played by ribonucleic acid (RNA) in translating the genetic message in DNA to an amino acid sequence in protein.

- comprehend the nature of the genetic codes in DNA and understand how various genetic sequences were established.

- conceptualize the processes of transcription and translation in the synthesis of protein as they occur in the cell.

- understand the need for gene control in a cell and summarize some negative and positive mechanisms by which that control comes about.

- explain the various types of regulatory mechanisms that contribute to the complex biochemistry by which a gene is expressed.

INTRODUCTION

At the beginning of this century scientists advanced human thought significantly when they realized that patterns of heredity can be explained by the activity of chromosomes (Chapter 1). That realization led not only to the science of genetics but also to great advances in medicine and agriculture. And it profoundly influenced how we think about ourselves because it removed the mystery from heredity and made our biological nature seem more approachable.

But the realization was not an end in itself. Indeed, the great discoveries concerning chromosomes also raised a question, one that would influence biological thinking for more than half a century: What exactly is the biochemical connection between chromosomes and the hereditary traits expressed by the chromosomes? Put simply, how do chromosomes work?

Modern biologists can partly answer that question. They know, for example, that the biochemical information in chromosomes is converted to information contained in a sequence of amino acids in proteins and that proteins so constructed compose our cells, tissues, organs, and organ systems. Hence, all parts of our anatomy reflect the activity of chromosomes. Moreover, the entire chemistry of our bodily activities (whether moving or thinking or digesting) is governed by biological catalysts called enzymes, and enzymes are proteins. Thus, the structure and chemistry of our bodies revolve around proteins, and chromosomes specify proteins.

The great deduction relating chromosomes to proteins was not arrived at in a sudden flash of light but was developed slowly by a succession of investigators working over many decades. It began with the realization that deoxyribonucleic acid (DNA) is the key substance of chromosomes (Chapter 1), and it continued with the explanation of DNA's structure (Chapter 2). This chapter will continue that story by focusing on the seminal experiments relating chromosomal DNA to protein. It will then consider the mechanisms by which the biochemical information in DNA converts to biochemical information in protein. By considering DNA in action, we are really studying how genes work; in doing so we are answering one of the great questions of biology—how do our chromosomes express our hereditary traits?

FOUNDATION STUDIES

In 1902, a prominent British physician named **Archibald Garrod** wrote that certain diseases seemed to occur time and again only in selected families. Garrod's survey of several generations of families convinced him that these disorders behave as though they are controlled by a single inherited factor. Mendel's work was in the process of being rediscovered at that time (Chapter 1), and it was acceptable to think of inherited traits in terms of Mendelian "markers" (or genes, as they would later be called). Garrod concluded that certain disorders could be inherited much as flower color was inherited in pea plants. He further postulated (quite remarkably for his time) that a genetic disease is caused by a change in an ancestor's genetic material, and the defect is passed along within the family.

One disease Garrod studied in detail was **alkaptonuria**. Patients with this disease expel urine that rapidly turns black on exposure to air. The color change takes place because the urine contains alkapton, a substance that darkens on exposure to oxygen. In normal individuals, alkapton (known chemically as homogentisic acid) is broken down to simpler substances in the body, but in persons with alkaptonuria, the body cannot make this transformation, and alkapton is excreted.

chromosome

a structural element of DNA and protein that serves on the repository of hereditary information in the cell.

alkaptonuria

al˝kap-to-nur´e-ah

a genetic disease in which the body fails to break down alkapton into simpler compounds because it lacks the necessary enzyme.

(a)   (b)

FIGURE 3.1

(a) George Beadle and (b) Edward Tatum, the biochemists who postulated the one gene–one enzyme theory and related gene activity to protein synthesis.

Although Garrod lacked a biochemist's background, he suggested that the body could not break down alkapton because it lacked the necessary enzyme. (The concept of an "enzyme" was also relatively new in the early 1900s, and relating a chemical reaction to an enzyme put one at the forefront of science. We will discuss enzymes presently.) Thus, the patient with alkaptonuria could have an enzyme defect, and if the defect was inherited, it followed that a gene defect could cause an enzyme defect. Other scientists accepted Garrod's insight, and the concept gradually developed that genes have something to do with enzyme production. This concept would be strengthened by the work of Beadle and Tatum, but the scientific community would have to wait forty years for their work to be performed.

enzyme

a protein that catalyzes a chemical reaction of metabolism while itself remaining unchanged.

THE ONE GENE–ONE ENZYME HYPOTHESIS

In the early 1940s, while Avery and his colleagues were attempting to identify the role of DNA as the transforming principle in bacteria (Chapter 1), other researchers were trying to clarify the chromosome's normal function in the living cell. Among the leaders in this field of chromosome research were **George Beadle** and **Edward Tatum** (Figure 3.1) of Stanford University. Beadle and Tatum performed experiments that were innovative and imaginative for the time: They mutated the DNA of an organism's chromosomes and studied the effects on the biochemistry of the organism.

For their experiments, Beadle and Tatum selected the common bread mold *Neurospora crassa*. This mold can be cultivated in the laboratory on a growth medium where the exact nature and amount of each component is known (e.g., 10 g of glucose per liter). Beadle and Tatum exposed *N. crassa* to X rays and mutated the cells so they could not synthesize a particular nutrient. The mutated organism would not grow on the medium unless the particular nutrient was supplied. With this approach, they were able to identify and isolate many mutated strains of *N. crassa*. Figure 3.2 displays their results.

Through their experiments, Beadle and Tatum noted that a mutated mold was able to pass on its **mutation** to the next generation. Furthermore, each mutation could be explained as damage inflicted on a cellular catalyst, known as an enzyme. **Enzymes** are protein molecules that bring about virtually all the chemical reactions in a cell. It appeared that each time a mutation developed in the cell, the ability to synthesize an important growth substance was lost, and the loss could be traced to an enzyme defect. Thus, two concepts emerged: mutations are traceable to the hereditary aspects of the cell (and thus, to its genes), and DNA within genes has something to do with the production of key enzymes.

Neurospora

new-ros'por-ah

mutation

a permanent change in the characteristics of an organism arising from a change in the genes of the organism.

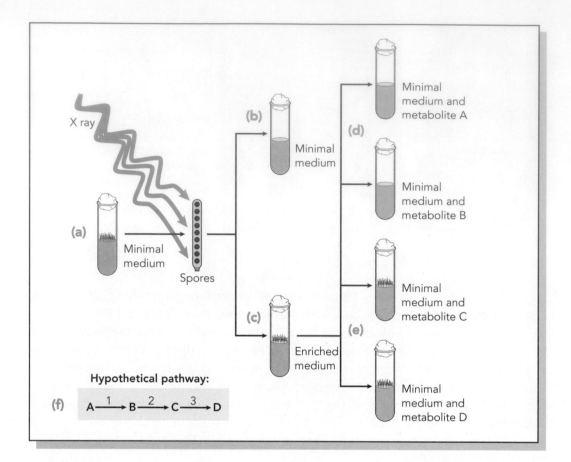

FIGURE 3.2

The Beadle-Tatum experiment, demonstrating the one gene—one enzyme hypothesis.
(a) Spores of the fungus *Neurospora crassa* are irradiated with X rays or ultraviolet light.
(b) Then the spores are transferred to a minimal medium lacking certain nutrients (metabolites)
or (c) an enriched medium containing various nutrients. (d) The fungi do not grow on the
minimal medium but (e) they do grow on the media containing metabolites C and D. Because
they grow only where C and D are present, the fungi must be missing enzyme 2 from the
hypothetical pathway of metabolism shown in (f). Thus, the irradiation must have eliminated
enzyme 2. Beadle and Tatum postulated that the X rays affected the gene needed to produce
this enzyme.

On these bases, Beadle and Tatum postulated that the genes of a cell influence the
production of cellular enzymes. They called their hypothesis the "**one gene–one
enzyme hypothesis.**" Beadle and Tatum guessed that an enzyme alteration was
disrupting at least one event in a series of chemical reactions.

An important implication of the one gene–one enzyme hypothesis related to the
chemical nature of enzymes. In the 1940s biochemists agreed that enzymes are
composed of proteins. They understood equally well that enzymes are catalysts for the
synthesis of all an organism's structural parts (e.g., proteins, carbohydrates, lipids,
nucleic acids, blood components, hormones, antibodies, hair fiber, muscle proteins,
and on and on). It stood to reason that if genes affect the production of enzymes and if
enzymes are composed of protein, genes must affect protein production.

In retrospect, it is easy to understand why scientific interest in DNA was growing in
the 1940s. Avery's group was in the process of identifying DNA as the substance of
bacterial transformation; Hershey and Chase were working to show that DNA was the
determining molecule in viral replication (Chapter 1); and Beadle and Tatum were

catalyst

a substance that facilitates
a chemical reaction.

focusing attention on genes for their ability to transmit mutations between generations and for their significance in the synthesis of proteins. The implications of DNA activity were capturing the imaginations of molecular biologists, and deservedly so because the activity manifested by DNA stood at the very roots of molecular biology.

PROTEINS AND CODES

Proteins are generally considered the working and structural molecules of all cells, be they microbial, plant, animal, or other cells. Chemically, proteins are composed of building-block units called **amino acids** (somewhat analogous to the nucleotide building blocks of nucleic acids). Only twenty different amino acids, shown in Figure 3.3, are used to produce the countless combinations found in the proteins of cells. For example,

protein

a processed chain of amino acids that performs structural, enzyme, transport, signal, or other functions in the cell.

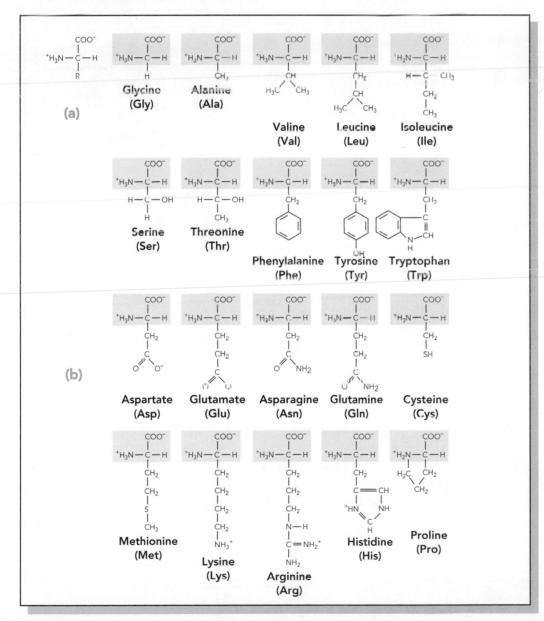

FIGURE 3.3

The twenty amino acids making up most of the proteins of living things. (a) The general structure of the amino acid shows the amino group with an added proton (NH$_3^+$) and the carboxyl (acid) group missing a proton (–COO⁻). (b) Note that the shaded portion is identical in all the amino acids.

amino acid

a chemical compound that contains at least one amino group and one acid group and is used in the construction of a protein.

biochemists estimate that there are roughly 10,000 different enzymes operating in a living cell. This implies 10,000 different combinations of the twenty amino acids. Moreover, by broad estimate, there are about 100,000 different proteins in a human cell. There must be 100,000 different combinations of the amino acids in these proteins.

One protein differs from another in the sequence of amino acids in the protein, not the nature of the amino acids. For example, two proteins may each have thirty-seven amino acids linked together, and the thirty-seven amino acids may be identical, but if the sequences of the amino acids in the proteins are different, the proteins are different. Think for a moment how many words can be composed from a twenty-six–letter alphabet. For proteins, the alphabet is not composed of twenty-six letters

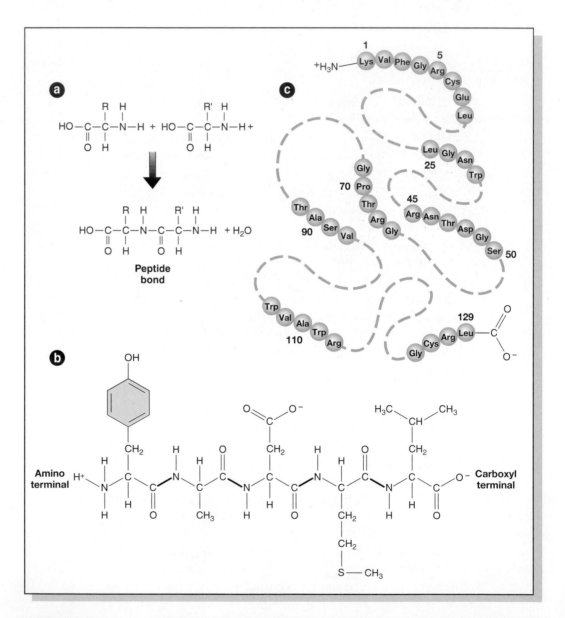

FIGURE 3.4

The linkage of amino acids in a protein. (a) The amino portion of one amino acid and the acid portion of another amino acid join together in a dehydration synthesis reaction to form a peptide bond. A dipeptide and a water molecule result from the reaction. (b) Additions at either the amino terminal or carboxyl (acid) terminal ends lengthens the protein. In the illustration, there are five amino acids in the protein chain. Can you identify them? (c) A protein (the enzyme lysozyme) with 129 amino acids.

but of twenty amino acids. Amino acid linkages in proteins are shown in Figure 3.4.

One of the first to suggest a cellular coding mechanism for proteins was the German physicist **Erwin Schrödinger**. In his 1944 book *What is Life?*, Schrödinger suggested that the cellular chromosomes might contain a "code-script." He wrote that variations of atoms in the chromosome could produce a Morse code-like packet of information. However, the prevailing wisdom of the day was that chromosomes were composed of protein, and so DNA did not enter Schrödinger's thinking.

Shortly after the 1953 announcement of the Watson-Crick model of DNA (Chapter 2), another physicist named **George Gamow** introduced the coding issue once again. Writing to Watson and Crick, Gamow proposed that the arrangement of bases along the double helix might produce a series of chemical "holes." Each hole, according to Gamow, could have a different structure, depending on the nature of the bases that bordered the hole. Different amino acids could then fit into different holes and link up to form a protein. Indeed, Gamow had developed a strong interest in another nucleic acid, ribonucleic acid (RNA). In 1954, he organized an "RNA Club" to investigate the role of RNA in protein synthesis. Members wore a tie with a sinuous lime-green curl of nucleic acid flanked by boxy yellow outlines of purines and pyrimidines on a black background.

Although chemical analysis showed that the amino acids would not fit the chemistry or spatial arrangements of the bases, Gamow's theory encouraged researchers to think in terms of a coding mechanism. They questioned, for instance, how could a string of bases along a double helix be translated into a string of amino acids along a protein?

The importance of the amino acid sequence in protein was highlighted by pioneering work performed in 1956 by **Vernon Ingram** and his group at Cambridge University. Ingram studied the molecular basis of **sickle cell anemia**. This disease is an inheritable defect in the protein of the red blood cells (RBCs). Ingram's experiments indicated that sickle cell anemia is caused by an incorrect amino acid at a single position in the **hemoglobin** protein of RBCs. (Figure 3.5 shows this aberration.) Hemoglobin carries oxygen in the RBC. With the incorrect amino acid in place, the defective protein collapses (and the RBC does likewise), thereby yielding a distorted RBC that vaguely resembles a sickle (hence, the name sickle cell anemia). Persons so affected cannot easily transport oxygen to their cells and that, together with the clogging induced by sickled cells, often results in early death.

Because sickle cell anemia is known to be an inheritable disease, there was no question that genes were involved. Ingram's research pointed to the possibility that the genes for normal hemoglobin and sickle cell hemoglobin differ in how they specify a single amino acid in the hemoglobin protein. In this case, a defective gene can miscode a single amino acid in a protein with devastating results.

As the 1950s progressed, evidence continued to accumulate that DNA is the material of heredity and that DNA expresses its genetic message through proteins. Many scientists were also supporting the concept that some sort of genetic code exists in DNA, a code that specifies the sequence of amino acids in protein. But what was the code and how was the code translated into an amino acid sequence? Eventually the answers to these questions would come, but only with considerable guesswork and experimentation coupled with a better grasp of the biochemical realities of living things (Questionline 3.1).

■ THE INTERMEDIARY

To understand the process of protein synthesis, biochemists had to know whether the chemical information in DNA passes directly to the cytoplasm or whether some intermediary substance is required. Scientists were aware that protein synthesis takes

---

Schrödinger
shray'ding-er

ribonucleic acid
a nucleic acid containing ribose rather than deoxyribose in DNA and uracil rather than thymine.

sickle cell anemia
a genetically linked disease in which the RBCs assume a collapsed sicklelike appearance.

hemoglobin
the red pigment of the RBCs that is responsible for transporting oxygen in the body.

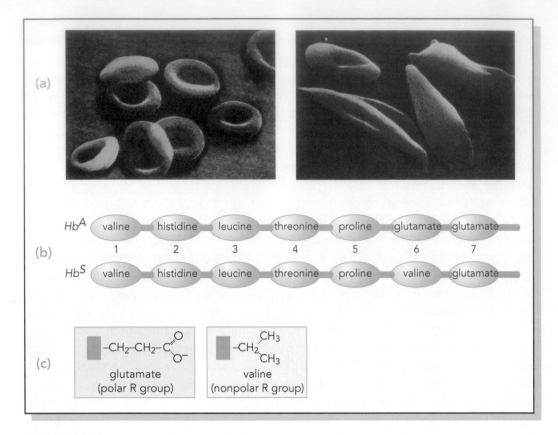

FIGURE 3.5

Sickle cell anemia and protein molecules. (a) Two scanning electron micrographs of human RBCs. (1) Normal cells. (2) Distorted cells from a person with sickle cell anemia. The cells are collapsed because of a change in the amino acid content of the hemoglobin protein within the cells. (b) The amino acid sequences of normal (HbA) and sickle-cell (HbS) hemoglobin. Of the 146 amino acids in the chain, only number 6 is different. Where there should be glutamic acid (Glu) in normal hemoglobin there is valine (Val). One erroneous amino acid in the sequence is responsible for the distorted RBCs. (c) The nonpolar group of valine is believed to be the source of the problem.

QUESTIONLINE 3.1

1. **Q.** **I've heard of DNA; but what is RNA?**

   A. RNA stands for ribonucleic acid. It is an organic compound acting as an intermediary in the synthesis of protein directed by DNA. RNA has ribose and uracil, whereas DNA has deoxyribose and thymine. Also, RNA is usually single-stranded, whereas DNA is usually double-stranded.

2. **Q.** **Exactly what is the genetic code?**

   A. The genetic code is a sequence of nitrogenous bases on a DNA molecule. The sequence provides a code for the correct placement of amino acids in a cellular protein. It acts much like a Morse code (e.g., dot-dot-dash) that can be translated to a word. Multiple words then make up a sentence.

3. **Q.** **How many bases specify an amino acid in a genetic code?**

   A. Nitrogenous bases act in blocks of three. A sequence of three bases (e.g., G-C-A) specifies a single amino acid (e.g., alanine). In the Morse code, by comparison, there are three dots or dashes for a letter.

place in the cell's cytoplasm, and because DNA was acknowledged to be largely confined to the cell's nucleus, a direct passage of information was unlikely. Scientists assumed that a relay system was operating in the cell.

In the 1940s, biochemists reported that cells undergoing protein synthesis possess an unusually large amount of **ribonucleic acid (RNA)**, a close relative of DNA. Liver cells and pancreas cells, for example, are active producers of protein and are also rich in RNA. Because RNA is lacking in cells that do not secrete many proteins (e.g., muscle and kidney cells), there seemed to be a definite correlation between protein synthesis and RNA. Moreover it had long been known that RNA, unlike DNA, occurs primarily in the cytoplasm of cells, where protein synthesis occurs. For example, embryologists had noted that embryonic cells, known for their high rate of protein synthesis, have equally high levels of RNA in their cytoplasm. All the lines of evidence pointed to the possibility that RNA is the intermediary in protein synthesis.

Although RNA and DNA are similar compounds, they differ in three important ways: (1) the carbohydrate portion of RNA is ribose, whereas in DNA it is deoxyribose; (2) RNA has the nitrogenous base uracil, whereas DNA has thymine; and (3) RNA is usually single-stranded, whereas DNA is almost always double-stranded. Figure 3.6 illustrates an RNA molecule.

nitrogenous

ni-troj′en-us

Despite the differences, it was clear that DNA could act as a model for RNA synthesis much as it does for the synthesis of new DNA (Chapter 2). The exceptions would be that uracil is positioned in the RNA instead of thymine, and ribose is used in the nucleic acid backbone instead of deoxyribose. Biochemists theorized that the synthesis could occur in the nucleus, then the RNA could travel to the cytoplasm, where it would determine the

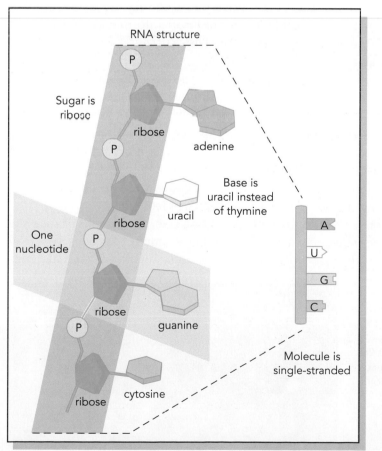

FIGURE 3.6

The structure of RNA. Like DNA, RNA is a polymer of nucleotides. However, three important differences distinguish RNA from DNA: the carbohydrate in RNA is ribose, whereas in DNA it is deoxyribose; RNA contains uracil as one of its four nitrogenous bases, whereas DNA has thymine instead of uracil; and RNA is usually single-stranded, whereas DNA is usually double-stranded.

amino acid sequence in the protein. Indeed, radioactive tracer experiments were indicating that RNA molecules move from the cell's nucleus to its cytoplasm.

Additional evidence for RNA's role was suggested by evidence from experiments with **viruses**, such as shown in Figure 3.7. Researchers found that when bacteria are infected by DNA-containing bacteriophages (Chapter 2), the bacteria synthesize RNA before they begin synthesizing protein. Hence, RNA appeared to be the intermediary compound between DNA and protein. Moreover, certain viruses, such as the tobacco mosaic virus, contain RNA rather than DNA. When the RNA enters a tobacco leaf cell, the cell produces new viruses, including the protein portion of the virus. Scientists guessed that RNA has the information necessary for synthesizing protein.

So scientists were drawn to RNA to explain how DNA's genetic information is translated to an amino acid sequence. As evidence mounted in favor of RNA's role, new questions were surfacing about the nature of the genetic information itself. We will consider those questions next.

■ THE GENETIC CODE

As of 1961, the evidence that DNA carries genetic information was compelling. As of that year, biochemists had identified the double helix structure of DNA; they had pinpointed the energy sources and enzymes used to link amino acids together in proteins; and they had accepted the concept of information flow from DNA to RNA to protein. But a basic question remained: How is the genetic message of DNA transmitted to RNA and how does RNA assist protein synthesis? In more specific terms, how does the order of nitrogenous bases in DNA encode the order of amino acids in a protein? What, essentially, is the genetic code?

Answers to these questions came in a series of experiments reported in 1961 by **Francis H. C. Crick** and his colleagues. Crick's group reasoned that the genetic code of DNA probably consists of a series of blocks of chemical information, each block corresponding to an amino acid in the protein. They further hypothesized (for mathematical reasons that will be discussed presently) that within a single block, a sequence of three nitrogenous bases specifies an amino acid. To verify these hypotheses, Crick's group performed experiments with viral DNAs, altering their nitrogenous bases to see what, if anything, would happen.

The experiments were elegant. Crick's group altered viral DNAs with acridine compounds. These compounds insert or delete DNA's nucleotides (and hence, its nitrogenous bases). For example, a DNA sequence might have the following sequence of bases AGGCATGCAATG. After acridine treatment, this sequence could be changed to AGGATGCAATG. In this case, the fourth base (C) has been deleted.

The Crick group sought to find out whether deleting one base would change the amino acids in the protein synthesized. If, in the preceding example, the codes were read in sequences of three bases, the first code would remain stable but the succeeding codes

bacteriophage

bak-te′re-o-faj″

a virus that replicates within a bacterium or integrates its genetic material into that of the bacterium.

tobacco mosaic virus

a virus that attacks tobacco plants and causes the leaves to assume a shriveled appearance because of the infection.

acridine

ak′ri-den

FIGURE 3.7

An electron micrograph of tobacco mosaic viruses. The viruses are the thin rods containing RNA. Latex beads 264 nm (0.264 um) in diameter have been added to the preparation for comparison purposes. Research on these viruses provided evidence that RNA has a role in protein synthesis.

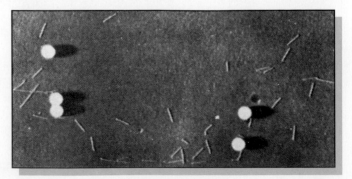

would change. Hence, the first amino acid would be in its proper position in the protein, but the remaining amino acids would change because new codes had arisen:

AGG CAT GCA TTG ➡ **(delete C)** ➡ **AGG ATG CAT TG**

And so the amino acids did change. In Crick's experiments, the first amino acid occurred in its correct position in the protein, but all succeeding amino acids were different.

Now Crick and his colleagues deleted two bases to see whether the amino acids would change their nature:

AGG CAT GCA TTG ➡ **(delete C and A)** ➡ **AGG TGC ATT G**

As you can see, deleting two bases at a fixed point preserves the three-base codes before that point, but the deletion changes the three-base codes after the deletion. In the protein, the amino acid encoded by AGG occurs in the correct position, but all succeeding amino acids are different because all the following codes (e.g., TGC and ATT) are different.

Finally, the key experiment was performed. Crick and his colleagues removed three bases to see the effect on the amino acids in the protein:

AGG CAT GCA TTG ➡ **(delete C, A, and T)** ➡ **AGG GCA TTG**

In the protein produced, the first amino acid occurs correctly because the AGG code remains. Also, and significantly, the succeeding amino acids also occur correctly because their base codes have been preserved by removing three bases. Of course, the amino acid specified by CAT is deleted, but the following codes are as originally constituted. The key to preserving the codes is eliminating three bases. Thus, by analyzing the nature and sequence of amino acids in the protein, the researchers could conclude that a nucleotide block for a single amino acid consists of three nitrogenous bases. The magic number for the genetic code was three (Figure 3.8a).

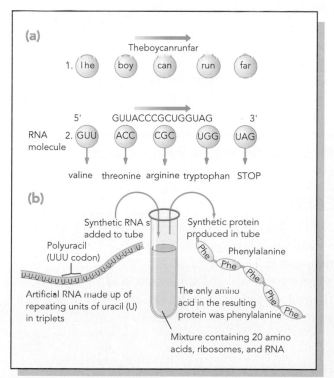

FIGURE 3.8

(a) The message in (1) is written in three-letter words reading from left to right. Each three-letter word has meaning in the message. In (2), we have a biochemical message written in three-letter bases reading from left to right. Each three-letter "word" (or codon) specifies an amino acid in the protein and thus has biochemical meaning to the cell. The codon at the far right is a stop signal. (b) How the genetic code was broken for the amino acid phenylalanine. Researchers synthesized a strand of RNA containing only uracil (U-U-U-U-U-U-U-U-etc). Then they placed the RNA into a tube containing twenty different amino acids and all the materials needed to synthesize protein. The system generated a protein consisting only of phenylalanine. Thus, U-U-U had to be the code for phenylalanine.

A bit of simple mathematics also pointed to the **three-base code**. Investigators knew that there are twenty amino acids from which virtually all proteins are made. They reasoned that if a single nitrogenous base encodes a single amino acid, there would only be enough information for four amino acids because there are only four bases (e.g., [A] adenine for amino acid No. 1, [G] guanine for amino acid No. 2, and so on). If two bases code for a single amino acid, only sixteen codes could be made and only sixteen amino acids could be encoded (e.g., [AG] for amino acid No. 1; [AT] for amino acid No. 2; [AG] for amino acid No. 3, and so on). Clearly this did not seem likely because there are twenty amino acids to encode. Now, if three nitrogenous bases comprise a code, sixty-four possible codes could exist in DNA (e.g., [ATA], [GAC], [TCG], [CAT], and so on). The experimental evidence provided by the Crick group pointed to this possibility and helped establish that a three-base sequence (a triplet) is the foundation for the genetic code for an amino acid.

The codes themselves were soon worked out, the first results coming in 1962. To determine the codes, biochemists combined synthetic RNA molecules with enzymes, amino acids, and other essential materials to see what protein would be made. (Figure 3.8b shows these experiments.) They found, for example, that a synthetic RNA molecule consisting of only uracil (U-U-U-U-U-U-U-U-U) would encode a protein consisting only of the amino acid phenylalanine. Thus, the RNA code for phenylalanine was U-U-U. Working backward another step, the DNA code was reasoned as A-A-A. Hence, the genetic code for phenylalanine is A-A-A.

The research to learn the genetic codes for all the amino acids was led by **Marshall Nirenberg** and **Heinrich Matthaei**, and independently, **Har Gobind Khorana**. Having worked out the code for phenylalanine, they continued this pattern of experiments and soon learned the codes for all twenty amino acids. These codes, occurring over and over again in a DNA molecule, specify which amino acid will occur in which position in the developing protein, as Figure 3.9 illustrates. Most amino acids have more than one code; and three codes are terminator, or "stop" signals, as we will see presently. For their work in explaining the nature of the genetic codes, Nirenberg and Khorana were awarded the 1968 Nobel Prize in Physiology or Medicine.

In the ensuing years, biochemists demonstrated that the **genetic code** is nearly universal: the same three-base codes specify the same amino acids regardless of whether the organism is bacterium, bee, or buttercup. The essential difference among species of organisms is not the nature of the nitrogenous bases but the sequence in which they occur in the DNA molecule. Different sequences of bases in DNA specify different sequences of bases in RNA, and the sequence of bases in RNA specifies the sequence of amino acids in proteins. This is the so-called **central dogma** of protein synthesis. And as the protein varies, so does the species of organism.

PROTEIN SYNTHESIS

The pieces were now in place: The DNA in genes encodes an amino acid sequence in proteins; RNA is the intermediary between a nitrogenous base sequence in DNA and the amino acid sequence in protein; and the genetic code consists of blocks of three nitrogenous bases, each block ultimately specifying an amino acid in the protein. Biochemists were ready to fit the pieces together to formulate a process for the synthesis of protein. The first step of that process would be transcription.

TRANSCRIPTION

As evidence was accumulating on the nature of the genetic code, an equally

phenylalanine
fen″il-al′i-nene

transcription
the biochemical process in which an enzyme catalyzes the synthesis of an RNA molecule by using the base sequence of DNA as a template.

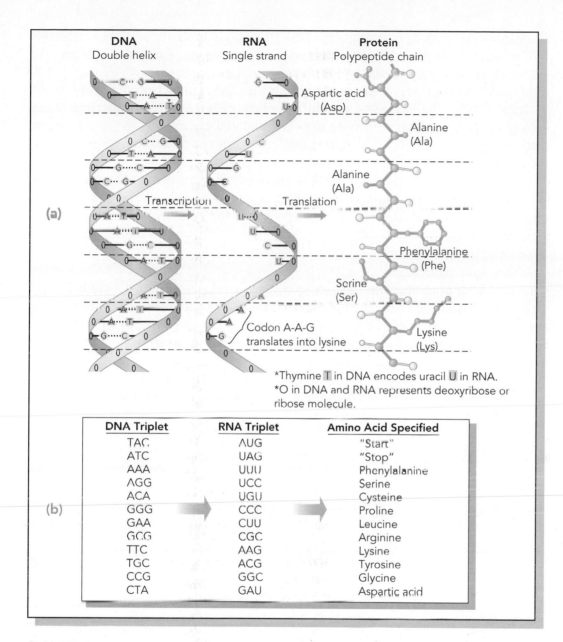

FIGURE 3.9

Gene expression and protein synthesis. (a) The base code in DNA is used to formulate a base code in RNA by the process of transcription. The RNA molecule is then used in translation to encode an amino acid sequence in a protein. (b) Some selected triplet codes in DNA and RNA and the amino acid specified in the protein. Note that the RNA code (known as a codon) is the complement of the DNA code and that certain codons are "start" or "stop" signals. The remainder of the sixty-four DNA codes and their amino acids can be found in most biochemistry textbooks

persuasive body of knowledge was being generated on the role of RNA in protein synthesis. RNA was an intermediary in the information transfer system (Questionline 3.2), but the place of RNA turned out to be even more considerable than originally suspected, and several types of RNA are now known to participate in protein synthesis. We will examine each next.

Modern biochemists agree that the process of protein synthesis is initiated by an uncoiling of the DNA double helix and an uncoupling of the two strands of DNA. A functional region of DNA, the **gene**, is thereby exposed. Using the sequence of nitrogenous bases along only one of the DNA strands, molecules of RNA are synthesized with complementary bases. Component nucleotides stored in the region are used for the synthesis, and an enzyme called **RNA polymerase** binds the nucleotides together to form the RNA molecule. The production of RNA is called **transcription**, a word coined by Crick in 1956. The fragments so constructed are known as **RNA transcripts**. These RNA molecules, together with ribosomal proteins and enzymes, constitute a system that carries out the task of reading the genetic message and producing the protein that the genetic message specifies.

At least three types of RNA transcripts are constructed for use in protein synthesis. One type is called **messenger RNA (mRNA)**. Each mRNA molecule is a long, single strand of RNA that passes out of the nucleus into the cell's cytoplasm carrying the genetic message (hence, its name "messenger RNA"). This molecule specifies which amino acid is to occupy which position in the protein.

Transcription of the mRNA by RNA polymerase begins at a specific DNA site on the gene called the **promoter site**. The promoter site is a sequence of nitrogenous bases. For a given gene, a promoter site exists on one DNA strand but not the other. The strand having the promoter site will transcribe its message to mRNA and is called the **sense strand**. The other strand remains nonfunctional and is termed the **nonsense strand**. We will discuss how this transcription occurs shortly.

The second type of RNA transcript is **ribosomal RNA (rRNA)**. Ribosomal RNA combines with protein to form ribosomes, the ultramicroscopic bodies existing along the cell's internal membranes and pictured in Figure 3.10. Functioning as the cell's structural "workbenches" where proteins are formed, the ribosomes are sites where enzymes assemble amino acids to proteins according to instructions delivered by mRNA molecules. Each ribosome has two subunits: a smaller one (called the 30s subunit), which binds the ribosome to mRNA molecules; and a larger one (the 50s subunit), where the enzymes for linking amino acids together are located. In living cells, the subunits

RNA polymerase

pol-im´er-ase

the enzyme that functions in transcription and synthesizes an RNA molecule with bases complementary to those in DNA.

promoter site

a sequence of bases on a DNA molecule that signals transcription to begin.

ultramicroscopic

below the ability of the light microscope to visualize.

QUESTIONLINE 3.2

1. **Q.** What do biochemists mean by transcription?

   A. Transcription is the biochemical process during which the base sequence of DNA is used as a model (or template) to synthesize an RNA molecule with a complementary base sequence. The RNA molecule so-formed is called an RNA transcript.

2. **Q.** How does the RNA transcript specify an amino acid chain in a protein?

   A. The RNA transcript (known as messenger RNA) moves through the cell to the ribosome, where it is met by transfer RNA (tRNA) molecules, each attached to an amino acid. The three-base code of a tRNA molecule matches to a complementary three-base code on the mRNA molecule, and an amino acid is thereby brought into position for attachment to the growing amino acid chain.

3. **Q.** How does amino acid "X" wind up in the correct position on the amino acid chain?

   A. The key is in the base codes. A three-base code on the mRNA molecule attracts a complementary code on the tRNA molecule, and because amino acid "X" is attached to that tRNA, it tags along. Thus, by matching the mRNA code and the tRNA code, the correct amino acid is brought into position.

combine with one another when the 30s subunit fits into a cleft in the 50s subunit to form the 70s ribosome. (The "s" stands for Svedberg unit, a measurement of the weight of the ribosome subunit as measured in the centrifuge.) The units are not cumulative in the complete ribosome because of the molecular weights and shape of the final molecule. Figure 3.11 depicts the assembly.

The third type of RNA transcript is **transfer RNA (tRNA)**. Transfer RNA molecules exist in dozens of different types and float freely in the cell's cytoplasm, where they bind to amino acids. Then they deliver the amino acids to the ribosome, as first postulated in the 1950 by **Paul Zamecnik** and his research group. (In 1996 Zamecnik was awarded the Albert Lasker Award for Achievement in Medical Science.) At least one form of tRNA exists for each of the twenty amino acids found in protein. The amino acid alanine, for example, has its own unique tRNA; glycine has its tRNA; and each of the other 18 amino acids has a tRNA that binds to it alone. The binding requires a specific enzyme and energy obtained from an energy-rich molecule called adenosine triphosphate (ATP).

ribosome

an ultramicroscopic cellular body where amino acids are enzymatically bonded to form a protein.

zamecnik

zam' chek

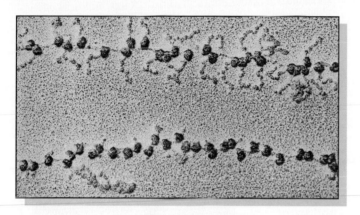

FIGURE 3.10

An electron micrograph of ribosomes with a messenger RNA molecule attached. The dark bodies are the ribosomes, and the threadlike strand is the mRNA. The strand begins at the lower right, extends to the left, then goes off the photograph. It reenters the photo at the top left and ends at the top right.

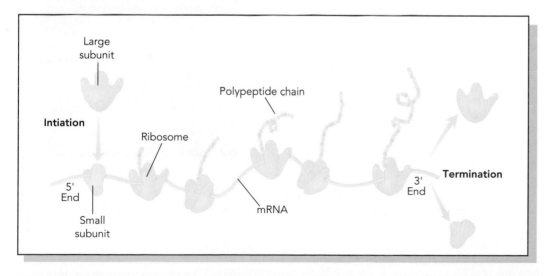

FIGURE 3.11

Details of the ribosome. In eukaryotic organisms such as animal and plants, the ribosome is composed of two subunits a smaller (30s) subunit and a larger (50s) subunit. The 30s subunit fits into a depression on the surface of the 50s subunit to form the complete (70s) ribosome. During the translation phase of protein synthesis, the ribosome moves along the mRNA molecule, serving as a workbench where amino acids are combined to form a polypeptide chain. The movement begins at the 5' end of the mRNA and terminates at the 3' end.

transfer RNA

a molecule of RNA that carries an amino acid molecule to the ribosome for use in protein synthesis.

translation

the biochemical process involving ribosomes in which an mRNA molecule provides a nitrogenous base code for the placement of amino acids in the synthesis of a protein.

Transfer RNA molecules exist in a form that vaguely resembles a cloverleaf shown in Figure 3.12. At one end of the molecule a sequence of three nitrogenous bases (a triplet) complements a triplet of bases somewhere on the mRNA molecule (for example, CGA of a tRNA molecule complements the GCU triplet on an mRNA molecule). The complementary triplets are an essential factor in correctly placing an amino acid into a protein chain. The triplets translate the genetic code of DNA into an amino acid sequence in proteins.

■ TRANSLATION

DNA is often called the "blueprint for life," which suggests that DNA is a set of chemical construction plans for building an organism. Perhaps this concept is oversimplified because reading the DNA code for a body part would hardly help one understand what that body part looks like. (For example, reading C-G-G-A-T-T-A-C-G-T-A-C would give us scant insight to how a nose looks or what it does). It would probably be more correct to consider DNA the starting point for a profoundly important chemical process in which a body part is constructed from available amino acids. In a context

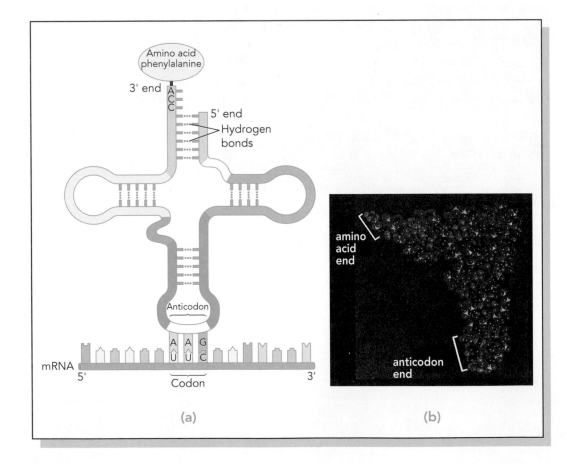

(a)   (b)

FIGURE 3.12

Structure of the transfer RNA (tRNA) molecule. (a) The tRNA molecule vaguely resembles a cloverleaf in two dimensions, although it is twisted in its active three-dimensional form. A chain of nucleotides, each with its nitrogenous base, makes up the molecules. Weak chemical bonds, called hydrogen bonds (shown by the dotted lines), connect various portions of the molecule. At the upper end of the molecule at the 3′ end is the amino acid attachment site. At the lower portion of the molecule, there is the key recognition site called the anticodon. During protein synthesis, the anticodon pairs with its complementary codon on a strand of mRNA. The pairing brings that amino acid into position. (b) A space-filling molecule of the tRNA molecule.

such as this, the process has more significance than the players. In essence, the language of the nitrogenous base code of DNA will be "translated" into the language of an amino acid sequence in a protein. In the paragraphs to follow, we will see what this process entails.

For organizational purposes, the actual mechanism of protein synthesis can be envisioned as beginning in the cell nucleus with the production of a molecule of mRNA. As noted previously, the enzyme RNA polymerase catalyzes mRNA formation from the sense strand of DNA in the transcription process. Figure 3.13 demonstrates this process. In transcription, the genetic code of DNA is used as a template, and the three-base triplets in DNA serve as a model to construct an mRNA molecule with complementary triplets. For example, a three-base C-G-T sequence in DNA would induce the enzyme to place the three-base G-C-A sequence in RNA. Each three-base sequence in RNA (e.g., G-C-A) is called a **codon** because it encodes a particular amino acid. The mRNA molecule is a series of codons formed as RNA polymerase sweeps down the DNA template. Figure 3.14 illustrates the complementary placement of bases.

But the mRNA is not yet ready to carry the genetic message. In complex (eukaryotic) cells, such as plant, animal, and human cells, the primary mRNA is processed by biochemically snipping out certain portions of "genetic gibberish." These segments of RNA apparently are not used for protein synthesis and are discarded by the cell. The segments are called **introns** because they are intervening sequences in the RNA (the name was first suggested by biochemist Walter Gilbert of Harvard University). The remaining mRNA contains the segments to be translated to an amino acid sequence. These segments are called **exons** because the codons will be expressed, as Figure 3.15 displays. The exons are chemically spliced together to form the "mature" mRNA molecule.

codon

a three-base sequence on an mRNA molecule that specifies a particular amino acid.

eukaryotic

u″kar-e-ot′ik

pertaining to a complex organism whose cells have a nucleus and organelles, multiply by mitosis, and have other features that separate them from more simple prokaryotic cells.

intron

an intervening section of mRNA that is removed before the production of the final mRNA molecule

exon

a section of mRNA that specifies an amino acid sequence and that is retained during the production of the final mRNA molecule.

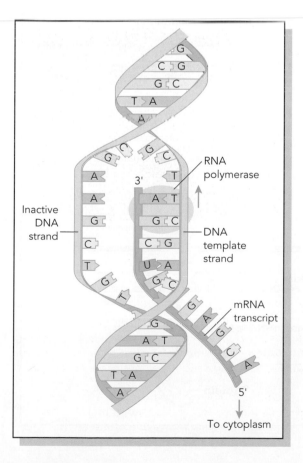

Inactive DNA strand

RNA polymerase

DNA template strand

mRNA transcript

3′

5′

To cytoplasm

FIGURE 3.13

The process of transcription. The enzyme RNA polymerase moves along one strand (the template strand) of the DNA molecule and synthesizes a complementary mRNA molecule (the RNA transcript), using the base code of the DNA strand as a guide. A nucleotide sequence called the promoter (not shown) signals the start of the transcription, and a sequence called the terminator signals its completion. The synthesis proceeds from the 5′ end of the RNA to its 3′ end. Note that the inactive strand of DNA is not transcribed. The mRNA molecule will move toward the cytoplasm after processing.

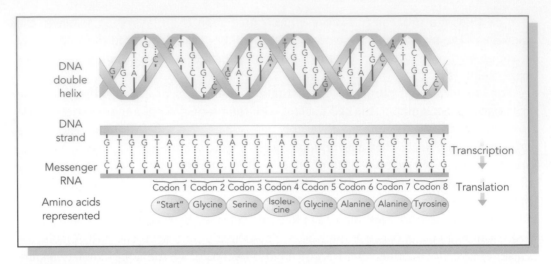

FIGURE 3.14

A broad view of protein synthesis. The DNA molecule unwinds, and the coding strand of DNA is transcribed to messenger RNA. The mRNA then operates as a series of codons, each codon containing three bases. During translation, a particular codon specifies a certain amino acid for placement in the protein chain during translation. Because codons 6 and 7 are identical, alanine molecules occur next to one another in the protein. Codons 2 and 5 (identical) encode glycine.

FIGURE 3.15

The formation of mRNA. A gene consists of exons, the parts of the gene expressed as protein, and introns, the intervening sequences between the exons. In the formation of mRNA, the gene is transcribed to a preliminary mRNA molecule. Then the introns are removed biochemically and the exons are spliced together. This activity results in the functional mRNA molecule, which is then ready for translation. This type of processing does not occur in mRNA production in prokaryotic cells such as bacterial cells; it occurs only in eukaryotic cells such as plant animal, and human cells.

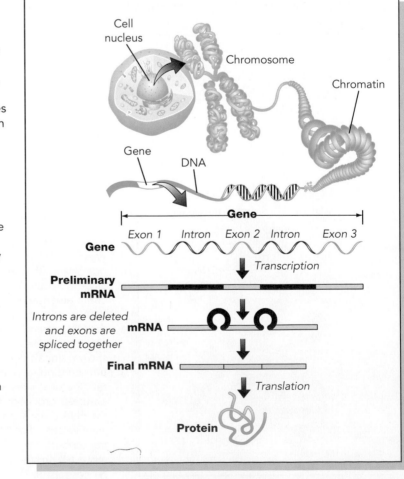

The existence of introns and exons was a novel observation in DNA biochemistry because it showed that genes can be split over a broad area of the DNA molecule. The concept of split genes was first demonstrated in 1977 by Richard Roberts and his colleagues at Cold Spring Harbor Laboratory and by Philip Sharp and his coworkers at Massachusetts Institute of Technology. For their work, the two researchers shared the 1993 Nobel Prize in Physiology or Medicine.

In simple (prokaryotic) cells, there are no split genes and no intron-exon processing. Here, the initial mRNA encoded by the DNA is identical to the final mRNA. We will encounter this biochemistry again in discussions of the production of pharmaceutical products (Chapters 5 and 6) and in the description of DNA fingerprinting (Chapter 9).

Still other modifications concern the nucleotide sequences in mRNA. For example, when the mRNA is only about twenty nucleotides long, a "**cap**" is added to the molecule's front end. This cap consists of an unusual nucleotide called 7-methylguanosine. (Figure 3.16 shows its structure.) The cap is essential for mRNA to bind correctly to the ribosome.

Another modification occurs at the end of transcription as the mRNA is being released from the DNA template. At this point, special enzymes add a string of adenine-containing nucleotides to the tail of the mRNA molecule. This added segment is called the **poly-A tail**. It may contain up to 200 nucleotides containing adenine. Although its function is not certain, the poly-A tail has value to the DNA technologist because it provides a method for isolating mRNA molecules from a complex mixture. This procedure is discussed in Chapter 5.

prokaryotic

pro″kar-e-ot″ik

pertaining to a more simple organism whose cells lack a nucleus or organelles, multiply by simple fission, and have other features that separate them from more complex eukaryotic cells; bacteria are prokaryotic organisms (prokaryotes).

poly-A tail

a sequence of adenine-containing nucleotides at the end of an mRNA molecule.

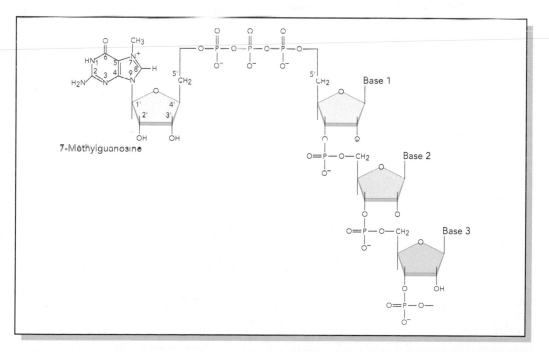

FIGURE 3.16

The cap of mRNA. At the beginning of transcription a "cap" is biochemically placed at the front end of the mRNA molecule. The cap consists of a molecule of 7-methylguanosine. The molecule resembles a guanine-containing nucleotide, except that a methyl group occurs at the No. 7 position of the guanine portion. Three phosphate groups instead of the usual one connect the cap to the remainder of the mRNA molecule. When the molecule of mRNA is complete, an enzyme adds a "tail" at the far end consisting of several adenine-containing nucleotides (the "poly-A tail").

In a eukaryotic cell, the mRNA molecule now moves through a pore in the nuclear membrane into the cellular cytoplasm. Here it complexes with one or more ribosomes. Because the mRNA molecule often contains thousands of nucleotides, only a small portion of the molecule contacts the ribosome at a given time, as Figure 3.17 depicts. Furthermore, several ribosomes may be using the genetic message in one mRNA molecule simultaneously, and several molecules of a certain protein may be forming at the same time.

While the mRNA-ribosome complex is evolving, activity is taking place elsewhere in the cytoplasm as different amino acids join with their specific tRNA molecules. These unions require considerable chemical energy and a series of highly specific activating enzymes, one for each amino acid. Each of the twenty different tRNA molecules has a distinctive three-base sequence at one end of the cloverleaf pattern. This three-base sequence (e.g., C-G-U) is called an **anticodon** because it will complex with a codon complementary to it in the mRNA.

Once bound to their amino acids, the different tRNA molecules transport their biochemical cargo to the ribosome, where mRNA is stationed. Now one portion of the mRNA molecule attaches to the 30s subunit, and a tRNA molecule with its amino acid attaches to the 50s subunit. In this carefully orchestrated step, the codon of the mRNA attracts a complementary anticodon on the tRNA. The codon-anticodon matching brings a specified amino acid into position. The matching thus denotes the amino acid's location in the protein chain. At this precise moment, the genetic code of DNA is

anticodon

a three-base code on a tRNA molecule that complements the codon on a mRNA molecule..

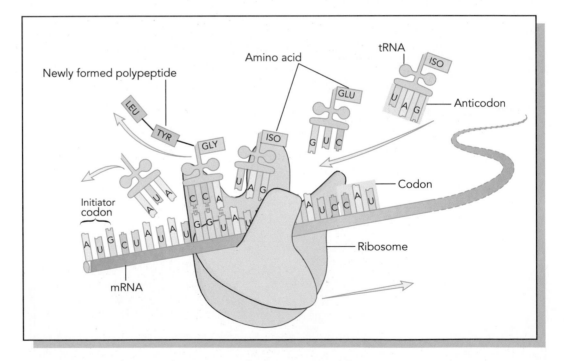

FIGURE 3.17

Translation. The ribosome moves along the mRNA molecule from codon to codon, as tRNA molecules bring their amino acids into position. Each amino acid is then attached to the growing protein chain. Once it has given up its amino acid, the tRNA molecule moves back into the cytoplasm to unite with another amino acid molecule. Meanwhile, the ribosome moves to the next codon and receives the next tRNA with its amino acid. Note that the codon in mRNA and anticodon in tRNA contain complementary bases. This complementary pairing specifies which amino acid is slotted into which position in the growing protein chain. Also note that tRNA molecules with different anticodons unite with different amino acids specified by their abbreviations.

expressed as the location of an amino acid in a protein chain.

After pairing with mRNA, the tRNA-amino acid is held in a viselike grip on the ribosome's larger subunit. The ribosome then moves along the mRNA to a new location. Here, a second tRNA with its amino acid approaches the ribosome and pairs its anticodon with the second codon on the mRNA molecule. Thus, two tRNA molecules and their amino acids stand next to one another on the mRNA. In a millisecond, an enzyme from the 50s subunit of the ribosome joins the amino acids together to form a **dipeptide** (two amino acids in a chain). The first tRNA is now free of its amino acid, and it moves back to the cytoplasm, leaving its amino acid behind and joined to the second amino acid.

Now the ribosome moves to a third location at the third codon of the mRNA. A new tRNA with its amino acid enters the picture. It matches its anticodon with the complementary codon on the mRNA, thereby ferrying its amino acid into position. Once again, an enzyme binds the new amino acid to the first two, thereby forming a **tripeptide** (three amino acids in a chain). The second tRNA is freed of its amino acid, then discharged back to the cytoplasm. Codon after codon is used as the ribosome moves along the mRNA, and amino acid after amino acid is joined to the growing chain, now called a **polypeptide**. By a one-by-one addition, the peptide increases in length. The translation process is now in full swing.

The final one or two codons of the mRNA are chain terminator or **"stop" signals**. As these codons (UAA, UAG, or UGA) are reached, no complementary tRNA molecules exist, and no amino acids are added to the chain. Instead, the stop signals activate release factors to discharge the polypeptide chain from the ribosome. Now the polypeptide will coil to yield the functional protein, or the protein may consist of several polypeptide chains such as in hemoglobin, the oxygen-carrying protein of RBCs. Whatever the nature of the protein, the message of the DNA has now been expressed in an amino acid sequence in the protein, and the process is complete. Figure 3.18 shows the overall process.

peptide

a sequence of amino acids that can be further processed or configured to yield a protein molecule; a dipeptide contains two amino acid molecules.

polypeptide

a peptide containing numerous amino acids.

■ GENE CONTROL

It should be obvious to even the most casual observer that all the genes are not operating all the time. There is little point for a digestive cell to produce a digestive enzyme when the target of the enzyme's activity (e.g., a particular food) is not present. Shutting down the gene and halting enzyme production save energy and eliminate the need for storing the gene product (Questionline 3.3).

In the broader sense, the growth and development of an organism entails a long series of intricate biochemical steps, each delicately tuned to achieve a precise effect. Specific enzyme activities are often called into play to catalyze a specific biochemical change. Once the change has taken place, the enzyme flow ceases and other activities ensue. Gene transcription occurs only for a specified period of time.

In most cases, the control of gene expression involves control of transcription at the level of the gene. The site of control is usually a sequence of nucleotides called the **regulatory site**. In many cases, a specific regulatory protein, a so-called **repressor protein**, will bind to the regulatory site and exert its controlling influence. This process is called **repression**.

When a repressor protein reacts with a regulatory site to inhibit transcription, the control mechanism is referred to as a **negative control**. Negative control often takes place between points where RNA polymerase binds and the gene for transcription begins (Questionline 3.3). By binding to the site, the repressor protein prevents the movement of RNA polymerase toward the gene. Without RNA polymerase, the gene cannot be transcribed. Placing a log across a railroad track would have a similar effect.

regulatory site

a sequence of nitrogenous bases where gene expression can be controlled by reaction with repressor or activator proteins.

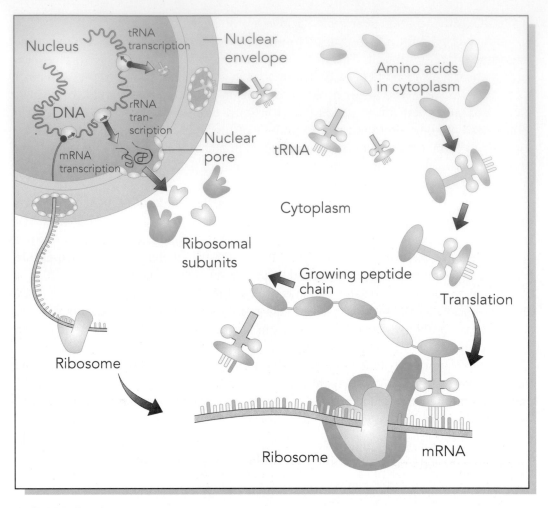

FIGURE 3.18

Gene expression. A summary of the events taking place during the expression of the gene in the synthesis of protein in the cell. This diagram summarizes many of the previous diagrams in this chapter. How many different processes can you see occurring?

activator protein

a protein that reacts with a regulatory site and encourages gene expression.

Gene expression can also be controlled by a type of **positive control**. In this case, the regulatory protein encourages (rather than inhibits) gene transcription. The regulatory protein is called an **activator protein**, and the process is called **activation**. Activation takes place when the activator protein binds to the regulatory site and stimulates unwinding of the DNA helix to encourage mRNA formation. As usual, the mRNA formation is directed by RNA polymerase, but the enzyme operates more efficiently once the DNA has been unwound. Thus, the process has been "activated."

Understanding activation and repression give one a clue to how genes can be turned on and off, but it does not show how cells control which genes are activated or which are repressed. Research evidence indicates that this level of control is regulated by the shape of the regulatory protein. The change in shape can enhance or destroy the ability of a regulatory protein to bind to the regulatory site. For example, in its new shape, the regulatory protein may recognize a binding site not recognized previously, or the newly configured protein may be unable to bind to a site where binding was previously possible.

Various levels of control are necessary for a cell to avoid metabolic anarchy. Genes must be tuned into the intracellular environment and must be switched on and off as conditions demand. Even a simple bacterium may have as many as 2500 genes, and

QUESTIONLINE 3.3

1. **Q.** Why is it important for cells to control the expression of their genes?

   A. Humans have the same 100,000 genes in each cell of their bodies. But the cells perform different functions, and certain genes are operative whereas others are dormant at any one time. This is why controls are necessary. Also, there are times when gene products are not required. For example, there is little sense to having digestive enzymes available if we are not eating.

2. **Q.** Are there different kinds of genes?

   A. Genes tend to be quite complex. There are structural genes that specify a protein. Adjacent to these are segments of DNA called promoter and operator sites, which initiate the activity of structural genes and regulate their performance, respectively. Somewhat distant from the structural genes is a regulator gene. This gene controls the operator site through regulatory proteins it encodes.

3. **Q.** I've occasionally heard the word operon used in connection with gene activity. What is an operon?

   A. An operon is a complex set of operational genes. It involves the myriad components of gene expression and regulation that lead to the synthesis of a protein. The *lac* operon in bacteria represents a well-studied operon.

precise control of their expression is an essential feature of the organism's life. Consider, for example, how dramatically the living conditions change from moment to moment for a bacterium living in the human gut: The bacterium must adapt constantly (for example, to what we eat), and controlling its gene activity helps it adapt and survive.

■ GENE COMPLEXITY

As this chapter has amply demonstrated, gene activity is an intricate series of recognitions, controls, regulations, and biochemical changes that eventually lead to the synthesis of a protein. Before DNA technology could emerge as an skill unto itself, scientists had to understand the gene and study its complexity. This was no easy matter, however, because the gene is more than a simple coding mechanism. A specific example of a single gene system will illustrate its complexity.

Bacteria produce **beta-galactosidase**, an enzyme that acts on the carbohydrate lactose and breaks it down to constituent parts (glucose and galactose). Production of the enzyme is directed by a gene system that has been studied since the 1950s and about which much is known. The accumulated knowledge regarding this gene indicates that at least five different regions make up the gene complex. The total gene complex is called the **operon** because the genes "operate" together (in this case, the operon is the *lac* operon because it functions in lactose digestion).

We will begin with the genes that actually encode beta-galactosidase, the enzyme. These genes are called **structural genes** because they encode the structure of the enzyme protein. In bacteria there are three structural genes for three enzymes in the enzyme system involved (we will treat them as a unit and refer to a single enzyme). All the genes are located next to one another on the bacterial chromosome, as Figure 3.19 shows. This circumstance is not found in all cells. Indeed, in complex eukaryotic cells, the structural genes are usually found on different chromosomes or, if on the same

galactosidase

ga``lac-tos´i-das

lactose

a disaccharide composed of molecules of glucose and galactose.

operon

the complex of genes and other regulatory regions of DNA that function together to bring about gene expression.

ribosomal binding site

a series of nitrogenous bases in the operon that specify the base code in mRNA to permit the mRNA to bind to the ribosome.

chromosome, there may be intervening sequences of nucleotides.

Next to the structural genes is a series of nitrogenous bases responsible for binding mRNA to the ribosome. Known as the **ribosomal binding site**, this base sequence encodes the base sequence in mRNA that ensures its ability to unite correctly with the ribosome. No genetic message for protein is present at this site.

The next site encountered as we move further away from the structural genes is the recognition site for RNA polymerase. This site is called the **promoter** because binding to the site promotes transcription. RNA polymerase binds to the promoter then moves to the right until it encounters a special "start transcription" signal (the sequence TAC) at the beginning of the structural genes. This action was mentioned earlier in the chapter.

Lying between the promoter and the structural genes, we encounter the **operator**. This is the regulatory site for repression or activation of the gene. An important level of control takes place here because RNA polymerase must pass through the operator region to get to the structural genes. For instance, when the base sequence of the

ribosomal binding site

a series of nitrogenous bases in the operon that specify the base code in mRNA to permit the mRNA to bind to the ribosome.

operator

a series of nitrogenous bases in the operon where the regulation of gene expression by the repressor or activator proteins occurs.

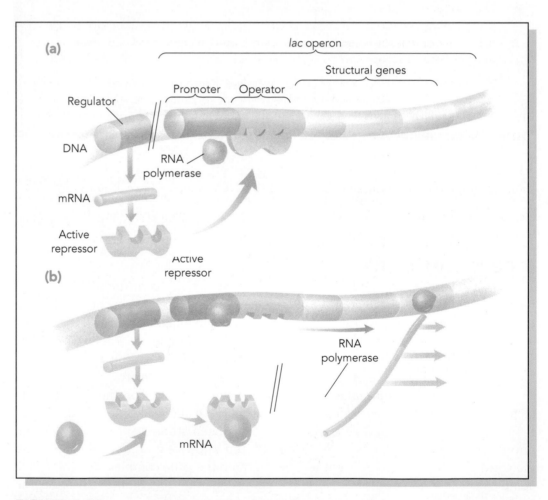

FIGURE 3.19

An example of control of gene expression as exemplified by the *lac* operon. (a) A regulator gene encodes an mRNA molecule, whose code is translated to a repressor protein. The repressor protein binds to the operator gene in the regulatory region of a gene. The enzyme RNA polymerase normally passes through the regulatory region, but with the repressor protein in place, the enzyme cannot move through, and gene transcription is halted. (b) Certain molecules called inducers can remove the repression by uniting with the repressor protein. In the *lac* operon, a lactose molecule unites with the repressor protein. This union inactivates the repressor and removes it from the regulatory site. Now the RNA polymerase is free to begin transcription of the gene, and enzyme molecules result.

operator binds with repressor protein, the RNA polymerase cannot pass the blockage, and the structural genes do not function. Still another promoter is located in front of the operator in the bacterial gene complex. This promoter facilitates unwinding of the DNA before RNA polymerase acts.

Another level of complexity occurs in the gene that produces the repressor protein. This gene is spatially removed from the operon at a different location on the chromosome. The gene encodes a **repressor protein**, which binds to the operator (the regulatory site) and shuts off the production of enzyme. When lactose appears in the environment, however, the repressor protein unites with lactose molecules (the signal molecules), and the repressor protein takes on a new shape. Now it is unable to hold onto the DNA sequence in the operator, and the repression is lifted. There follows a burst of synthesis of beta-galactosidase. Thus, the appearance of lactose molecules encourages the synthesis of the enzyme necessary for the digestion. When the lactose is used up, the repressor protein changes back to its original shape. It then complexes to the operator and biochemically shuts off the production of beta-galactosidase.

repressor protein

a protein that reacts with a regulatory site and restricts expression of the gene by inhibiting transcription.

Although involved and quite complex to the beginning student of DNA activity, the operation of the gene system in bacteria (a prokaryotic organism) is relatively simple compared with the gene systems in complex eukaryotic cells. One reason is that many more genes exist in eukaryotic cells and more regulation of expression is required. A human cell, for example, is estimated to have more than 100,000 genes (vs. about 2500 for a bacterium). Also a eukaryotic organism will probably have multiple cell types and, consequently, vastly different patterns of gene expression. For example, human liver cells and human brain cells function quite differently. Moreover, a eukaryotic organism undergoes many forms of development (such as embryonic, fetal, adolescent, and adult development), and the different cells at different stages of life require regulation of gene expression at varying stages of life.

Biochemists have discovered that eukaryotic organisms have noncoding, nucleotide sequences called **enhancers**, which have a powerful influence on gene expression. Enhancers can be thousands of bases away ("upstream") from the promoter region. Special proteins called **transcription factors** attach to the enhancers and appear to induce the DNA to bend into a loop. This bending brings the enhancer into contact with the promoter further down ("downstream") on the DNA molecule. Different transcription proteins keyed to different enhancer sequences could activate different genes during the development of the cell. Hundreds of transcription factors have been discovered thus far.

enhancer

a series of nitrogenous bases that encourages DNA activity at a distant promoter site.

Nor is regulation of gene expression confined to control at the time of transcription (Figure 3.20). Researchers have established that regulation can occur at multiple levels of gene expression. It can occur, for instance, in how mRNA molecules are processed before leaving the nucleus, in the transport of mRNA molecules out of the nucleus, where mRNA molecules bind to the ribosome, and in the activity and stability of the protein products of gene expression

These multiple and complex levels of regulation often confound scientists as they attempt to reproduce in a test tube the gene expression taking place in the cell. Regardless of their complexity, the regulatory mechanisms must be considered, understood, and resolved if DNA technology is to move ahead. We tend to relegate the control and regulatory aspects of DNA action to secondary importance; yet they are as important to gene activity as transcription and translation. Grappling with the mechanics of gene control remains a major task that DNA technologists must confront.

FIGURE 3.20

Levels of regulation of gene expression. Gene expression can be regulated at multiple points during the process of protein synthesis. (1) DNA transcriptional controls (e.g., repression of gene activity); (2) Processing of RNA transcripts (e.g., removal of introns and preservation of exons). (3) Transport of RNA molecules out of the nucleus. (4) Digestion of the used mRNA molecule for reuse of the components. (5) Translation of the mRNA code to an amino acid sequence in protein. (6) Modification of the protein to its final functional form.

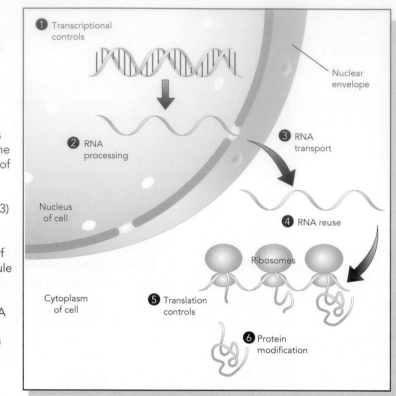

SUMMARY

The realization that chromosomes are involved in patterns of heredity led scientists to investigate the role of DNA in living systems and to study how that role is performed. Garrod's studies indicated that enzymes were related to gene activity and the Beadle-Tatum experiments indicated that a gene regulates the production of an enzyme. This concept was broadened by Ingram's experiments, which showed that genes regulate protein production and that a gene mutation affects the placement of a single amino acid in the protein. Evidence also accumulated that RNA is the intermediary between the biochemical message in DNA and the protein's amino acid sequence. The work of Crick's group was instrumental in identifying a three-base sequence in DNA as the genetic code.

Modern biochemists now recognize that DNA encodes three forms of RNA: messenger RNA (mRNA), whose codons transport the genetic message; ribosomal RNA (rRNA), which is used to construct ribosomes; and transfer RNA (tRNA), which transports amino acids to the ribosomes and whose anticodons complement the codons in mRNA. Transcription is the synthesis of mRNA using the genetic message in DNA as a template, whereas translation is the transfer of the genetic message in mRNA to a carefully orchestrated amino acid sequence in protein.

Gene expression is controlled by regulatory proteins and is required to preserve cell energy and chemical resources. Negative control can occur when a regulatory protein binds to a regulatory site and inhibits transcription. Positive control develops when an activator protein stimulates transcription to take place, such as by encouraging the unwinding of the DNA helix.

The complexity of DNA in action is illustrated by the activity of a cluster of genes called an operon. Studied in bacteria, the operon consists of structural genes that encode an enzyme, a nitrogenous base series that controls ribosome binding, a promoter site for RNA polymerase binding, and an operator gene for repressing or activating the structural gene. This complex setup of genes is further involved because the repressor or activator protein is encoded by a gene at a distant site. Compared with human systems the bacterial system is relatively simple. Such systems must be understood and controlled if experiments in DNA technology are to move ahead.

REVIEW

These pages have focused on the mechanisms by which the genetic message in DNA is expressed in a protein. To assess your understanding of the discussions, answer the following questions:

1. Briefly summarize the insight of Garrod and the work of Beadle and Tatum and show why both were important to the evolving interest in DNA.

2. What did Ingram's experiments indicate and why were his results significant?

3. What evidence pointed scientists to RNA's role in protein synthesis, and how did Crick's experiments verify the three-base coding sequence in the DNA molecule?

4. Name three types of RNA that participate in protein synthesis and specify the function of each type.

5. Describe the synthesis of RNA taking place in transcription, and specify how mRNA is modified before its release from the nucleus of a eukaryotic cell.

6. Explain the structure and function of transfer RNA molecules as they occur in the cell's cytoplasm.

7. Summarize the activities of translation taking place at the ribosome once the mRNA molecule and the tRNA molecules have arrived.

8. Compare negative and positive controls that may influence the expression of the gene. Use the processes of repression and activation as guides for your comparison.

9. Using the gene for lactose-digesting enzyme as an example, explain the various levels of complexity that must be considered to understand gene expression.

10. Describe in broad terms why gene controls are necessary in a cell and why regulations of expression are more complex in human cells than in bacterial cells.

FOR ADDITIONAL READING

Beardsely, T. "Smart Genes." *Scientific American* (August, 1991).

Becker, W. M., et al. *The World of the Cell*. Menlo Park, CA: The Benjamin/Cummings Publishing Co., 1996.

Cooper, G. *The Cell: A Molecular Approach*. Washington, DC: ASM Press, 1997.

Fox, M. "Breaking the Genetic Code in a Letter to Max Delbruck." *Journal of College Science Teaching* (March/April, 1996).

Freedman, D. H. "Life's Off-switch." *Discover* (July, 1991).

Judson, H. F. *The Eighth Day of Creation*. New York: Simon and Shuster, 1979.

Watson, J., and Tooze, J. *The DNA Story*. San Francisco: W. H. Freeman and Co., 1981.

Foundations of DNA Technology

 CHAPTER OUTLINE

LOOKING AHEAD

This chapter focuses on the discoveries and developments that laid the foundations for the practical applications of DNA technology. When you have completed this chapter, you should be able to:

- understand the nature of the biochemical tools used in DNA technology and recognize why microorganisms occupy an important place in the experiments of DNA technology.

- discuss the key discoveries and experiments that formed the basis for modern DNA technology.

- identify some of the individuals whose work laid the foundations for DNA technology.

- conceptualize the basic processes used in gene splicing and DNA technology and prepare for the chapters ahead.

- discuss briefly some of the theoretical and practical applications of DNA technology and acquire an overview of the implications of this skill.

- recognize some important terms and concepts used in DNA technology and broaden your vocabulary of this science.

INTRODUCTION

During the 1950s and 1960s, scientists made substantial gains in molecular biology as they clarified the role of DNA in the biochemistry of protein synthesis and explained the intricate details of this process (Chapter 3). In the 1970s, they set aside their tendency to observe DNA in action and devised methods to cut and splice DNA fragments to produce recombined DNA molecules (as we will see in this chapter). Next, they induced the recombined DNA molecules to do their bidding, and thus, they introduced the era of DNA technology.

With the advent of DNA technology, there was a new frontier to explore, a frontier where the wildest imaginations could be accommodated. Researchers discovered, for example, that they could transplant animal genes into bacterial cells and coax the genes to function in this new environment; they kindled hopes that plants, noted for their ability to produce carbohydrate, could be transformed to produce proteins; and they laid the foundations for the study of treatments for genetic diseases. In addition, the future held promise for inexpensive sources of bioenergy, mass production of pharmaceuticals, and novel types of vaccines. There appeared to be no limit to what could be accomplished.

Today, in the twenty-first century, much of what scientists predicted has come to pass. The public has already witnessed how gene-altered bacteria can produce insulin for human needs (Chapter 6) and how human hemoglobin can be manufactured in animals (Chapter 11). The first gene therapy experiments have begun (Chapter 8), and significant progress in agricultural biotechnology has been achieved (Chapter 10). DNA technology and the general field of molecular biology continue to have staggering potential for improving pharmaceutical yields, diagnosing and treating human disease, and creating new products for home and industry.

At the same time, DNA technology has triggered a certain degree of public concern and confusion. It has created dilemmas for governmental and regulatory agencies, and it has raised fears that genetically altered organisms could pose a threat to human welfare. We will discuss many of these problems in the chapters ahead as we consider the foundations for DNA technology and the practical applications of this awesome skill. For the present, we will explore the thought processes and discoveries that laid the foundations for modern DNA technology. The fruits of this technology are the subject matter for the remainder of this book.

THE TOOLS OF GENETIC ENGINEERING

Humans have guided the flow of genes for thousands of years without realizing it. They have bred useful agricultural plants from wild ancestral stock and have invented new organisms by crossing ones that ordinarily do not mate. A nectarine, for example, is a cross between a plum and a peach, and a mule is derived from breeding a horse with a donkey. Figure 4.1 shows several other products of human ingenuity.

But the genetic alterations of DNA technology go well beyond those of the past because scientists can now intervene directly in the genetic fate of organisms. Contemporary experiments are much more exotic, and importantly, they can be performed in test tubes. By taking DNA fragments from different species and splicing the fragments together, researchers can produce novel forms of DNA and change the character of organisms so profoundly that new species emerge. For example, a gene-altered, insulin-producing bacterium is no longer identical to its native species; it could be considered a new species of organism.

---

recombined DNA

a DNA molecule altered with a segment of foreign DNA.

insulin

a pancreatic hormone that facilitates the passage of glucose molecules from the bloodstream into the body's cells.

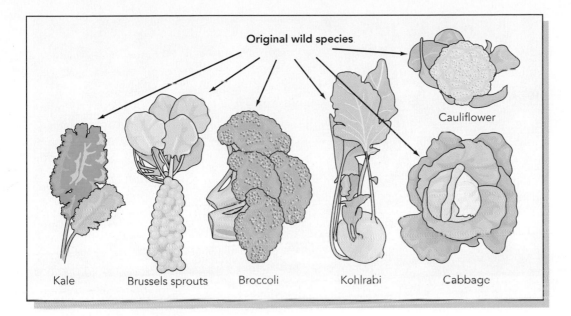

FIGURE 4.1

Guiding the flow of genes. Through the centuries, humans used a cabbagelike wild species of plant to breed all these modern plants. Depending on the climate of cultivation and the desires of the local population, some varieties were bred to be hard-headed as in a modern cabbage, some formed masses of flower buds as in broccoli and cauliflower, and some made clusters of leaf buds as in Brussels sprouts.

During the 1960s and 1970s the new technology involving DNA required new approaches to science, new insights, and new materials. Ways of cutting and splicing DNA fragments had to be found, organisms had to be enlisted to carry the new fragments, and biochemical methods had to be developed to permit expression of the DNA fragments. The times called for flamboyance and breadth of imagination. In the pages ahead we will explore some products of that imagination as we survey the breakthroughs that formed the bases for the new science of DNA technology.

■ THE ORGANISMS

The new DNA technology has been made possible by seminal work with viruses and bacteria, such as that pictured in Figure 4.2. These organisms can be cultivated easily, and their biochemistry can be studied in convenient test tubes. By far the most popular bacterium used in DNA technology has been ***Escherichia coli***. The widespread use of this microscopic rod is probably an accident of history. Having studied the organism for decades, molecular biologists know more about its biochemistry, morphology, physiology, and genetics than any other organism (including humans). *E. coli* has become the workhorse for experiments in DNA technology because it is easy to cultivate and, except for certain strains, it is not regarded as a serious human pathogen (Questionline 4.1) .

The **bacterial chromosome** is particularly well suited for DNA experiments because it consists solely of DNA. In the more complex eukaryotic cells (such as animal cells), a major portion of the chromosome is protein, which is intimately combined with the DNA. Moreover, because there is only one chromosome in a bacterium, its genes can express themselves without being affected by a second chromosome. (Eukaryotic chromosomes, by comparison, occur in pairs.)

Another factor contributing to the use of the bacterial chromosome is that it usually

Escherichia
esh′er-ik′e-ah

eukaryotic
u′′kar-e-ot′ik

pertaining to a complex organism whose cells have a nucleus and organelles and multiply by mitosis and have other features that separate them from more simple prokaryotic cells.

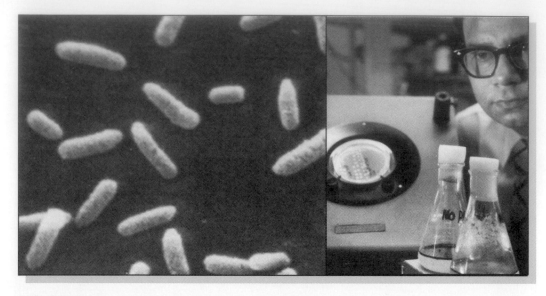

FIGURE 4.2

Bacteria in service to humanity. This is an electron micrographic view of the bacterium *Pseudomonas*, a bacterial rod. The investigator pictured in the insert, Anand Chakrabarty, has engineered *Pseudomonas* to produce enzymes to decompose oil and assist environmental cleanups of oil spills.

genetic code

the sequence of nitrogenous bases on the DNA molecule that specifies an amino acid sequence in a protein.

ultramicroscopic

below the ability of the light microscope to visualize; requiring an electron microscope.

genome

the total DNA component of all the genes of a cell or a single-celled microorganism; includes DNA from the nucleus and cytoplasmic organelles.

floats freely in the cytoplasm and is easy to reach. Eukaryotic chromosomes, by contrast, are enclosed in the nucleus of the cell within a nuclear membrane. Because the genetic code is nearly universal (Chapter 3), the bacterium can express foreign DNA from any organism as long as the foreign DNA is attached to the bacterial DNA. Indeed, a foreign gene attached to a bacterial chromosome is replicated and transcribed in exactly the same way as the native bacterial DNA.

Viruses are equally useful for the experiments in DNA technology. These ultramicroscopic particles consist of minuscule fragments of RNA or DNA enclosed in a protein coat and, in some cases, a surrounding envelope of lipid. (Note: Figure 4.3 displays several viral particles.) Viruses replicate only in living cells, as Chapter 1 explains. They shed their protein coats and use the cell's molecular machinery to produce new viruses. The DNA of a DNA virus acts as a gene and directs the synthesis of new viral parts; the RNA of an RNA virus often acts as a messenger RNA molecule to provide codons for enzyme proteins and viral structures.

Sometimes the virus does not replicate itself immediately. Instead, it integrates itself into the host's chromosome and becomes part of the cell's genome. The DNA of a **herpesvirus**, for example, integrates into a nerve cell genome and remains with that cell for years, causing recurring herpes infections in the body. Another process occurs with the **human immunodeficiency virus (HIV)**. The RNA from this virus serves as a template for the synthesis of DNA. The DNA now inserts itself to the host cell genome as Figure 4.4 displays. (The individual is said to have HIV disease.) From this position, the DNA encodes new viral particles. The ability of viruses to insert to host cell genomes drew the attention of DNA technologists as a way of bringing genes into a cell, as Chapter 5 notes.

Numerous organisms are used in DNA technology, as well. They include yeast cells, various other species of bacteria and viruses, insect and mammalian tissues, and a number of plants. We will mention their names and significance as we encounter them in succeeding chapters.

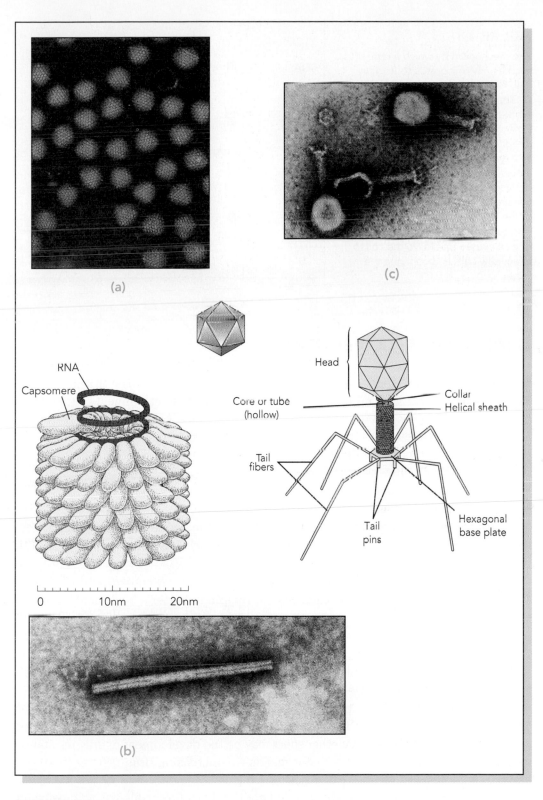

(a)

(c)

RNA

Capsomere

Core or tube
(hollow)

Head

Collar
Helical sheath

Tail
fibers

Tail
pins

Hexagonal
base plate

0   10nm   20nm

(b)

FIGURE 4.3

The structure of viruses. (a) A plant virus (turnip mosaic virus) displaying icosahedral symmetry
(with a model of the icosahedron). (b) Another plant virus (tobacco mosaic virus) displaying
helical symmetry, with a model of the virus. (c) A bacteriophage (bacterial virus) together with its
model of the virus.

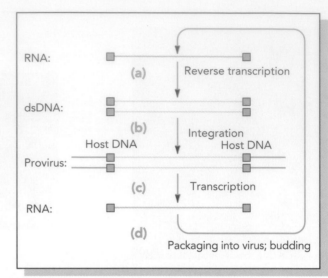

FIGURE 4.4

The mechanism by which a retrovirus such as HIV integrates to its host cell. Retroviral nucleic acid is RNA. (a) The enzyme reverse transcriptase synthesizes a linear double-stranded DNA molecule (dsDNA), which (b) integrates into the host cell genome as a provirus. (c) At some later point (hours, days, weeks, or longer), host-derived RNA polymerase transcribes the provirus to form RNA genomes of the retrovirus. (d) Once packaged with protein, the genomes form a set of new viruses that leave the cell to infect new cells and continue the cycle.

MICROBIAL RECOMBINATIONS

Research conducted in the 1950s demonstrated conclusively that genetic recombination occurs among bacteria and may even involve viruses. The story began with Griffith's experiments with bacteria in 1928 (Chapter 1). These experiments showed that genetic recombinations were possible and led to Avery's identification of DNA as the molecule responsible for recombination. Then, in the 1950s, bacteriologists discovered three forms of recombination by which bacteria change their genetic constitutions. We shall examine those three processes next.

One of the more remarkable observations of the 1950s was that it is not unusual for bacteria to take up genetic material from molecular debris in the environment. The phenomenon came to be known as **transformation**. Transformation takes place in less than one percent of a bacterial population, and it can bring about profound genetic changes.

During transformation, a number of **donor bacteria** break apart, and their DNA explodes out of the cells into fragments. Should **recipient bacteria** be present, segments of double-stranded DNA containing about 10 to 20 genes pass through their cell walls and membranes. Enzymes dissolve one strand of the DNA, and the remaining second strand displaces a segment of single-stranded DNA in the recipient's chromosome. To make the transformation complete, the foreign genes express themselves during protein synthesis. Figure 4.5 summarizes this process.

Under natural conditions, transformation takes place in organisms whose DNAs are very similar. One of the effects is to increase the pathogenicity of the recipient organism (as in Griffith's pneumococci). Another effect may be the development of drug resistance. Indeed, scientists believe that a major reason for increasing drug resistance among bacterial pathogens is the uptake of genes from the environment.

Continuing research in the 1950s pointed out that bacteria can also recombine by the process of **conjugation**. Seminal findings by **Joshua Lederberg**, **Francois Jacob**, and **Elie Wollman** contributed to knowledge of this process. During conjugation, two live bacteria, a donor and a recipient, come together and join by a cytoplasmic bridge (or pore), as the photograph in Figure 4.6 shows. Single-stranded DNA from the donor bacterium then crosses the cytoplasmic bridge to the recipient. Here it integrates to the recipient's chromosome (a rare event), or it remains in the recipient's cytoplasm as a free-

transformation

the genetic recombination process in which bacteria acquire fragments of DNA from the local environment and express the proteins encoded by the genes in those fragments.

conjugation

the genetic recombination process in which one live bacterium acquires fragments of DNA from another live bacterium and expresses the proteins encoded by the acquired DNA.

Jacob

zhah-kob´

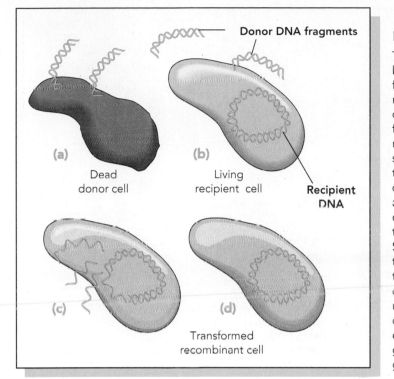

(a) Dead donor cell

(b) Living recipient cell

Donor DNA fragments

Recipient DNA

(c)

(d)

Transformed recombinant cell

FIGURE 4.5

Transformation in a bacterial cell. (a) A fragment of DNA is released from a dead donor cell. (b) The DNA fragment enters a living recipient cell. (c) One strand dissolves, and (d) the remaining strand displaces a segment from a strand in the recipient's chromosome and transforms the cell. Scientists studied transformation in depth in the 1950s and demonstrated that it occurs under laboratory conditions. The process is essential to the uptake of genes by a cell during genetic alteration.

QUESTIONLINE 4.1

1. **Q.** Exactly what is the *E. coli* I hear so much about?

A. *E. coli* is short for *Escherichia* (esh-er-ik'e-a) *coli*. It is a bacterium normally found in the human and animal intestine. *E coli* has been used for decades in university, research, and industrial laboratories and is generally considered harmless, although pathogenic strains have been isolated in recent years. *E. coli* has a single chromosome, which makes it an appealing organism to work with.

2. **Q.** Can bacteria like *E. coli* take on new genetic characteristics?

A. Yes, it has been demonstrated that *E. coli* and other bacteria can acquire genes from their local environment and incorporate the genes to their chromosome. They then synthesize the proteins encoded by the new genes. Humans, animals, or plants have no such ability.

3. **Q.** Can viruses carry genes into bacteria?

A. Viruses are at the threshold of life. They are packets of nucleic acid encased in a protein coat. Viruses are able to enter bacterial cells and attach their nucleic acid fragments to the bacterial chromosome. As such, they can carry their own genes into a bacterium, as well as any other genes that might be experimentally attached to their nucleic acids.

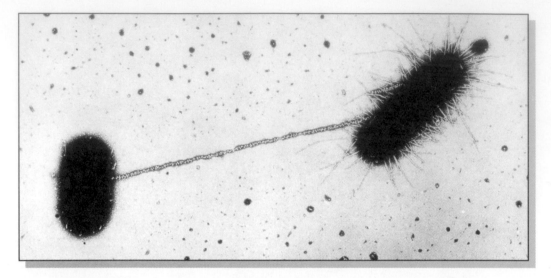

FIGURE 4.6

Conjugation in bacteria. Two *Escherichia coli* cells are joined by a cytoplasmic bridge (or pore) and are engaged in conjugation. The cell on the right has surface appendages called pili and is the donor cell. The cell on the left is the recipient. Conjugation provides an opportunity for the transfer of genetic material from one cell to another. It is one method by which bacteria acquire the genes that allow drug resistance.

plasmid

a closed loop of DNA containing about a dozen genes and existing in multiple copies in the cytoplasm of bacterial cells.

floating loop of DNA (what molecular biologists call a **plasmid**). The newly acquired genes then express themselves.

Conjugation has been demonstrated among cells from different genera of bacteria such as *Salmonella* and *Shigella* cells. (By contrast, transformation occurs only among cells of the same bacterial genus.) For this reason, conjugation accounts for the passage of genes among unrelated bacteria and explains recombinations among widely divergent species. Genes for antibiotic resistance are believed to move among bacteria in this manner.

In the late 1950s, Joshua Lederberg worked with **Norton Zinder** to discover a third form of recombination. This form is unlike the previous two because viruses are involved as transfer mechanisms among bacterial cells. The process came to be known as **transduction**.

transduction

the genetic recombination process in which bacterial viruses acquire fragments of bacterial DNA during viral replication and transport those fragments into another live bacterium, where the DNA is expressed.

In transduction, the virus involved is called a **bacteriophage** (or simply, **phage**). It attaches to and enters a bacterial cell (Figure 4.7), then integrates its DNA to the bacterial chromosome. At some later time, the virus disengages from the chromosome and carries along a tiny fragment of bacterial DNA. When the virus replicates, it reproduces its own DNA, as well as the bacterial DNA, and packages the DNA into new phages. Later, when the virus enters a new bacterium, it attaches its DNA and the bacterial DNA to the new bacterial chromosome. The new bacterium has been recombined, or transduced.

bacteriophage

bak-te´re-o-faj

a DNA-containing virus that replicates within bacteria.

Transduction is an infrequent event in bacterial cells, and examples are difficult to identify. However, the potential for transduction is great because many viruses are known to integrate into host cells (as the phage does in transduction). The **diphtheria bacterium**, for example, harbors a bacteriophage that encodes a physiologically destructive toxin. Also, the ***Salmonella*** that cause food-borne infections are known carriers of bacteriophages. Among human cells, the herpesviruses integrate into nerve cells, as noted earlier. Chickenpox viruses, infectious mononucleosis viruses, and HIV are other viruses that integrate their DNA to human cells.

diphtheria

dif-the´re-ah

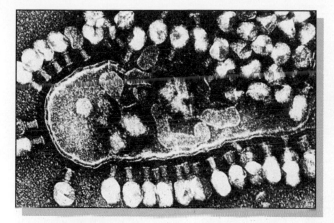

FIGURE 4.7

An electron micrograph of the bacteriophage and its host bacterium *Escherichia coli*. Numerous bacteriophages can be seen attached to the surface of the bacterial cell. Note the somewhat hexagonal heads and tails of the phages. After attachment, the nucleic acid of the virus enters the bacterium and encodes the protein materials necessary for viral replication.

The processes of transformation, conjugation, and transduction permit a bacterium to acquire new bits of DNA and assume new genetic characteristics In the 1950s, molecular biologists first began using the word "recombined" to refer to genetically altered bacteria and their new DNA. Gradually, the term "**recombinant DNA**" crept into sthe biochemical lexicon. Then, as today, recombinant DNA refers to a DNA molecule with its natural DNA component and some foreign DNA permanently attached to the original molecule. The laboratory experiments performed in the 1950s demonstrated that bacterial recombination is possible in nature. Scientists began wondering whether they could perform similar recombinations in the laboratory.

recombinant DNA

a molecule of DNA that contains "foreign" DNA spliced into the molecule.

■ RESTRICTION ENZYMES

Interest in recombinant DNA work heightened considerably in the 1950s when **Salvador Luria** and his colleagues observed that *Escherichia coli* (*E. coli*) could resist destruction by bacteriophages by "restricting" replication of the virus. By 1962, **Werner Arber** and his research group showed that an enzyme system retards replication by cleaving the viral DNA before it reaches the bacterial cytoplasm. They isolated the DNA-cleaving enzyme from *E. coli* and called it an **endonuclease**. The enzyme cuts viral DNA but not the host bacterial DNA because it first protects and modifies the bacterial DNA by adding methyl groups to it. Because the endonuclease "restricts" viral replication (and because it operated at a "restricted" site on the DNA molecule), it eventually came to be known as a **restriction enzyme**.

Arber's group found that their restriction enzyme had limited practical value for manipulating DNA because it cut the DNA at too many sites. Then in 1970, **Hamilton Smith** and his colleagues isolated a new restriction enzyme from the bacterium *Haemophilus influenzae*. This enzyme cuts the DNA molecule at a more predictable point, so its activity is more precise than Arber's enzyme. **Daniel Nathans** and his coworkers then used the bacterial enzyme to cut DNA from the simian virus SV40 at the same precise location (Figure 4.8). For their work, Arber, Smith, and Nathans were awarded the Nobel Prize in Physiology or Medicine in 1978.

As the years unfolded, DNA technologists found they could use restriction enzymes to split a DNA molecule on demand, regardless of the source of the DNA. For example, a restriction enzyme from a bacterium could be used to cleave a DNA molecule at point X, regardless of whether the DNA was from plant, animal, or human. And one enzyme could scissor the DNAs of various viruses. Eventually, more than 1200 restriction enzymes were isolated and purified from bacteria. In this group are enzymes that act at many different nucleotide sequences; about 75 of the enzymes are now commercially available (Questionline 4.2).

endonuclease

an enzyme that cleaves a DNA molecule within a living cell.

methyl

a chemical combining group composed of one carbon atom and three hydrogen atoms.

Haemophilus

he-mof'i-lus

simian

referring to a monkey or monkeylike animal.

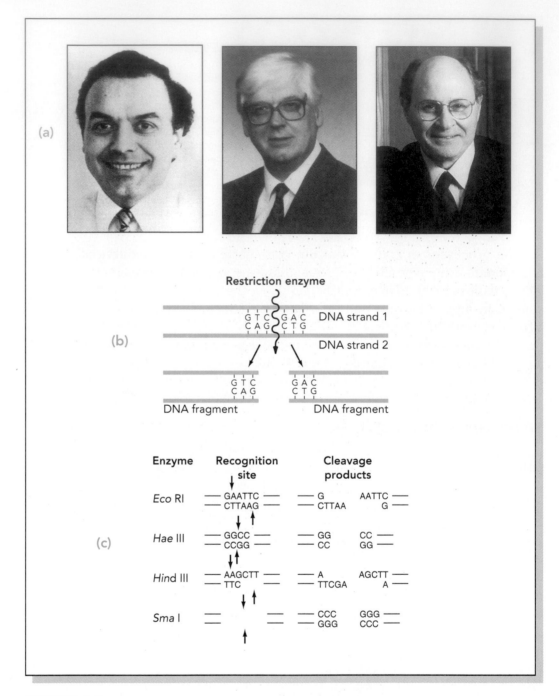

FIGURE 4.8

Restriction enzymes. (a) Werner Arber, Hamilton Smith, and Daniel Nathans, winners of the 1978 Nobel Prize in Physiology or Medicine for their work with restriction enzymes. (b) Activity of a restriction enzyme (also called an endonuclease). A restriction enzyme cuts through the two strands of a DNA molecule to produce two fragments. The restriction enzyme shown leaves blunt ends in the DNA fragments. (c) The recognition sites of several restriction enzymes. Arrows show where the different restriction enzymes function. Note that at a particular recognition site the sequences of nucleotides of the two DNA strains run in opposite directions. This symmetry is called twofold rotational symmetry. Also, note that two enzymes HaeIII and SmaI produce fragments with blunt ends, whereas enzymes EcoRI and HindIII leave ends that dangle. These dangling ends are more desirable and are essential to gene-combining techniques.

QUESTIONLINE 4.2

1. **Q.** What is the difference between a restriction enzyme and an endonuclease?

   A. There is no difference. Both terms refer to an enzyme that digests a DNA molecule. A restriction enzyme is given that name because it operates at an identifiable "restricted" site on a DNA molecule and because it "restricts" viral invasion of a bacterium. An endonuclease is an enzyme that works on a nucleic acid *within* a cell (as opposed to an exonuclease that works *outside* the cell).

2. **Q.** How do endonucleases and ligases compare with one another?

   A. Endonucleases and ligases have opposite modes of activity. Endonucleases are enzymes that cut open DNA molecules, whereas ligases are enzymes that chemically stitch together two fragments of DNA (to ligate is to join together). Both types enzymes are commonly formed within living cells.

3. **Q.** Is there any difference between bacterial plasmids and chromosomes?

   A. In bacteria, the chromosome is a closed-loop molecule of DNA containing about 4000 genes and encoding all the essential proteins of the bacterium. Plasmids are a series of small loops of DNA present in the bacterial cytoplasm. They contain only a few genes and do not appear to encode proteins essential for the organism's livelihood. Plasmids are not present in human, animal, or plant cells.

Restriction enzymes, such as those listed in Table 4.1, are named for the bacterium from which they have been isolated. The first letter used for the enzyme is the first letter of the bacterium's genus name (in italics), then comes the first two letters of its species modifier (also in italics), next is the strain of the organism, last is a Roman numeral signifying the order of discovery. Consider, for example, the following names:

*Eco*RI from *Escherichia (E)*
        coli (co)
        strain RY13 (R)
        first endonuclease (I)
*Bam*HI from *Bacillus (B)*
        amyloliquefaciens (am)
        strain H (H)
        first endonuclease (I)
*Hind*III from *Haemophilus (H)*
        influenzae (in)
        strain Rd (d)
        third endonuclease (III)

How a restriction operates is interesting. The enzyme scans a DNA molecule and stops when it recognizes a sequence of four or six nucleotides. This **recognition sequence**, as it is called, is the place where the DNA is scissored. The recognition sequence displays a twofold rotational symmetry; that is, the nucleotides at one end of the sequence are complementary to those at the other end (Figure 4.8c). Essentially, the two strands of the double helix have the same nucleotide sequence running in opposite directions for the length of the sequence. Many restriction enzymes leave

recognition sequence

a sequence of nitrogenous bases on a DNA molecule that is chemically recognized by a restriction enzyme as its location of activity.

blunt ends at the cutting site, but others leave single-stranded dangling ends.

The unique nucleotide arrangement in the two strands has two effects: (1) Because the same recognition sequence occurs on both strands of the DNA (although running in opposite directions), the restriction enzyme recognizes and cleaves both DNA strands, thereby cutting both strands of the double helix. (2) The cut sites are often offset from one another, thereby leaving a few nucleotides dangling from the ends. Notably, the bases of the dangling nucleotides are complementary to one another, as shown in Figure 4.9.

It is important to remember that every cleavage performed by a restriction enzyme will occur at the same recognition site regardless of the source of the DNA. It is equally important that the recognition site will probably occur at least once in any available sample of DNA, so that any source of DNA can be used. Indeed, the fewer bases in the recognition site, the more probable that the site will occur in a given genome. Each of the fragments will have dangling nucleotides, the so-called **sticky ends**, resulting from the restriction enzyme's activity. Because the single-stranded ends are complementary,

complementary

being the opposite member of a well-recognized pair; also, an adjective describing the bases in the double-stranded DNA molecule.

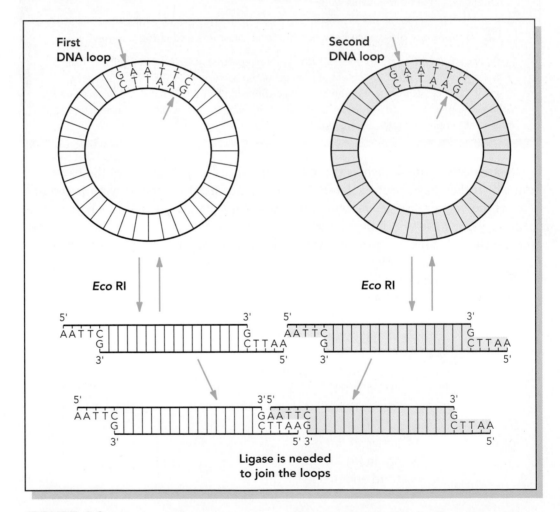

FIGURE 4.9

Construction of a recombinant DNA molecule from two unrelated plasmids. The restriction enzyme *Eco*RI cleaves the DNA of both plasmids because, even though they are unrelated, the plasmids have the identical recognition sites in their DNA. The circular plasmids now become linear. At this point the exposed ends join with each other, and the two linear molecules can rejoin with one another to form a single, long DNA molecule. Also, the ends can join to one another to form a large loop. The union is catalyzed by DNA ligase. Unions such as these are the bases for synthetic genetic recombinations.

TABLE 4.1.
Some Restriction Enzymes, Their Sources, and Their Recognition Sites

ENZYME	MICROBIAL SOURCE	SEQUENCE*
AluI	Arthrobacter luteus	5′—A—G—C—T—3′ 3′—T—C—G—A—5′
BamHI	Bacillus amyloliquefaciens H	5′—G—G—A—T—C—C—3′ 3′—C—C—T—A—G—G—5′
EcoRI	Escherichia coli	5′—G—A—A—T—T—C—3′ 3′—C—T—T—A—A—G—5′
EcoRII	Escherichia coli	5′—C—C—T—G—G—3′ 3′—G—G—A—C—C—5′
HaeIII	Haemophilus aegyptius	5′—G—G—C—C—3′ 3′—C—C—G—G—5′
HindIII	Haemophilus influenzae b	5′—A—A—G—C—T—T—3′ 3′—T—T—C—G—A—A—5′
PstI	Providencia stuartii	5′—C—T—G—C—A—G—3′ 3′—G—A—C—G—T—C—5′
SalI	Streptomyces albus	5′—G—T—C—G—A—C—3′ 3′—C—A—G—C—T—G—5′

*Note: The arrows indicate the sites of cleavage on each strand.

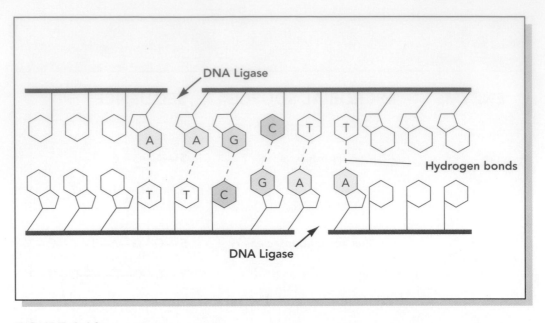

FIGURE 4.10

The bonds that hold DNA fragments together in a recombinant DNA molecule. Hydrogen bonds form weak bonds between complementary bases, whereas a strong link is forged by DNA ligase. The latter joins the backbones of the DNA strands.

they can pair with each other or with any other DNA fragment having complementary sticky ends. This latter property makes restriction enzymes the invaluable tools of genetic engineering. The source of DNA (bacterium, bird, or buttercup) is immaterial as long as the sticky ends of the DNA fragment are complementary.

▮ LIGASES

Bringing together complementary fragments of DNA does not of itself forge a bond between the fragments. Although hydrogen bonds will form between the complementary bases of the dangling nucleotides, these bonds are not strong enough to hold the ends together indefinitely, especially at physiological temperatures.

To forge a permanent link between the ends, an enzyme called **DNA ligase** is used. First isolated from bacteriophage T4, DNA ligase joins (or ligates) the biochemical backbones of DNA strands by forming a new chemical bond between the phosphate group at the 5' carbon of one deoxyribose molecule and the 3' carbon of the next deoxyribose molecule, as Figure 4.10 indicates. This bond, known as a phosphodiester bond, is a covalent bond much stronger than the hydrogen bonds between opposite nitrogenous bases. The **phosphodiester bond** exists between all nucleotides of the DNA strands. Its formation between the fragments seals the fragments together and completes the new molecule (Figure 4.11).

DNA ligase was initially isolated in the late 1960s. It exists not only in viruses but also in eukaryotic cells and in *E. coli*, where it stitches together fragments of DNA and works with other enzymes in the synthesis of DNA. The enzyme also assists DNA repair by combining broken pieces when a cell is recovering from injury. In DNA technology it functions as the final link in the chain of events leading to a recombinant DNA molecule.

DNA ligase

li´gaze

an enzyme that joins fragments of DNA together by linking the phosphate group of one nucleotide to the deoxyribose molecule of the next nucleotide

phosphodiester

fos´´fo-di-es´ter

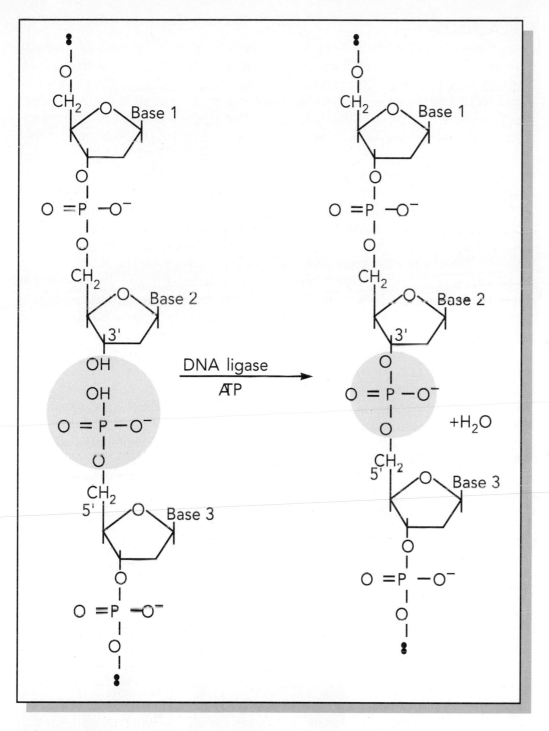

FIGURE 4.11

The activity of DNA ligase. DNA ligase links DNA fragments to one another by joining the phosphate group at the 5′ carbon of one deoxyribose molecule to the 3′ carbon of the next deoxyribose molecule. This type of linkage is called a phosphodiester bond. It is considerably stronger than a hydrogen bond forming between opposing nucleotides.

■ PLASMIDS

In the early 1970s, **Paul Berg** and his collaborators at Stanford University investigated the possibility of using restriction enzymes to alter the structure of DNA. Berg's group isolated the chromosome from *E. coli* and split open the DNA with a restriction enzyme. However, the DNA had blunt ends, and it was exceedingly difficult to attach the foreign DNA from a virus. (It was somewhat akin to attaching a brick to the end of a wall.) They succeeded, however, and in so doing, produced the first recombinant DNA molecule. Berg shared the 1980 Nobel Prize in Chemistry for his feat.

At about the same time, two events were taking place to revolutionize DNA technology. One was **Herbert Boyer's** isolation of *Eco*RI, the restriction enzyme that leaves mortiselike dangling ends. As noted previously, the dangling ends (or sticky ends) permit DNA fragments to overlap and attach easily (the proper way to attach a brick to a wall). The second event was taking place at **Stanley Cohen's** laboratory at Stanford University. Cohen was accumulating data on small loops (closed circles) of DNA found in the cytoplasm of bacteria but not in more complex organisms. The loops are known as **plasmids**. With functions not entirely understood, plasmids exist apart from the bacterial chromosome. They contain as few as a dozen or as many as several hundred genes (compared with several thousand in the bacterial chromosome), and they represent about two percent of the organism's genetic information. Plasmids are apparently not essential for bacterial growth; they can be lost without significant harm to the bacterium.

In 1972, Cohen constructed a new plasmid beginning with DNA isolated from *E. coli.* The new plasmid had three important features: (1) it had only a single recognition site where *Eco*RI would cut the molecule, thus identifying the spot where the plasmid would open; (2) it had a sequence of nucleotides called an **origin of replication**, which is needed to initiate DNA replication; such a sequence stimulates plasmids to multiply in a host organism; (3) it contained a gene for resistance to the antibiotic tetracycline; thus, bacteria having the plasmid could resist the effects of tetracycline, whereas bacteria lacking the plasmid would be killed by the tetracycline. Cohen named the plasmid pSC101 ("SC" for Stanley Cohen).

Another key characteristic of Cohen's plasmid was the ease of **inserting** it to a host cell. Cohen, shown in a Figure 4.12, found that he could insert plasmids to fresh bacteria by suspending the latter in cold calcium chloride, then rapidly heating the bacteria to 42° C. So treated, the bacterial wall and plasma membrane open to permit the plasmids to pass through and into the cytoplasm.

origin of replication

a sequence of nitrogenous bases on a DNA molecule that signals the beginning point of replication of the DNA molecule.

tetracycline

an antibiotic consisting of four benzene rings and associated chemical groups; inhibits protein synthesis in certain groups of bacteria and thereby destroys them.

FIGURE 4.12

Gene engineers. (a) Paul Berg, the American biochemist who synthesized the first recombinant DNA molecule and won the 1980 Nobel Prize in Chemistry. (b) Stanley Cohen, who developed the use of plasmids in DNA technology. Cohen was also a Nobel laureate.

(a)   (b)

Once inside the bacterial cell, a single plasmid **multiplies** to form a few dozen identical replicas among the normal plasmids. If the plasmid contains a foreign gene (such as a human gene), that gene is copied along with the rest of the molecule. Because the bacterium harboring the plasmid is also multiplying—as often as once every twenty minutes—each new bacterium obtains a few new plasmids. Before long, a single bacterium breeds millions of descendants. Such a population derived from a single parent cell is called a **clone**. All cells in the clone have the identical plasmids. Likewise, millions of copies of the foreign gene now exist. The gene has also been cloned.

It did not take too much imagination to realize that plasmids were the ideal carriers, or **vectors**, to carry foreign genes in DNA technology experiments. Thoughts even turned to carrying human genes. The stage was set for the recombinant DNA experiments that would change the course of biochemical history.

clone

a group of organisms, cells, molecules, or other objects all originating from a single individual; synonymous with colony.

vector

an organism or molecule used to transport foreign genes into a recipient organism.

RECOMBINANT DNA EXPERIMENTS

To some historians, the discipline of DNA technology came into being over pastrami sandwiches at a local delicatessen in Waikiki Beach. The year was 1972. It happened that Herbert Boyer was at a scientific conference in Hawaii speaking about the *Eco*RI restriction enzyme. Stanley Cohen was in the audience. After the presentation, Cohen invited Boyer to lunch to explore their possible collaboration on a series of experiments. The two researchers sat and considered the experiments that would send DNA technology into a new era.

Cohen had been conducting fruitful experiments with plasmids, but he was experiencing difficulty cutting open the plasmids. Boyer's *Eco*RI enzyme seemed to be the ideal solution, so Cohen suggested they join forces. They would use his plasmids and Boyer's enzyme to recombine the plasmid DNA. First they would try to recombine two plasmids to form a single plasmid. If successful, they would attempt to bring DNA from a foreign species into the plasmid to produce a recombinant DNA molecule. Boyer agreed, and the bargain was struck.

The experiments of Boyer and Cohen were performed in 1973, twenty years after Watson and Crick published their historic article on DNA's structure. The plasmid that Boyer and Cohen used was pSC101, shown in Figure 4.13. In their first experiments, Boyer and Cohen successfully recombined pSC101 with plasmid pSC102 and cloned it within *E. coli* cells. Their approach was interesting: Plasmid pSC101 contained a gene that imparts resistance to the antibiotic tetracycline; plasmid pSC102 had a gene for

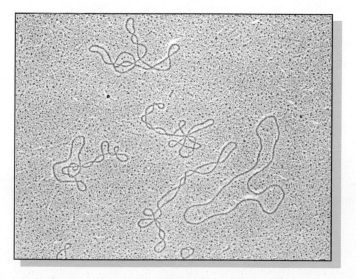

FIGURE 4.13

An electron micrograph of bacterial plasmids. Note that the DNA occurs in a closed loop. Some loops are relaxed, whereas others are tightly coiled. In eukaryotic cells, such as from animals or plants, the DNA is in a linear form.

kanamycin

an aminoglycoside antibiotic inhibitory to species of Gram(-) bacteria.

Xenopus laevis

zen´o-pus la´vis

chimera

ki-mer´ah

Staphylococcus

staf´´i-lo-kok´us

resistance to another antibiotic, kanamycin. When the recombined plasmids were successfully incorporated into bacteria, the cells displayed their new genetic capabilities by resisting the effects of both tetracycline and kanamycin.

Then came the second set of experiments, the ones involving foreign DNA, as represented in Figure 4.14. Boyer and Cohen obtained a gene from cells of the African clawed frog *Xenopus laevis*. This gene encodes a segment of protein used to form ribosomal RNA. They used *Eco*RI to cut open the frog DNA, then used the same enzyme to open plasmid pSC101. Because *Eco*RI was used for both DNAs, the dangling ends were similar in the plasmids. Frog DNA fragments were then mixed with the plasmid DNA, and pairing occurred between the complementary bases of the two DNA fragments. DNA ligase was added and the circular plasmid reformed, except that now it contained the inserted DNA from the frog. The researchers called the recombined plasmid a **chimera**, from the mythical lion-goat-serpent of Greek mythology.

Next it was time to see whether the foreign DNA could encode protein. Using standardized techniques for introducing plasmids to bacteria, Cohen coaxed the plasmids into *E. coli* cells and placed the bacteria in a growth medium. Within minutes, the bacteria and their recombined plasmids were replicating, and before long, the *E. coli* cells were producing an extra protein, the protein normally formed by frog cells. Here was one species of organism (*E. coli*) producing its own proteins plus the protein of another species (*X. laevis*). The genes for *X. laevis* protein had been successfully transplanted to another organism.

The scientific media and the national press trumpeted the success of Boyer and Cohen because the two scientists had breached the barriers separating biological species and launched the era of DNA technology. As one observer noted: "Biotechnology used to be BBC (before Boyer-Cohen); now it is ABC (after Boyer-Cohen)." Many years later, in 1996, the two scientists were honored with the prestigious Lemelson-MIT prize and shared the $500,000 award.

Molecular biologists were quick to see the implications of recombinant DNA technology, and they rushed to perform their own gene manipulation experiments. Within weeks, scientists were investigating other gene transfers and were trying to transcend other species barriers. In one experiment, genes from *Staphylococcus aureus*, the common skin bacterium, were transferred to *E. coli* cells. Other scientists attempted to isolate human genes and insert them to bacterial plasmids. And still others began to speculate on the overpowering implications of experiments in DNA technology. They reasoned, for example, that genes could be inserted into living cells to relieve genetic deficiencies and that rare pharmaceutical proteins could be produced in mass quantities. They dreamed of inexpensive sources of bioenergy and infants free of birth defects. There seemed to be no limit to what the new technology could accomplish. Figure 4.15 summarizes some possibilities.

■ THE SAFETY ISSUE

Despite the enthusiasm for the new DNA technology, many scientists warned of its dangerous consequences and suggested a more cautious approach to the research. There was alarm, for example, when Paul Berg's group proposed to insert genes from a cancer virus into *E. coli* cells. His colleagues emphatically pointed out that if the recombined *E. coli* cells escape from the laboratory, they could enter the human intestine (where they normally live) and express the cancer genes. Berg considered their objections and canceled the experiment.

But the safety issue was would not go away, and through the next few months,

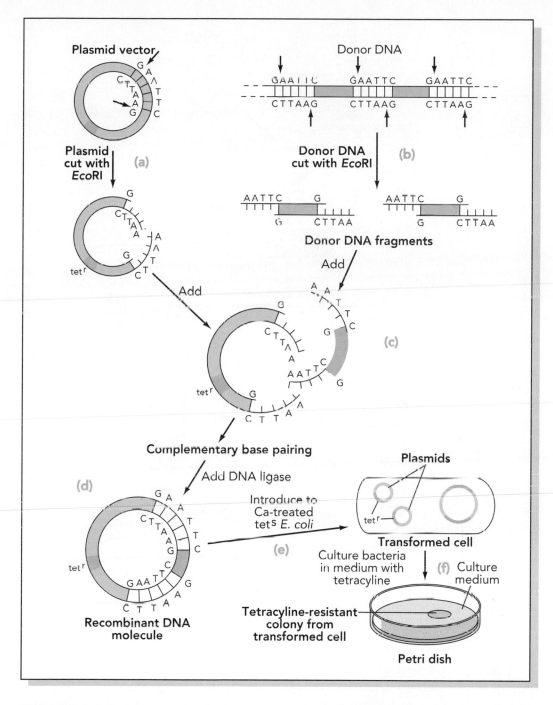

FIGURE 4.14

The Boyer-Cohen experiment of 1973. (a) Plasmid pSC101 is opened with the restriction enzyme *Eco*RI at the points shown by the arrows. The plasmid contains the gene for tetracycline resistance designated tetr. (b) Donor DNA from the toad *Xenopus laevis* is also treated with *Eco*RI, and fragments of foreign DNA are obtained. Note that *Eco*RI acts at the same recognition sites in the plasmid vector and the donor DNA. (c) Fragments of donor DNA are combined with the opened plasmids, and complementary base pairing takes place. (d) When DNA ligase is added, the plasmid closes and a recombinant DNA molecule is formed. (e) The plasmids are introduced to tetracycline-sensitive (tet^s) *E. coli* cells by treating the cells with calcium compounds. The *E. coli* cells are transformed. The plasmids multiply within the cells and encode the proteins specified by the frog DNA. (f) When the bacteria are cultivated in medium containing tetracycline, the cells containing recombined plasmids grow and form colonies. This is because they posses the gene for resisting tetracycline. Bacteria containing the normal plasmids do not possess the tetracycline-resistant genes and fail to form colonies in the medium.

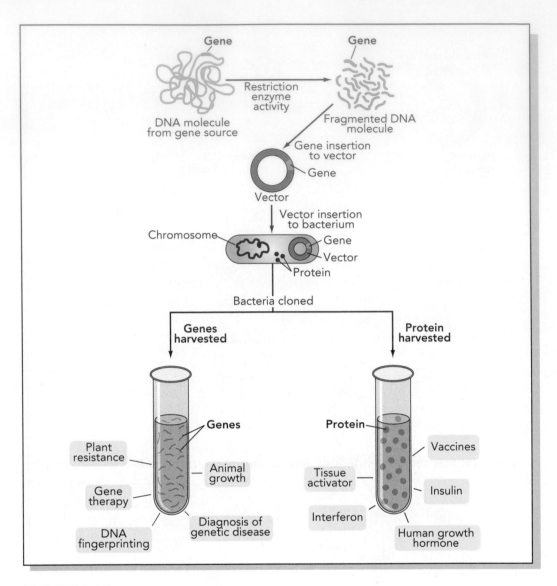

FIGURE 4.15

A review of DNA technology and genetic engineering showing the two major objectives of the process: the cloning of human genes and the large-scale production of proteins encoded by the genes.

National Academy of Sciences

a prestigious scientific organization in the United States to which members are elected for their achievements.

small groups of molecular biologists began to discuss the safety implication of DNA technology in meetings held under the auspices of the National Academy of Sciences. Then, in 1974 an unprecedented event took place: a letter signed by Paul Berg and nine other scientists appeared simultaneously in *Science, Nature,* and *Proceedings of the National Academy of Sciences,* perhaps the three most prestigious journals in the English-speaking world. The letter pointed out the potential danger:

> Recent advances in techniques for the isolation and rejoining of segments of DNA now permit construction of biologically active recombinant DNA molecules in vitro. Although such experiments are likely to facilitate the solution of important theoretical and practical biological problems, they would also result in the creation of novel types of DNA elements whose biological properties cannot be completely predicted.... There is a serious concern that some of these DNA molecules could prove biologically hazardous.

The letter went on to ask molecular biologists worldwide to "...voluntarily defer..." certain types of DNA experiments until an international conference could be held to discuss the possible dangers and the necessary safeguards. Cautions were expressed for experiments involving the introduction of antibiotic-resistant genes to plasmids and the linking of cancer virus genes to any DNA carrier molecule. This was the first time that scientists in any field had restricted research even though it had not yet proven dangerous. To ensure that the public knew what was going on, a news conference was arranged to coincide with the publication of the letter.

Although the recommendations did not carry the force of law, scientists around the world subscribed to its wishes. Then, in February 1975, a group of 139 researchers from 17 nations assembled for a four-day meeting at Asilomar, a conference center in Pacific Grove, California, to develop guidelines and recommendations for conducting experiments in DNA technology (Figure 4.16). One of the first things the attendees did was reassure an uneasy public that the bacteria used for recombinant DNA experiments were specifically bred and "disarmed" so they could not survive outside the laboratory. They also wrestled with the central dilemma: evidence was lacking either supporting or rejecting the hazardous nature of DNA work.

Asilomar
ah-sil′o-mar

In the end, the researchers opted for caution and set about policing their own work. They assigned recombinant DNA experiments to low-risk, medium-risk, or high-risk categories and decided that certain very risky experiments (such as those proposed by Berg) would not be conducted at all until better containment methods could be developed. The containment strategies would be physical, as well as biological; that is, a high priority would be given to developing strains of bacteria unable to survive outside the laboratory.

The Asilomar recommendations led to the formation of a **Recombinant DNA Advisory Committee (RAC)** of the National Institutes of Health, which issued a set of stringent guidelines that paralleled the Asilomar recommendations. As the years passed, it became clear to the public that DNA researchers were not about to set loose the celebrated "Andromeda strain" of microorganism, and there was a gradual relaxation of the guidelines and periodic revisions of the regulations (Questionline 4.3). The Recombinant DNA Advisory Committee continues in existence today, and it remains the watchdog over the DNA technology experiments that we will consider in

RAC
the Recombinant DNA Advisory Committee, a federally sponsored group of individuals who oversee research projects involving gene splicing and recombinant DNA.

FIGURE 4.16

Scientists at the Asilomar Conference of 1975 drawing up principles for containment facilities for DNA technology experiments.

QUESTIONLINE 4.3

1. **Q.** **When was the first recombinant DNA molecule synthesized?**

    A. Credit for the first recombinant DNA molecule goes to Paul Berg and his coworkers at Stanford University. In the early 1970s, Berg's group succeeded in attaching a fragment of DNA to a chromosome from *E. coli*. The process was exceedingly difficult, however, and a much improved method was pioneered by Stanley Cohen and Herbert Boyer in 1973, using plasmids and the *Eco*RI restriction enzyme.

2. **Q.** **Are recombinant DNA experiments safe?**

    A. Although there was much concern in the 1970s that recombinant DNA experiments could yield strange and dangerous new forms of life, the current outlook is that the experiments are safe and capable of bearing substantial fruit.

3. **Q.** **Who oversees recombinant DNA experiments?**

    A. There are many levels of supervision over recombinant DNA experiments. Institutional review boards and special working groups pass judgment on proposed experiments. The National Institutes of Health has established the Recombinant DNA Advisory Board to consider novel experiments, especially those involving human subjects (e.g., gene therapy). There are also a series of public watchdog groups that offer comment on laboratory experiments and their applications.

chapters ahead.

■ THE FUTURE

The experiments performed in 1972 profoundly affected the way geneticists think. Before the successful gene-splicing experiments, geneticists inferred all that they knew about genes by studying their properties and expression. Suddenly they could manipulate the genes. Soon, DNA technology permitted geneticists to understand the structure, function, and regulation of genes. They could now begin working toward cures for diseases such as inherited disorders and cancers, where the workings of genes somehow go awry. Science entered an Age of Biochemical Genetics that some observers paralleled to the Neolithic period when humans learned how to domesticate animals and raise crops.

The successes in DNA technology also gave rise to the discipline of **biotechnology** (known also as molecular genetics). Biotechnology is a vast and entirely new discipline in which molecular biology is used to solve problems associated with pollution, food production, energy production, and synthesis of new medicines. Biochemists soon came to regard bacteria as the chemical factories of the future and reasoned that they could be programmed with DNA to produce any number of substances having industrial, economic, and medical significance.

Through the 1970s, some of the promise was fulfilled, and numerous corporations began applying the techniques of DNA science for manufacturing useful products. By 1980, for example, one company was harvesting insulin from bacteria engineered with genes from human pancreas cells (Chapter 5). Another company was using bacterial cells to produce interferon, a viral-inhibiting substance normally produced by human cells. Still other companies had recombined the DNA in bacteria so they could

Neolithic

referring to a late Stone Age period of human history.

interferon

in´ter-fer´on

produce such things as human growth hormone, various vaccines, and enzymes to dissolve oil spills.

During the 1980s, many other products of DNA technology were either realistically forecast or already available. Recombined bacteria were used for disposing of toxic waste and dissolving hair in sinks and drains; bacteria were producing a human clot-dissolving enzyme named urokinase and a kidney hormone erythropoietin (Chapter 5). Hundreds of companies worldwide were working on the industrial applications of DNA technology, and many scientists were speaking optimistically of a future in which fertilizers were obsolete, plants could use microbial toxins to drive off pests, and crops were cultivated without danger of frost damage (Chapter 9).

Today, as we embark on the new century, researchers have gone well beyond inserting foreign DNA to bacterial cells and have developed DNA technologies in myriad fields, as Figure 4.17 displays. For instance, a new form of forensic science called **DNA fingerprinting** has won acceptance in the court system (Chapter 9); in a process called **gene therapy** humans have been infused with cells containing foreign genes to treat or cure disease (Chapter 8); animals have been engineered to possess a human immune system (Chapter 11); and the project to learn the sequence of every base in every human gene is now under way (Chapter 12).

erythropoietin

e-rith″ro-poi′e-tin

(a) (b)

(c) (d)

FIGURE 4.17

The many facets of DNA technology. (a) Copper leach dumps in New Mexico that use bacteria to extract copper from the ore. (b) Giant fermenters used in the synthesis of antibiotics by genetically engineered organisms. (c) Novel and imaginative pharmaceutical products such as interleukin. (d) New directions in biomedical and bioagricultural research.

More than any of these possibilities, DNA technology will give humans the opportunity to become masters of the molecules they are made of. The most optimistic scientists envision that humans will come to understand the human body and all things that creep or swim or run or fly. Hardly any facet of the human existence will remain untouched by DNA technology, and the political questions raised will require solutions on a global scale. For example, will DNA technology lead to gene tampering and efforts to "improve the race?" It is clear that ethical and moral dilemmas will abound at every step of the genetic path along which scientists travel.

Discussions of these and numerous other topics form the basis for the remainder of this book. As we will see in the pages ahead, DNA technology has entered the mainstream of human life and has become one of the most eloquent applications of scientific research. Once used with caution, the word "breathtaking" is now heard more and more to describe the implications and fruits of DNA technology.

SUMMARY

Genetic alterations using DNA technology permit scientists to intervene directly in the fate of organisms. The technology has been made possible in part by knowledge about microorganisms gleaned over the decades. For example, the biochemistry and genetics of the bacterium *Escherichia coli* are well known, and the replication patterns of viruses have been studied in depth. Knowledge of the genetic recombinations in bacteria, including transformation, conjugation, and transduction, also aided the research in new DNA technology.

Interest in DNA technology heightened considerably with the discovery of restriction enzymes. These enzymes catalyze the cutting of a DNA molecule at a "restricted" point regardless of the source of the DNA. Moreover, some restriction enzymes such as *Eco*RI leave dangling ends of single-stranded DNA at the point where the DNA is scissored. Biochemists use this property to unite the DNA fragments from different sources and to construct a recombinant DNA molecule. They use an enzyme called DNA ligase to forge a permanent link between the dangling ends of the DNA molecules at the point of union.

Another key development has been the use of plasmids as carriers for foreign DNA fragments. Plasmids are closed loops of DNA derived from bacteria. In 1973, Stanley Cohen and Herbert Boyer performed several experiments in which they spliced foreign DNA from a frog into plasmids, then inserted the plasmids to fresh bacteria. The bacteria proceeded to form bacterial proteins, as well as frog proteins, and the era of DNA technology was launched.

Safety concerns surfaced during the 1970s, and biochemists gathered at the Asilomar conference center to help establish guidelines for conducting the experiments in DNA technology. The Recombinant DNA Advisory Committee, a governing body for DNA technology, is a modern outgrowth of that conference. As postulated in the 1970s, the gene-splicing technology has had profound consequences in pure, as well as practical, research. Pharmaceutical products, agricultural advances, gene therapy, and the novel approaches to forensic science reflect these consequences. In the fundamental sense, DNA technology also gives humans mastery over the molecules they are made of.

REVIEW

This chapter's concentration has been the research discoveries of the 1960s and 1970s that formed the basis on which DNA technology could be developed. To test your knowledge of the chapter's contents, consider the following review questions.

1. Describe some of the attributes of a bacterium and its DNA that make it a useful organism for studying DNA technology.

2. What are viruses and how are they useful to DNA technology experiments?

3. Explain three forms of genetic recombination that can take place in bacteria. What is the significance of these recombinations in DNA technology?

4. Discuss the mechanism of the activity and the source, advantages of use, and nomenclature for restriction enzymes.

5. What are ligases? What functions do they perform? How are they used in DNA technology?

6. Describe some of the characteristics of plasmids, including their source, composition, method for insertion into bacteria, and value to DNA technology.

7. Summarize the experiments performed by Boyer and Cohen that set the foundations for DNA technology and gave impetus to gene-splicing experiments.

8. Describe an example of how an experiment in DNA technology can be dangerous and explain how scientists addressed the issue of safety in the 1970s.

9. What are the functions of the Recombinant Advisory Board (RAC) of the National Institutes of Health?

10. List some of the possible uses of DNA technology for resolving human problems of a practical nature and indicate some of the theoretical implications of DNA technology.

FOR ADDITIONAL READING

Alberts, B., et al. *Essential Cell Biology*. New York: Garland Publishing Co., 1998.

Alcamo, I. E. *Fundamentals of Microbiology*, 5th ed. Menlo Park, CA: The Benjamin/Cummings Publishing Co., 1997.

Elmert-Dewitt, P. "The Genetic Revolution." *Time* (January 17, 1994).

Karp, G. *Cell and Molecular Biology*. New York: John Wiley and Sons, Inc., 1999.

Malacinsk, G. M., and Freifelder, D. *Essentials of Molecular Biology*. Sudbury, MA: Jones and Bartlett Publishing, Inc., 1998.

Miller, R. V. "Bacterial Gene Swapping in Nature." *Scientific American* (January, 1998).

Raven, P. H., and Johnson, G. B. *Biology*, 4th ed. Dubuque, IA: W. C. Brown/McGraw Hill Publishing Co., 1996.

Methods of DNA Technology

 **5**

 CHAPTER OUTLINE

 LOOKING AHEAD

The pages of this chapter describe the methods used by DNA technologists to recombine cells with foreign genes and stimulate those cells to produce proteins having practical use. On completing the chapter, you should be able to:

■ understand how genes are isolated from cells or synthesized in the laboratory for use in DNA technology.

■ list some criteria for selecting vectors to carry foreign genes and host cells to produce the protein encoded by those genes.

■ conceptualize the biochemistry involved in gene expression, understand the problems that can arise in the process, and learn how DNA technologists resolve those problems.

■ explain the concept of the gene library, describe the process for establishing two types of gene libraries, and discuss how genes are recovered from the library.

INTRODUCTION

The methods for developing the products of DNA technology differ considerably from the pharmaceutical methods used during past generations. Consider, for example, the methods used for developing drugs. In previous decades, development was largely a matter of intelligent guesswork: Researchers would synthesize a new drug, test it for therapeutic value, and hope for the best. Serendipity often intervened, and the research often bore fruit. Unfortunately, the research sometimes came to a unhappy end, and another search began.

DNA technology differs substantially. It encourages researchers to pinpoint, isolate, and clone the gene that encodes a certain drug. Then the researchers attempt to splice the gene into a living cell and stimulate that cell to produce the drug. The drug is purified and concentrated, and it is ready for use in the human body. Here it may be used to replace a missing protein as insulin does for diabetic patients; or it may provide relief from physiological stress as tissue plasminogen factor (TPA) does for heart attack patients; or it may be useful as a vaccine such as the hepatitis B vaccine.

Each of these examples is explored in Chapter 6 of this book. Before discussing the processes for industrial manufacture, however, it is important to appreciate the basic methods and biochemistry of DNA technology. That way, the methods can be applied to specific circumstances and modified to fit the scientist's needs. This chapter focuses on those methods and provides an introduction to practical DNA technologies described in following chapters.

THE BIOCHEMISTRY OF GENE EXPRESSION

On the surface, the DNA-based methods for producing proteins and other biomolecules appear so simple that we might wonder why it took scientists so long to develop them. For instance it appears that to manufacture insulin all one has to do is identify the right gene, plug it into a bacterium, and stand back while the bacterium synthesizes all the insulin needed. In our naiveté, we might even suggest many other applications of DNA technology. How often, for instance, have we heard our contemporaries suggest "Why don't they just...?"

But it is not that simple. In the nitty-gritty world of the research and development laboratory, DNA technology is a highly sophisticated blend of biology, chemistry, physics, and mathematics. The technology is available to the relatively few individuals who have the educational background to use it and the monetary support to make it work (Figure 5.1 shows some necessary machinery). Producing proteins and manufacturing biomolecules are among the most elegant endeavors of our society, and the processes are far more complex than oversimplified explanations make them appear.

To begin, the DNA technologist must be familiar with **microorganisms** (or "microbes") because they are the biochemical factories commonly used for engineering pharmaceuticals. Fortunately, humans have had enormous amounts of experience in cultivating microbes cheaply and efficiently on a large scale. Brewers and bakers, for example, have been using yeasts for centuries to manufacture products for human consumption (e.g., breads, alcoholic products, yeast tablets), and since the 1940s, antibiotics have been isolated from the waste products of microorganisms grown by the ton. By contrast, obtaining products from laboratory-cultivated animal or plant cells can be extremely difficult. (Ironically, insulin is obtained much more easily

insulin

a pancreatic hormone that facilitates the passage of glucose from the bloodstream into the body cells.

biomolecules

chemical molecules that function in living things.

FIGURE 5.1

An industrial model fermentation tank used to cultivate massive quantities of microorganisms. To manufacture the products of DNA technology in useful amounts, laboratory experiments must be "scaled up" in tanks such as this.

from genetically engineered bacterial cells than from pancreas cells cultivated in test tubes.)

At its most basic level, the biochemistry of DNA technology includes gene insertions into a microbial organism to endow it with novel biochemical capabilities. The methods for accomplishing this goal can vary depending on which gene is transferred and which cell or organism receives the DNA. The personal preference of the scientist may enter the picture as well. For example, a scientist may feel more comfortable working with yeast cells than mouse tissue cells.

Several important procedures contribute to the success of an experiment in DNA technology: the gene must be isolated from the original cell or organism; then the correct carrier (vector) of the gene must be selected; a suitable host cell or organism should be available; the recipient cell or organisms must be induced to produce the protein encoded by the gene; and finally, the gene product should be in a state where it can be collected from the carrier (Figure 5.2). We shall consider each of these criteria in the pages ahead as we consider the biochemistry of how genes are expressed and how pharmaceuticals are manufactured through DNA technology.

vector

an entity that transports foreign DNA into a cell or organism where the DNA is expressed; a plasmid, cosmid, or virus can serve as a vector.

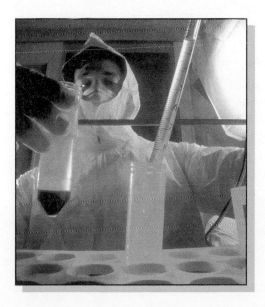

FIGURE 5.2

A successful experiment in DNA technology requires attention to gene isolation, vector selection, gene expression, and isolation of the gene product. The laboratory has become a highly sophisticated environment in which skilled researchers usually wear this type of special clothing for personal protection and to guard against contamination of the materials.

■ OBTAINING THE GENE

genome

the set of all genes possessed by a cell.

restriction enzyme

an enzyme that catalytically opens a DNA molecule at a certain point.

messenger RNA

a small RNA molecule that receives and carries the genetic message from DNA.

poly-A tail

a stretch of adenine-containg nucleotides at the end of an mRNA molecule.

The cells of a living organism contain an incredible array of genes numbering into the tens of thousands or more. A human cell, for example, has an estimated 100,000 genes in its genome, and searching for a single gene is a formidable task. (Indeed, the DNA from a single set of human chromosomes is about a meter in length. Of this amount, the genes occupy roughly a centimeter in length.) Fortunately the search has been somewhat simplified by the use of **restriction enzymes**, a set of enzymes that scissor the DNA molecule at specific locations, thus enabling the biochemist to isolate specific DNA segments. (Chapter 4 explores these enzymes at length.) The use of several different restriction enzymes increases the possibility that the target gene can be found within the DNA fragments.

But it may not be necessary to hunt for a specific gene among the array of DNA fragments. A useful alternative is to search for specific molecules of **messenger RNA (mRNA)**. When a gene is expressing itself in the cell, it is actively encoding mRNA molecules, as explained in Chapter 3. Thus, if the biochemist wishes to isolate the gene for protein X, it is useful to identify cells actively producing protein X and isolate from those cells the mRNA molecules contained in their cytoplasm. The mRNA will be in abundance because the cells are actively producing the protein.

Once the correct cells have been identified, the task of isolating the mRNA begins. The cells are disrupted and the cellular contents are subjected to an exhaustive series of chemical and physical treatments to exclude all other proteins, fats, carbohydrates, and nucleic acids. Then, the mRNA molecules are collected by zeroing in on the **poly-A tails** they all possess. These are chains of 150 to 200 nucleotides containing the base adenine (hence the name "poly-A"). The mass of mRNA is combined with cellulose particles containing on their surfaces a series of nucleic acid segments having thymine

---

QUESTIONLINE 5.1

1. **Q.** Are plasmids the only vectors used in DNA technology?

    A. Plasmids are important vectors for carrying foreign genes into cells, but they are not the only ones used. Other vectors include cosmids and viruses. Cosmids are stretches of viral DNA whose ends can be joined together to form a plasmidlike vector. Viruses used as vectors include bacteriophages, which infect bacteria and can be engineered to carry foreign bits of DNA.

2. **Q.** How is reverse transcriptase used in DNA technology?

    A. Reverse transcriptase is an enzyme that uses the nitrogenous base code in RNA as a model to synthesize a DNA molecule with a complementary base code. The enzyme can be used to construct a gene by obtaining the messenger RNA (mRNA) molecule for a particular protein and using the mRNA as a model for the complementary DNA. The DNA resulting from reverse transcriptase activity is the gene.

3. **Q.** Is there any difference between DNA and cDNA?

    A. Functionally, there is no difference between DNA and cDNA. The only difference is the origin. DNA is the naturally occurring genetic material of a gene, whereas cDNA is the product of reverse transcriptase activity. The enzyme uses RNA as a template to synthesize "complementary" DNA or cDNA. Thus synthesized, cDNA will not contain introns; natural DNA may have introns.

("poly-T"). The poly-A section of the mRNA binds tightly to the poly-T molecules, and the remaining debris is washed away. Now the mRNA molecules can be collected from the cellulose particles to yield concentrated mRNA.

After the mRNA has been purified, its biochemical information can be converted back to DNA information, as Figure 5.3 describes. To perform this step, the sequence of nitrogenous bases in mRNA is used as a template (or model) to synthesize a complementary sequence of bases in DNA. An enzyme called **reverse transcriptase** helps perform this synthesis. Discovered by Nobel Laureates Howard Temin and David Baltimore in the 1970s, reverse transcriptase uses RNA as a template to synthesize a DNA molecule with complementary bases. The enzyme is named reverse transcriptase because it reverses the usual process of transcription in which DNA is used as a template for the synthesis of mRNA.

The use of reverse transcriptase to synthesize a DNA molecule requires a **primer**, or starting nucleotide sequence. The primer consists of a string of thymidine nucleotides (poly-T), which binds to (or hybridizes to) the poly-A tail of the mRNA and acts as an initiation site for DNA production. The reverse transcriptase then moves along the mRNA molecule, encoding a DNA molecule complementary to the mRNA. This new DNA molecule is referred to as **complementary DNA**, or **cDNA**. At the end of the cDNA chain, the enzyme inserts a short hairpin loop of nucleotides, for reasons not yet understood. If single-stranded cDNA is desired, this loop of nucleotides is removed by a nuclease enzyme. Then, using degradation techniques in alkaline solutions, the cDNA molecule is cleaved away from the mRNA template molecule and isolated in pure form.

To produce a gene for insertion to a carrier, a double-stranded DNA molecule must be synthesized. To accomplish this, the hairpin loop of nucleotides is left in place and an enzyme called **DNA polymerase** is used. This enzyme uses the loop nucleotides as a primer

reverse transcriptase

an enzyme that uses the base sequence in an RNA molecule as a model for synthesizing a complementary DNA molecule.

cDNA

the DNA molecule that complements the base sequence in RNA and results from reverse transcriptase activity.

polymerase

po-lim′er-ase

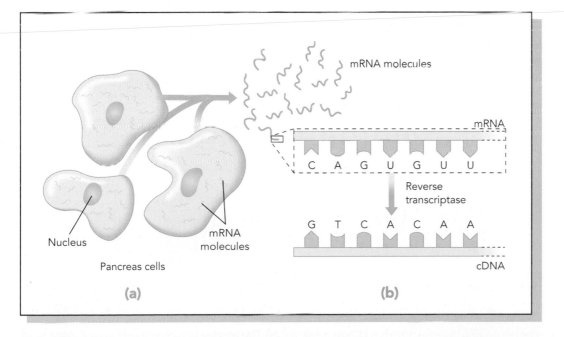

FIGURE 5.3

The general method for producing complementary DNA (cDNA). (a) Protein-synthesizing cells such as a pancreas cells are secured, and messenger RNA (mRNA) molecules are extracted from the cells. (b) The mRNA molecules are treated with the enzyme reverse transcriptase. This enzyme uses the base code in RNA to synthesize a complementary strand of DNA. The cDNA can later be used to synthesize a DNA strand complementary to the first strand.

and moves down the DNA molecule, using the molecule as a template to synthesize a complementary DNA molecule. Now the hairpin loop is removed by a nuclease, leaving a double-stranded cDNA molecule. The molecule is similar to the gene that originally encoded the mRNA molecule. The cDNA molecule, however, does not have introns, the "genetic gibberish" contained in the original genetic material (assuming the gene came from a eukaryotic cell such as a pancreas cell). It consists solely of exons, the encoding genes. Figure 5.4 summarizes the process.

For cDNA to function in a recipient cell, the molecule must duplicate itself during reproduction processes. Under normal circumstances random DNA fragments do not replicate inside a living cell because DNA polymerase, the enzyme functioning in DNA replication, will not operate unless a specific sequence of nitrogenous bases, the **origin of replication**, is present. The origin of replication can be provided by a carrier DNA molecule, to which the cDNA is attached. The carrier is commonly referred to as a vector. In the next section, we will discuss how the **vector** is selected for gene expression.

eukaryotic

applying to organisms such as animals, plants, fungi, and protozoa whose cells have a nucleus, organelles, and other complex features.

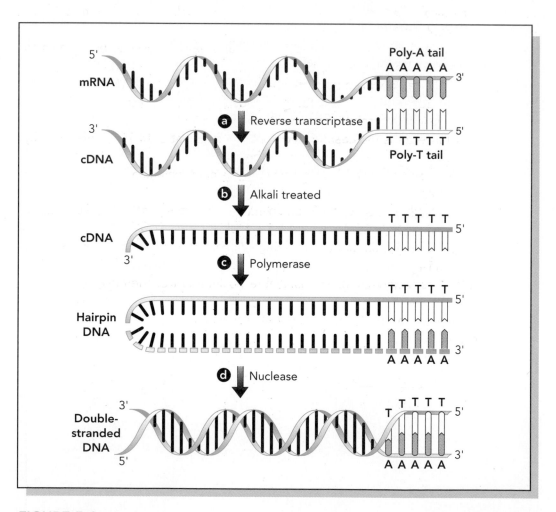

FIGURE 5.4

Formation of a double-stranded cDNA molecule. (a) The mature RNA is isolated from tissue cells with the poly-A tail intact. Reverse transcriptase and deoxynucleotides are used to synthesize complementary DNA (cDNA), which contains a complementary poly-T tail. (b) The RNA molecule is destroyed with alkali treatment, leaving the cDNA with its hairpin loop at the 3′ end. (c) The cDNA serves as both template and primer for synthesis of a strand of complementary cDNA. (d) The hairpin loop is cleaved away with a nuclease to leave a double-stranded cDNA suitable for insertion to a vector. The final cDNA consists solely of exons.

■ SELECTING THE VECTOR

For experiments in DNA technology, the vector (carrier) commonly used is the plasmid. A **plasmid** is a closed loop of DNA lying outside the chromosome of a bacterium, as shown in Figure 5.5. The more complex eukaryotic cells, such as animal and plant cells, do not have them. In bacteria, the function of plasmids is uncertain, but scientists are reasonably sure that plasmids are not essential for bacterial survival. In the 1970s, Stanley Cohen and his coworkers discovered that plasmids could be opened and connected to fragments of DNA to form recombined plasmids, or **chimeras**, as Chapter 4 describes. Chimeras contain the origin of replication required by DNA polymerase.

The process of adding a cDNA fragment to a plasmid uses a series of restriction enzymes. A restriction enzyme scans the plasmid's double helix until it recognizes a particular sequence of bases. Then it opens the plasmid in a staggered fashion, leaving four bases dangling from each cut strand. The four bases link up loosely with the complementary bases on the open cDNA molecule. Next, the enzyme DNA ligase joins the sugar-phosphate backbones of the plasmid and cDNA. DNA ligase produces a permanent chemical bond and stabilizes the chimera for insertion to fresh recipient cells. Chapter 4 explores this process in more detail.

For use in DNA technology a plasmid must be as small as possible because small plasmids are less easily damaged than large plasmids during the isolation techniques. Also a small plasmid can be taken up more efficiently by the host cell. The limitation on plasmid size implies that a cDNA molecule inserted into the plasmid cannot be unreasonably large for successful insertion (Questionline 5.1).

Another useful vector is the **cosmid**. A cosmid is a fragment of DNA manufactured by inserting a cDNA molecule between cohesive end-sites (COS) at the ends of a DNA molecule. The DNA molecule used for constructing a cosmid is obtained from a virus, often the bacteriophage lambda, which is known to replicate in bacteria. Viruses such as bacteriophages contain the genetic codes for factors permitting host cell penetration. Once inside the host cell, the linear DNA circularizes to a plasmidlike loop as its cohesive end-sites combine. Using a cosmid is similar to threading a needle, then tying together the loose ends to form a plasmid. Cosmids ferry large cDNA segments into cells because they penetrate cells more easily than plasmids. Figure 5.6 summarizes their use.

A third possible vector is a **virus** engineered to carry foreign bits of DNA such as cDNA. In recent years, an RNA-containing virus known as a **retrovirus** has been used to carry genes into human cells for gene therapy (Chapter 9). Viruses have also been

chimera

a plasmid or other vector containing DNA not normally found there.

cosmid

a linear stretch of DNA whose ends can be biochemically stitched together to form a loop and that can be used as a vector.

virus

a particle consisting of nucleic acid surrounded by a protein coat; replicates in host cells and causes disease in some instances.

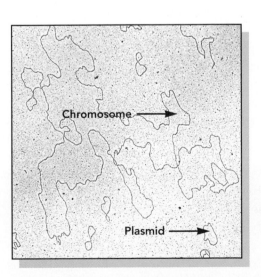

FIGURE 5.5

An electron micrograph of plasmids and a chromosomal DNA molecule showing the difference in size. In the photograph, the plasmids are the small loops. The DNA molecule, isolated from the chloroplast of a green plant, is the large loop in the center of the photograph.

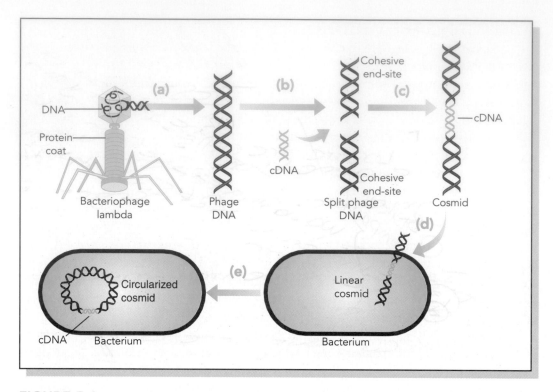

FIGURE 5.6

Synthesis and use of a cosmid. (a) DNA is obtained from the bacteriophage (phage) lambda. (b) The phage DNA is opened with a restriction enzyme, and the cDNA fragment is joined to it. (c) A cosmid results from the union. It contains base sequences for proteins that encourage entry into a bacterium. (d) The cosmid penetrates a bacterium. (e) Here it circularizes to form a plasmidlike body that carries the cDNA.

phage

faj

abbreviation for bacteriophage; a virus that replicates within bacteria or incorporates its nucleic acid to the bacterial genome.

marker gene

a gene whose activity helps identify the preserves of foreign enzymes in a cell.

used to transport genes into plant cells and have helped spark many innovative approaches to agriculture (Chapter 10). Viruses can even be engineered to transport the genetic codes of microbial fragments for use as immunizing agents. These **viral-vector vaccines**, as they are called, are a novel way of immunizing large populations of people. They portend of wave of optimism about future vaccines (Chapter 6).

A virus widely used as a vector is the **bacteriophage**, or simply, **phage**. The genome of the phage usually replicates itself within the bacterial cytoplasm, but this replication is not universal. In many cases, the phage genome unites with the bacterial chromosome and remains there permanently. The phage genome can therefore be used to deliver cDNA to a bacterial chromosome. An identical process occurs between certain plant viruses and animal viruses (e.g., retroviruses) and their host cells.

Like plasmids, viruses can be forced to accept short segments of foreign DNA into their genomes. Because they fail to replicate without a host cell, viruses are considered inert. They can usually incorporate much larger DNA segments than the plasmid and, therefore, they are often chosen for working with large genomes, such as genomes from human cells. Figure 5.7 shows one type of bacteriophage.

In addition to harboring an origin of replication, a vector should have other properties to increase its value. It should, for example, have only a single site where a particular restriction enzyme functions. If it has multiple sites, the enzyme will scissor the vector in many locations and fragment the gene.

The vector should also have an identifying **marker gene** so the vector can be located in cells after insertion. A gene that confers antibiotic resistance is a suitable marker because cells having the marker will live in the presence of an antibiotic, whereas

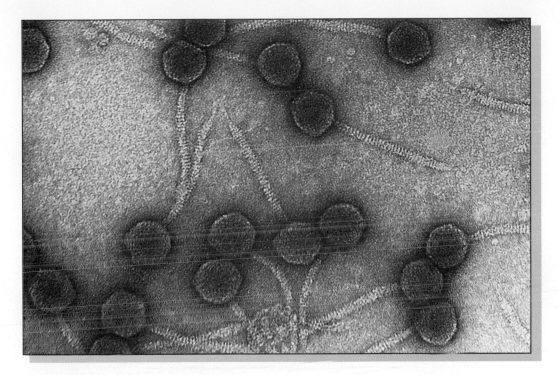

FIGURE 5.7

An electron micrograph of the bacteriophage lambda. Phages such as these penetrate bacterial cells, and their DNA cores attach to the chromosomes of bacteria. When engineered with foreign DNA, the phage serves as a vector for that DNA.

cells lacking the marker will die. Scientists assume that the desired gene has been inserted if the marker gene is displayed. Such a device has been used in plant technologies described in Chapter 10.

A marker gene of a different sort is one that stimulates a light flash. One example is a jellyfish gene that encodes a protein, which emits a green fluorescent light when illuminated with blue light. Scientists from Columbia University have used the marker gene to determine whether desired genes were inserted to *E. coli* cells during transformation experiments. When transformation was unsuccessful, the bacteria failed to glow when illuminated with blue light. However, when the transforming genes were successfully inserted, the marker gene was also inserted and the bacteria gave off a green glow on illumination.

Another factor increasing a vector's worth is its stability. Little is to be gained if a gene-altered plasmid places a physiological burden on host cells and causes them to die off in culture. It is desirable for plasmid-free cells to grow at a faster rate than plasmid-bearing cells to avoid overwhelming the plasmid-bearing cells and losing the valuable plasmids.

Vector stability can be achieved by placing the gene under strict control in the transcriptional process, such as during the period of most rapid cellular growth. To control expression, DNA technologists focus their attention on activity at the promoter site, a sequence of bases that triggers messenger RNA production. Stability can also be attained by use of chromosomes rather than plasmids as vectors. This strategy has the effect of commingling the foreign DNA with the organism's DNA and making the new gene less obvious than on the plasmid. The gene is effectively "hidden" within the morass of cellular genes.

To reduce the possibility of completely losing the plasmids, DNA technologists

promoter site

a sequence of nitrogenous bases that stimulates an appropriate enzyme to initiate transcription of a DNA molecule.

introduce to cells a large number of plasmids, a so-called **high copy number**. Loss of plasmid-bearing cells can occur during cell division if partitioning leaves certain cells with few or no plasmids and if these cells outpace the gene-altered cells. High copy numbers ensure a high probability that at least one plasmid molecule is distributed into daughter cells at cell division.

Another way to encourage plasmid retention is to make the cell dependent on the plasmid for its continued life. For example, the gene for an essential cell enzyme can be placed on the plasmid together with the desired gene. The plasmid is then implanted in cells deficient in the enzyme. To survive, the cells must have the plasmids (and the gene for producing the enzyme). Therefore, only the plasmid-bearing cells will survive when cell division takes place. It is also possible to incorporate a gene for antibiotic resistance to the plasmid. With this gene, the plasmid-bearing cells will resist the antibiotic and survive, and the cells lacking the plasmid will die off.

human artificial chromosome

a vector DNA molecule synthetically produced with qualities of a human chromosome.

A vector of a completely different type is the **human artificial chromosome (HAC)**. This vector, developed in 1997, carries human genes that are too long to fit a virus or plasmid, and it encodes a protein more consistent with human protein than do the other vectors. The HAC behaves like a natural human chromosome and passes from cell to cell during mitosis. It is a wholly synthetic, self-replicating human "microchromosome," about one-fifth to one-tenth the size of a normal human chromosome.

Formulation of the HAC was the work of **Huntington Willard** and his coworkers at Case Western Reserve University. The researchers fashioned the HAC with an origin of replication and a telomere, a protective cap of repeating DNA sequences at the end of the chromosome that prevents chromosome shortening during mitosis. They also stitched in a centromere, a knotlike stretch of DNA that joins chromatids during mitosis. However, they did not supply any histone protein with the HAC—the researchers hoped the living cell would supply this protein needed for support.

And indeed it did. When the researchers examined experimental white blood cells under the microscope, they found a minuscule but stable chromosome among the normal chromosomes (Figure 5.8). It was large enough to carry human genes and important regulatory genes. The newspapers trumpeted a significant advance in molecular genetics because the HAC enhances the possibility for introducing new genes to an individual with damaged genes, a form of gene therapy. This concept is explored further in Chapter 8. A second type of artificial chromosome, the yeast artificial

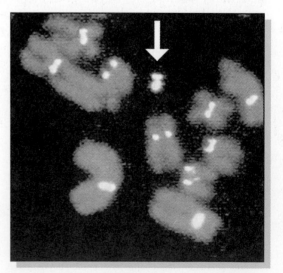

FIGURE 5.8
A human artificial chromosome (arrow) among normal chromosomes. Synthesized by researchers at Case Western Reserve University, the chromosome is a new vector for carrying foreign genes into host cells.

chromosome (YAC), is discussed in Chapter 12.

Before leaving the topic of vectors, it should be noted that these carriers constitute a type of gene delivery system known as **infection**. A second type of delivery system is transfection. **Transfection** is a physical process whereby genes are introduced to cells by such methods as liposomes, biolistics, microinjection methods, and electroporation. These methods are discussed in various other chapters together with the type of material introduced into cells.

infection

a gene delivery system using a vector organism or chromosome.

■ SELECTING THE HOST CELL

The nature of the vector is as important as the nature of the host cell or organism. Among other requirements, the host must be suitable for cultivation in the laboratory and should be able to incorporate the vector's genetic material. Also, the biochemist must be able to control gene expression within the host and collect the gene product.

One of the major vehicles for protein expression is the bacterium *Escherichia coli* displayed in Figure 5.9a. As noted in Chapter 4, this bacterium was used in the first experiments in DNA technology because so much was known about it. It had been used

Escherichia

esch″er-ik′e-a

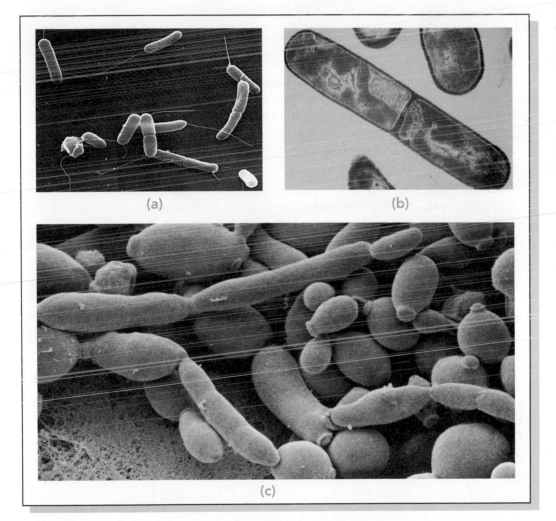

(a)

(b)

(c)

FIGURE 5.9

Some microbial hosts for DNA technology. (a) An electron micrograph of *Escherichia coli*, the common bacterium of the human and animal intestine. (b) *Bacillus subtilis*, the bacterium often isolated from the soil environment and generally regarded as a nonpathogen. (c) *Saccharomyces cerevisiae*, the nonpathogenic yeast commonly used in fermentation and baking processes. Each of these organisms has unique qualities that increase its appeal for DNA procedures.

in virus studies of the early 1950s; its genetics were established during experiments in transformation and conjugation in the 1960s; and it was used to decipher the process of protein synthesis during the 1960s and 1970s.

In addition, *E. coli* cells are widely used because they display an extraordinarily high reproduction rate. Under ideal growth conditions the bacteria double their numbers every twenty minutes. As the bacteria reproduce, their plasmids and inserted genes also reproduce. Within hours, the population contains millions of descendants of the bacteria, as well as millions of copies of the gene-altered plasmids. Such a population is called a **colony**, or **clone**. The same term, clone, is applied to the plasmids and genes. In the jargon of DNA technology, the gene "has been cloned." Gene cloning in animal or plant cells is a much slower process because these cells multiply with lower frequency than bacterial cells.

Although *E. coli* has been the workhorse of molecular genetics, the bacterium has a number of disadvantages. For instance, certain strains of *E. coli* have been related to infantile diarrhea and travelers' diarrhea (also known as "tourista" and Montezuma's revenge). Moreover, the *E. coli* cell wall contains endotoxins that can induce shock and circulatory collapse in the body; these endotoxins may be difficult to remove from pharmaceutical products. And *E. coli* is a relatively poor exporter of proteins, a factor that reduces its value in DNA technology.

An alternative host organism is the bacterium ***Bacillus subtilis*** shown in Figure 5.9b. This rod-shaped organism is nonpathogenic. In 1958, it was transformed in experiments similar to Griffith's, and since then, the genetics of the bacterium have been thoroughly studied. Strains of *B. subtilis* export proteins actively, and the organisms have been used in industry to produce antibiotics, insecticides, and industrial enzymes. The plasmids of *B. subtilis* have been scrutinized in detail, and the viruses that attack it have been studied.

In certain instances, it is desirable to use a **eukaryotic organism**, one whose cells have a nucleus, organelles, and a higher level of complexity than the prokaryotic bacteria. A suitable organism is the yeast ***Saccharomyces cerevisiae*** (Figure 5.9c). Yeasts are nonpathogenic, well-explored, and routinely used in brewing and baking. Use of a eukaryotic organism is preferred for producing human proteins (such as insulin) because complex proteins can be manufactured more easily by complex organisms than by simple organisms. Other eukaryotic microorganisms such as fungi have also been used in DNA technology, as we will see in other chapters.

A different eukaryotic organism currently in use is the **insect cell**. This cell is infected by the **baculovirus**, a rodlike particle that attacks more than 500 species of insects. The DNA of the virus encodes high levels of biologically active products, and the productivity in insect cells and larvae can exceed that in mammalian cells (e.g., mouse cells) by more than 10,000 times. The baculovirus system can carry hundreds of different genes for expression into insect cells, and the cells perform the important processing steps that *E. coli* cells cannot complete. And because baculoviruses do not infect humans, other vertebrates, or plants, there is a high safety factor. Chapter 10 discusses the vector and its host cell in more detail.

Certain experiments in DNA technology require **mammalian cells**, such as mouse cells. The use of mammalian cells has the advantage of avoiding toxic bacterial molecules that could enter a protein preparation. Biochemists have also found that certain proteins are simply too large and complex for bacteria to produce (one example, tissue plasminogen activator, is discussed in Chapter 6). Complex proteins tend to fold incorrectly in bacteria, and this defect can result in a biologically inactive product. Also, the bacterium may lack the enzyme systems for modifying the protein

clone

a mass of cells, organisms, or genes that have resulted by multiplication of a single cell, organism, or gene.

endotoxins

toxic substances located in the cell walls of certain bacteria.

Saccharomyces cerevisiae

sak"ah-ro-mi'sez ser"e-vis'e-a

baculovirus

bak'u-lo-vir-us

mammalian

applying to warm-blooded vertebrate animals that have mammary glands and hair; examples of mammals are rodents, barnyard animals, and primates, including humans into its final form (for example, by adding carbohydrate molecules).

On the negative side, working with mammalian cells is much more difficult and expensive than working with bacterial cells. Cultivating mammalian cells is very tedious, and inserting vector molecules to mammalian cells is extremely complex. Nevertheless, DNA technologists have devised ways to regulate protein synthesis in mammalian cells by working with promoter sequences, new vector systems, and novel methods for protein secretion.

Once the vector has been prepared, insertion to the recipient cell or organism is relatively easy. Plasmids and cosmids can be made to penetrate cells by alternate heating and cooling in the presence of calcium chloride, and viruses enter cells during their normal process of replication. Other transfection methods are also available as previously noted.

■ EXPRESSING THE GENE

Having prepared the vector and inserted it to the host cell or organism, the DNA technologist must be aware of certain biochemical considerations contributing to the success or failure of the process. These steps may be as critical as the previous ones because the biochemistry relates to the process of gene expression going on within the confines of the cell.

One such step depends on **strategic location**, that is, the precise place where the inserted gene is located on the plasmid, as Figure 5.10 shows. When RNA polymerase transcribes DNA to mRNA, it begins the process by finding a recognition site on the

RNA polymerase

an enzyme that synthesizes an RNA molecule from its component nucleotides.

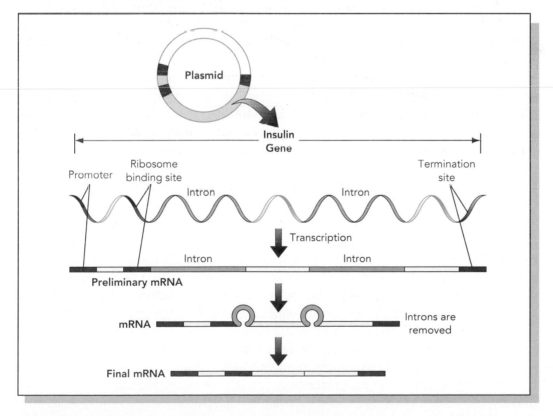

FIGURE 5.10

Essential features of a useful vector such as the plasmid that encodes insulin. The plasmid contains sites to initiate mRNA formation (the promoter); it also has sites to encouragebinding to the ribosome and to terminate mRNA production (the terminator). In eukaryotic cells, the preliminary mRNA molecule is modified by removing the introns to form the final mRNA molecule.

DNA molecule. This site, called the **promoter site**, is a sequence of bases that tells the enzyme to begin mRNA formation. The enzyme then moves along the DNA molecule and synthesizes mRNA until a termination base code is reached (Chapter 3). For gene expression to occur correctly, a promoter site must be placed at the correct spot on the vector relative to the gene.

Equally important to the promoter site is a **termination site**. This is a sequence of bases that signals the end to the transcription. It helps to place an efficient termination signal right after the gene so a readthrough into adjacent genes is prevented. To ensure proper transcription, the desired gene should be positioned correctly and between promoter and termination sites.

In addition to the promoter and termination sites, the vector must also contain a base sequence for a **ribosomal binding site**. This site is necessary because the mRNA molecule encoded by the gene must complex with the ribosome, and it must therefore have a base sequence complementary to that on a ribosome. Should binding fail to occur, the translation phase of protein synthesis will not take place and the gene will not be expressed.

Another consideration for controlling gene expression concerns the production of mRNA molecules. In bacteria, the entire DNA sequence is transcribed to an mRNA molecule, but this is not the case in eukaryotic animal and plant cells. In these cells, a number of intervening coding regions called **introns** are removed from preliminary mRNA molecules before final mRNA is prepared; the coding regions that remain are called **exons** because they are expressed. Thus the mRNA formed by eukaryotic cells is actually a collection of exons as Figure 5.10 illustrates. This biochemistry presents a problem for the DNA technologist because bacteria do not have enzymes to remove introns from mRNA molecules. Therefore, it could be difficult to use bacteria for the expression of human genes.

One way of circumventing this dilemma is to use complementary DNA (cDNA) derived from the activity of reverse transcriptase on mRNA, as discussed earlier. The cDNA formed by reverse transcriptase has no introns, and only exon DNA would be involved. The cDNA thus formed could then be used for insertion to the bacterium (Questionline 5.2).

Another way of dealing with the intron-exon problem is to synthesize an **artificial gene** by use of nucleotide building blocks. This method works backward from the standard biochemistry and uses the amino acid sequence of the protein as a basis for synthesizing DNA. It requires that the entire amino acid sequence of the protein be known. Then the genetic codes for the amino acids can be looked up and the proper nucleotides assembled. As of this writing, small artificial genes have been manufactured. They are called **oligonucleotides** ("oligo-" is derived from the Greek *oligos* meaning little, a reference to the small size of the gene). Oligonucleotides can be inserted into vectors and used for protein expression, but the methods are difficult. Among other challenges, promoter, termination, and ribosomal binding sites must be synthesized. Oligonucleotides have been used to produce antisense molecules, as described in Chapter 6.

Protein production rarely comes to an end with the linking of amino acids at the ribosome. The modification step represents another challenge that occupies the attention of DNA technologists. In the case of insulin, for example, a section of 35 amino acids must be removed from the "preliminary" insulin molecule. Then the remaining amino acids will form two chains and link together. Bacteria cannot perform this modification, so DNA technologists must engineer the bacteria to form the two chains separately. Then the two chains are linked together in a separate biochemical process outside the bacterium. (Chapter 6 explains this process in more detail.)

exon

a sequence of nitrogenous bases that encodes protein; it is retained in the formation of the final mRNA molecule.

intron

a sequence of nitrogenous bases that does not encode protein and is removed from the preliminary mRNA molecule in the formation of the final mRNA molecule.

oligonucleotides

ol´´i-go-new´cle-o-tide

small, synthetic fragments of RNA or DNA; a synthetic gene.

QUESTIONLINE 5.2

1. **Q.** Is *E. coli* the only organism used in DNA technology?

    A. Traditionally, *E. coli* has been the organism of choice for experiments in DNA technology because so much is known about its physiology, biochemistry, and genetics. In recent years, attention has shifted to the bacterium *B. subtilis*, the yeast *S. cerevisiae*, and cultivated mammalian cells. These cells can be used for the expression of foreign genes.

2. **Q.** What must be added to a structural gene to make it operational in a carrier cell or organism?

    A. A structural gene has the base sequence for a particular protein, but to make it work, other base sequences must be present. For example, there must be a promoter site that tells the enzyme to begin transcription and a termination site that brings transcription to an end. A ribosomal binding site is required to ensure that the mRNA unites with the ribosome, and it is helpful to have available a base sequence that signals the cell to release the protein after synthesis.

3. **Q.** What is the exon-intron problem that bacteria encounter with foreign genes?

    A. Introns are segments of DNA removed from mRNA in the formation of this molecule during transcription. Eukaryotic mRNAs have introns, but bacterial mRNAs do not have them, nor do they have the enzymes for dealing with introns. Thus, when an artificial gene from a eukaryote is placed into a bacterium, the latter becomes "confused" and finds it impossible to express the gene. This problem can be circumvented by use of a cDNA molecule as a gene or by use of an artificial gene; neither of these has any introns.

Another modification may involve adding a carbohydrate molecule to the protein chain to form a molecule called a **glycoprotein** (glyco- refers to carbohydrate). Two forms of interferon, the important antiviral substance, are glycoproteins. Bacteria are unable to add the carbohydrate portion to the protein encoded by the human gene, so interferon cannot be produced in bacteria. To resolve this problem, DNA technologists use animal cells (Chapter 6).

Regardless of the type of cell or organism that is used, it is always desirable to **increase the yield of product** (Figure 5.11). This step requires increasing the level of gene expression, for which there are several methods. It may be possible, for instance, to increase the number of plasmids in a bacterium (more plasmids produce more mRNA molecules and more protein). Also it may be feasible to increase the biochemical signal imparted by the promoter site to initiate gene expression or to increase the strength of mRNA binding to ribosomes by working with the gene for the ribosomal binding site. DNA technologists have investigated all these methods.

Another way to increase gene expression is to use the most advantageous set of codons for a particular protein in a particular organism. This approach may appear obvious, but many possible codons may exist for an individual amino acid, and it is important to learn which codon is favored by the organism. The less-preferred codons can then be deemphasized for the favored ones. Moreover, the protein-producing cells manufacture their product at the expense of other cellular activities (such as cell division), and the "producing" cells may be at a disadvantage when "ordinary" cells are present. Therefore, the smart DNA technologist will wait until the cell population is well

glycoprotein

a protein containing one or more carbohydrate molecules bonded to one or more of the amino acids in the protein.

codon

a three base code on an mRNA molecule specifying an amino acid in a polypeptide.

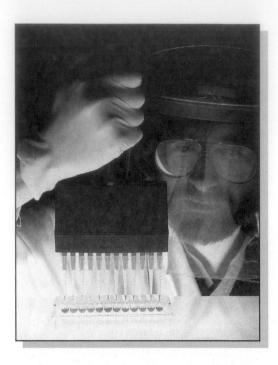

FIGURE 5.11

An example of the highly specialized apparatus used in DNA technology procedures. This apparatus is used for DNA extraction experiments.

established before attempting to maximize gene expression and will ensure that there are few competing cells. The promoter genes can be used to control the expression of structural genes until the time is right.

■ COLLECTING THE GENE PRODUCT

Once the biochemistry for gene expression has been stimulated, the gene product will begin to accumulate. Now comes the task of collecting the gene product. This process may not be as straightforward as it appears because a gene product is usually foreign to the cell producing it. Human insulin, for instance, is not a usual product of bacteria. Indeed, to deal with a nonbacterial protein such as this, a bacterium possesses enzymes known as **proteases** that recognize and degrade the protein. This enzyme activity can severely limit the ability to collect the insulin.

To address the potential destruction of the gene product, DNA technologists use strains of cells deficient in proteases. Such strains of *E. coli* are now in use. Of course, it would be best to develop strains that lack proteases completely, but these strains would probably not live without the defense protease lends. An alternative method of solving the problem is to fuse the gene product to proteins native to the cell. Such a route has been used for insulin production and human growth hormone. A final method is to use organisms where foreign proteins accumulate as aggregates in the cytoplasm and thereby avoid protease activity. Extraction of these proteins may be a problem, however, because the extraction methods may lead to protein breakdown.

Once the problem of protein loss has been solved, the expression of genes can continue. Gene products can accumulate inside the cellular cytoplasm or move to the outside environment. Export of gene products has the advantages of increasing the yield and ease of recovery and purification. Thus, considerable effort has been devoted to developing methods that result in exported products. Much of this work has focused on *Bacillus subtilis* because this bacterium normally secretes large quantities of extracellular proteins. DNA technologists have cleverly introduced to *B. subtilis* the **signal sequences** from other *Bacillus* species that secrete proteins actively. The bacteria hook onto the gene product a **signal peptide** that encourages secretion as Figure 5.12 explains. The signal

protease

a protein-digesting enzyme commonly found in bacteria.

signal peptide

a series of amino acids that encourages the release of a protein from the cellular interior to the extracellular environment.

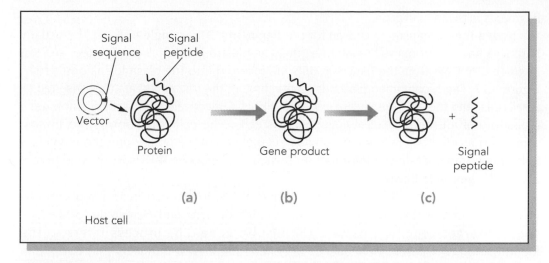

FIGURE 5.12

Use of a signal sequence and signal peptide. (a) The vector used in protein synthesis contains a signal sequence that encodes a signal peptide. The peptide is attached to the gene product. (b) The signal peptide enhances export of the gene product to the external environment. (c) Once exported, the signal peptide is biochemically removed, and the gene product is recovered.

peptide is removed after the gene product has been purified.

The use of signal sequences accelerates harvesting of the gene product while keeping the cells at their work of producing protein. The process has been adapted to the production of insulin and is explored further in Chapter 6.

GENE LIBRARIES

The DNA technologist would face a monumental task if genes had to be isolated from their original cells each time they were needed for an experiment. It makes sense, therefore, that genes should be saved in some repository once they have been isolated and purified from the mass of cellular DNA. The repositories are called **gene libraries**. Each library is a collection of living cells.

To begin the process of forming a gene library, a mass of cells containing the desired gene must be collected. Human pancreas cells, for example, contain the gene for insulin. The problem is that pancreas cells have approximately 100,000 genes (as do all human cells), and the 100,000 genes represent only about five to ten percent of the cell's DNA (the remaining ninety to ninety-five percent is nonfunctional "genetic gibberish"). The insulin gene must be isolated from this mass of DNA. We shall now see how this is accomplished.

ESTABLISHING A LIBRARY

To establish a cell line containing the insulin gene, the DNA technologist begins by obtaining the gene itself. To achieve this, a series of restriction enzymes is used to biochemically cut the cellular DNA of pancreas or other cells into fragments. Each fragment is slightly larger than an average-sized gene and contains about 4000 nitrogenous bases (or 4 kilobases [kb]). An average human chromosome has about 100,000 kb, so each chromosome is cut into about 25,000 fragments, each 4 kb long. Because there are twenty-three different chromosomes per human cell (twenty-four in males), the total number of 4 kb fragments is 575,000 fragments. Among these 575,000

gene library

a series of cells containing the genes from foreign cells for use in DNA technology experiments.

kilobase

a unit consisting of a thousand nitrogenous bases in a DNA or RNA molecule.

fragments is the insulin gene.

Now a host organism is chosen for the fragments. An example is *E. coli*, the usually harmless bacillus from the human intestine. The plasmids of *E. coli* are isolated, and the fragments of DNA from the pancreas cells are inserted into the plasmids as outlined in Figure 5.13. The recombined plasmids (containing all the fragments) are then inserted to *E. coli* cells, and the cells are encouraged to multiply. Collectively, the *E. coli* cells now are a repository of all the DNA of human pancreas cells. The cell collection is the library, whereas the cells themselves represent 575,000 books of the library, and the individual DNA fragments contain the information in each of the books. This sort of gene library is called a **genomic library**.

But which of the library books contains the informational gene for producing insulin? To answer this question, the library must be screened. Screening locates the specific target gene, in this case the insulin gene. The process refers to the identification of a particular sequence of genetic material within cells that have been genetically transformed.

■ SCREENING THE GENE LIBRARY

To screen a cell population in a gene library, certain materials and some information must be available. For example, when screening for cells containing insulin genes, a specific messenger RNA molecule is used. This mRNA molecule has a base sequence complementing the base sequence of the insulin gene. Acting as a type of genetic probe (Chapter 7), the mRNA unites specifically with the insulin gene. If the probe is linked to a radioactive isotope, the probe's radioactivity highlights when and where the reaction takes place.

To begin the screening process, the transformed *E. coli* are cultivated in the laboratory in Petri dishes containing a nutrient gel. Visible colonies resembling tiny buttons appear on the gel surface within a day or two. Ideally, each colony contains cells with a single insert of human DNA. However the cells could have normal plasmids (with no human genes) or they could have resisted invasion by any plasmids. Thus, each colony must be examined to see whether it is the one with the insulin gene insert.

To provide microbial "photocopies" of the plate for further work, the biochemist uses a technique called **replica plating**, as described in Figure 5.14. Developed in 1952 by Joshua and Esther Lederberg, the technique consists of pressing a piece of sterile velvet to the gel then pressing the velvet to a new Petri dish of nutrient gel to establish a "working" plate. By so doing, a replica of the original plate is made. Now, samples of different colonies from the working plate can be analyzed for the target gene.

The analysis can take a number of forms. For the insulin gene, the radioactive **mRNA probe** is used. First a sample from a suspected colony is secured and its DNA is fragmented and separated on nitrocellulose filter paper (special cellulose paper containing nitrate groups). The separation process is called electrophoresis. (Chapter 7 describes the process in detail.) Then the molecular probe is applied to see whether it will react with any of the DNA fragments from the cells. Like the right hand searching for the left hand, the probe mingles among the DNA seeking its complement. If no radioactivity accumulates, the technologist can be certain that no reaction with insulin mRNA took place and no insulin gene is present. However, if the radioactivity gathers at one site, that DNA is identified as the "library book" from which the information came. Figure 5.15 depicts the process.

At this point, the key colony has been located in the working plate. By checking backwards, the original colony in the original plate can be identified as the one

---

genomic library

a mass of cells that have incorporated into their plasmids all the genes derived from another cell such that the genes can be obtained in a relatively easy manner.

replica plating

a laboratory technique for making replicates of bacterial colonies.

nitrocellulose

ni″tro-cell′u-lose

electrophoresis

e-lek″tro-fo-re′sis

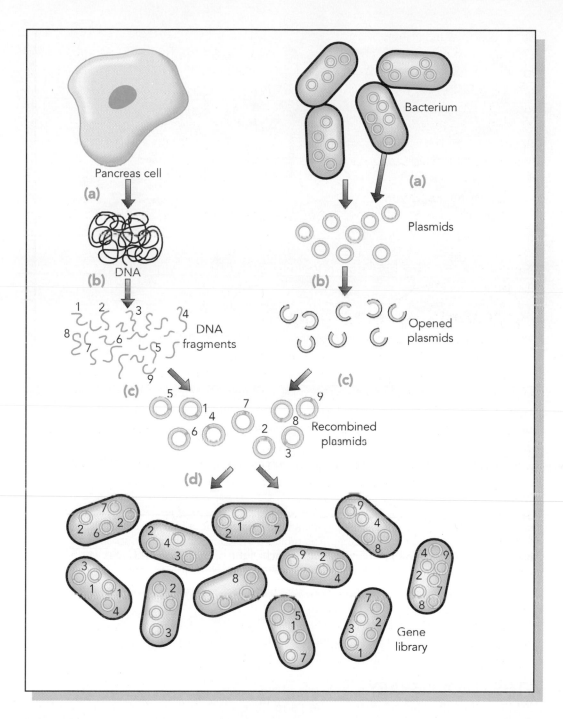

FIGURE 5.13

Establishing a gene library. (a) The DNA is isolated from tissue cells (such as pancreas cells), and plasmids are obtained from the host organism such as bacteria. (b) A particular restriction enzyme is used to digest the cell DNA, and the identical restriction enzyme is used to open the plasmids. The digestion results in open plasmids and multiple different fragments of DNA designated by numbers 1 through 9. (c) The DNA fragments are combined with the open plasmids and joined together to yield recombined plasmids. Note that different open plasmids acquire different numbered DNA fragments. (d) The recombined plasmids are inserted to fresh bacterial cells, and the plasmids take up residence within the cells and multiply as the cells multiply. The population of bacterial cells is the gene library. Different cells in the population contain different DNA fragments as noted by the numbers. Not all cells contain the same DNA fragments, and some cells have multiple copies of a particular recombined plasmid because of replication of the plasmid. The library can be screened according to procedures described in the next figure.

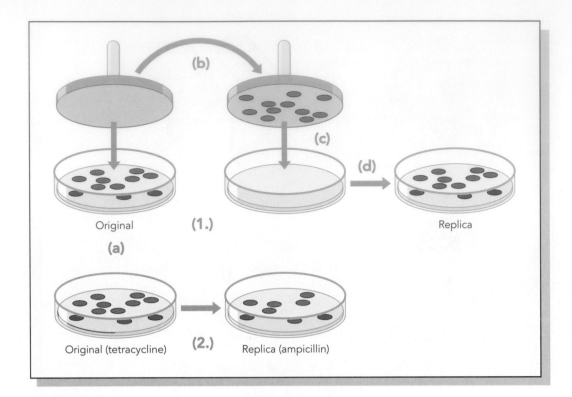

FIGURE 5.14

Use of the replica plating technique for screening bacterial colonies. (1.) Colonies of bacteria are cultivated on a Petri dish containing a nutrient gel. This is the original plate. (a) A pad of sterilized fabric such as velvet is placed against the colonies, and (b) some of the bacteria from each colony are transferred to the fabric. (c) Then the fabric is touched to a fresh Petri dish of nutrient gel. (d) The nutrient gel is incubated, and colonies appear, thereby making a "carbon copy," or replica, of the original plate. (2.) The technique can be used to screen for ampicillin-resistant bacteria (i.e., bacteria containing the ampicillin-resistant gene) growing on a plate containing tetracycline-resistant bacteria. The replica plate contains ampicillin, so only the ampicillin-resistant bacteria will grow. Reference back to the original place indicates which bacteria have the genes for both tetracycline resistance and ampicillin resistance.

containing cells with the insulin genes. Should the DNA technologist desire to work with these genes, they can be obtained at a later day by extracting them from the cells, a far easier task than returning to the original pancreas or other cells.

■ THE cDNA LIBRARY

Before leaving the concept of the gene library, we will examine a second form (Questionline 5.3). This form consists of complementary DNA (cDNA) molecules, synthesized as noted earlier in this chapter. A **cDNA library** consists of organisms (such as *E. coli*) that carry selected DNA fragments for specific cellular functions. (By contrast, the genomic library contains genes that encode more general cellular functions.) More specific genes are contained in the cDNA library because the cDNA library has been synthesized by the activity of reverse transcriptase from the specific information in mRNA.

To **screen a cDNA library**, a special type of molecular probe must be used. This probe is prepared by isolating plasmids with their cDNA inserts from the carrier organisms. The plasmid DNA is then treated chemically to separate the double helix into two single strands of DNA. The strands are attached to nitrocellulose filter paper, and a mixture of mRNA molecules is mixed with the filter paper. The mRNA molecules

cDNA library

a mass of cells that have incorporated into their plasmids a specific set of genes such that the genes can be obtained in a relatively easy manner.

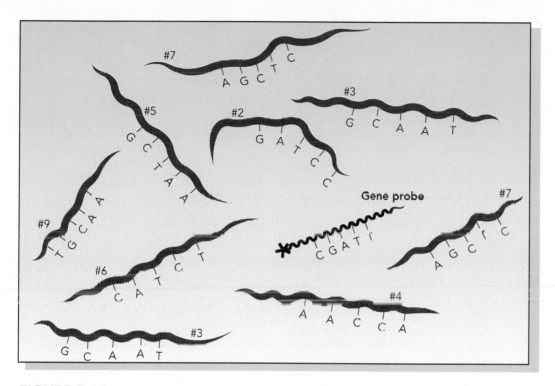

FIGURE 5.15

The use of a gene probe to locate the insulin gene. After a copy of the original gene library has been made (previous figure), the DNA from a bacterial colony is tested to determine whether the cells contain the desired insulin gene. To detect the gene, a gene probe is mixed with the DNA fragments. In this case, the gene probe has a base code complementary to the code of the insulin gene. When it unites with that gene, a radioactive signal is emitted to indicate that a union has occurred. In the figure, the gene probe would unite with fragment #5, which is the insulin gene.

are complementary to a multitude of gene sequences, so if any mRNA molecules are complementary to the DNA strands, they will stick to the DNA. All other mRNA molecules are then removed from the filter paper by washing.

What now remains on the filter paper is a population of similar mRNA molecules bound to single-stranded DNA molecules. The combinations are removed from the paper, and the RNA is separated. Next the mRNA is combined in a cell-free system with all the materials necessary for protein synthesis. The protein produced is then identified. Working backward, the protein gives information on the cDNA that supplies its genetic code. In so doing, the carrier cells are identified as the ones containing the cDNA for a particular protein.

Once the cDNA has been identified by this method, it can be used as a **gene probe** to screen a population of cells for that DNA (or RNA) sequence. For example, extracts from an unknown cell population can be combined on filter paper with known cDNA. If complementary binding takes place, the cDNA in the library of the unknown cell population is identified. If no binding occurs, other cDNA probes can be used until complementary binding occurs.

A **modern adaptation of the replica plating technique** can also be used to screen a genomic or cDNA library. In this process, colonies of the carrier organisms are cultivated on the gel surfaces of Petri dishes. But instead of sterile velvet, nitrocellulose filter paper is used to obtain replicas of the colonies. The nitrocellulose paper is pressed to the colonies on the surface to obtain a "photocopy" pattern, then the paper is soaked in a sodium hydroxide solution at a high temperature (65°C) to

gene probe

a molecule of DNA or RNA that unites with its complementary DNA or RNA molecule when introduced to a mass of foreign nucleic acid.

QUESTIONLINE 5.3

1. **Q.** **How are genes obtained for use in genetic engineering experiments?**

   A. Going to the source each time genes are needed would be a difficult task. For example, separating one human gene out from 100,000 genes in a human cell is a highly involved and biochemically difficult procedure. Instead, DNA technologists go to a gene library to obtain their gene. A gene library is a collection of host cells, one of which contains the gene they desire. The technologists screen the cells to locate those harboring the desired gene, then they extract the gene by biochemical methods and use it in their experiments.

2. **Q.** **How is a gene library screened to locate the desired gene?**

   A. There are several ways to screen a gene library. One method uses a genetic probe, a segment of mRNA that complements the base code on the desired gene. The library cells are broken open and the mRNA probe is added. Like a left hand searching for a right hand, the probe mingles among the mass of DNA strands and unites only with the complementary DNA molecule (assuming it is there). A radioactive signal indicates that a match has been made, and the population of cells is identified as containing the desired gene. If no signal is sent, no match was made, and a different population of cells is tested for the gene.

3. **Q.** **Are there different kinds of gene libraries?**

   A. Two major types of gene libraries exist: the genomic library and the cDNA library. The genomic library is a population of cells containing all the genes from a foreign organism. For example, a population of bacteria may have in their plasmids all the genes from a human pancreas cell. Of course, no single bacterium has all the human genes; different bacterial cells have different genes. A cDNA library consists of a population of bacterial cells containing complementary DNA molecules for controlling a specific function, such as insulin production. This sort of library is more valuable than the genomic library, but it is more difficult to prepare.

release the DNA from the cells and separate the strands (a process called denaturation). Now a radioactive gene probe is added to see where complementary binding will take place. A suitable incubation period under conditions that encourage binding (usually 24 hours) is permitted to pass.

After the incubation period, the filters are subjected to a wash to remove the excess gene probe and radioactivity. The filters are then sandwiched against sheets of X-ray film. Radioactivity in the form of beta particles is released from the radioactive probe into the X-ray film. After the film is developed, the radioactive areas show up as dark smudges. By comparing the location of colonies to those on the original plate, the DNA technologist can identify the colonies carrying the DNA sought. Figure 5.16 summarizes the process.

Still another method for screening a gene library is by taking advantage of **nutritional defects**. A "mutant" bacterium, for example, may be unable to grow in a culture medium unless the amino acid histidine is supplied. (The organism is a nutritional mutant because it cannot synthesize its own histidine.) If genes were inserted into a normal strain of the organism (i.e., one that can synthesize histidine), the gene-carrying bacteria could be selected out of a mixed population by culturing in a histidine-free medium. Under these conditions, the mutant bacteria could not grow because the medium lacks histidine, whereas the desirable, gene-carrying bacteria would form colonies (they synthesize their own histidine and therefore can grow). Still another method is to use

histidine
a heterocyclic amino acid used in the synthesis of protein by living organisms.

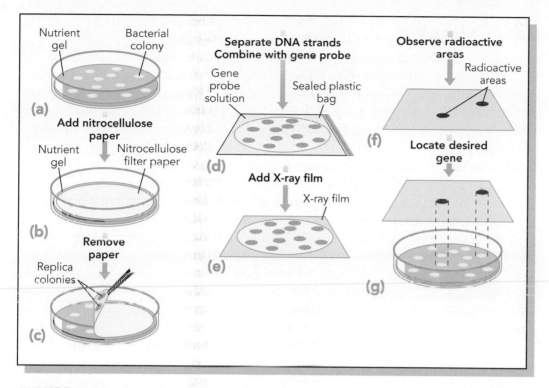

FIGURE 5.16

An alternative method for screening a DNA library. (a) Host carrier organisms containing the desired gene are cultivated on nutrient gel, where they form colonies (b) A pad of nitrocellulose filter paper is applied to the surface to obtain a replica of the colonies. (c) The nitrocellulose paper is peeled off the gel surface taking along bacteria from the different colonies. (d) The paper is treated to disrupt the cells and separate the strands of DNA. Then it is placed in a sealed plastic bag together with a solution containing a radioactive gene probe. The gene probe is specific for the desired gene and will react with its complementary DNA during the incubation period. (e) Now the paper is pressed against X-ray film. If radioactivity is emitted as a result of the gene-probe interaction, it will expose the X-ray film. (f) Dark areas appear on the film where radioactivity has been emitted and where an interaction has taken place. (g) The X-ray film is compared with the original plate to determine which colonies contained the DNA that emitted the radioactivity. These are colonies containing bacteria with the desired gene.

antibiotic-resistant genes as described in Chapter 7.

Like a regular library a gene library requires a regular **"refreshing"** to maintain its collection of information-containing cells. The library, either genomic or cDNA, is easily maintained by transferring its cells to fresh growth medium at regular intervals. The methods for cultivation and maintenance are well established for most carrier organisms. An alternative method is to freeze-dry the organisms, a process called **lyophilization**. So maintained, the organisms can be thawed and the genetic information retrieved at the whim of the DNA technologist. Storage at −70°C in liquid nitrogen is also possible. The gene library provides a regular source of genes and permits the DNA scientist to avoid the otherwise tedious task of returning to the gene source.

It should be clear from the foregoing that the many problems arising from DNA technology have been matched by equally inventive solutions. DNA technologists have manipulated the gene almost like a biochemical plaything and have succeeded in getting the gene to do their bidding. Their successful development of biochemical methods has led to a plethora of practical uses of DNA technology for improving the quality of human life. The remaining chapters of this book detail the nature of those successes.

lyophilization

li-of″i-li-za′shun

the process of freeze-drying.

SUMMARY

The methods of leading edge DNA technology are highly sophisticated and available only to individuals who have the educational background to understand them and monetary resources to investigate them. For example, stimulating a cell to manufacture a desired gene product is much more complex than might be imagined. The method begins with the isolation of a gene or its synthesis. The enzyme reverse transcriptase can be used for this task. The gene is then attached to a carrier molecule called a vector. A plasmid can be a vector, or this role can be filled by a cosmid or a virus such as a bacteriophage. The vector must have gene sequences for an origin of replication, a promoter, and a termination signal. Vector stability within a host cell is an important consideration, as is the nature of the host cell. The bacterium *E. coli* has been the traditional host cell of choice, but the bacterium *B. subtilis* and the yeast *S. cerevisiae* have also gained favor. Eukaryotic cells such as mammalian cells can be used, although cultivation difficulties must be addressed.

To produce the gene product inside the host cell, the DNA technologist must deal with such considerations as the proper gene signals on the vector, the ability of the mRNA to bind to ribosomes, and the presence of introns in the DNA molecule. Collecting the gene product may also require attention because the host cell will attempt to destroy foreign proteins within its cytoplasm. For the biochemist, exporting the gene product is more desirable than retaining it, so secretion is an important consideration.

For immediate use, the genes used in DNA technology are obtained from host cells or organisms called gene libraries. For example, the genes for insulin can be stored in *E. coli* cells. Screening the *E. coli* cells uses a detailed analysis involving a molecular probe that will bind specifically to mRNA molecules for insulin production. A library such as this is a genomic library compared with a cDNA library where genes for specific functions are stored. Probes are also used to screen cDNA libraries; the use of nutritional defects is an alternative screening method. The gene library can then be used to provide the gene for insertion to a vector during DNA technology experiments.

REVIEW

This chapter has surveyed the methods, problems, and solutions that attend the processes of DNA technology. To test your knowledge of the chapter, provide answers to the following questions:

1. What is reverse transcriptase and how can it be used to synthesize a gene?

2. Which enzyme is used to synthesize a double-stranded DNA molecule from a single-stranded molecule and how does it operate? Why is cDNA given that name?

3. Name three possible vectors for genes used in DNA technology and describe the advantages of using each vector.

4. How can vector stability and retention be achieved within a host cell or organism?

5. Name three microorganisms used as host organisms for vectors and describe the advantages of using each organism.

6. What is the intron-exon problem that surfaces during gene expression and how can it be resolved?

7. How do DNA technologists increase the amount of gene product during protein synthesis, and how is the problem of protease production circumvented?

8. What is a gene library, and why is it important to have genes deposited in a gene library?

9. Using the insulin gene as an example explain how a genomic library is formed. Describe how the gene library can be screened for the insulin gene.

10. Explain how a cDNA library is formed and how this library can be screened using a gene probe.

FOR ADDITIONAL READING

Berns, M. W. "Laser Scissors and Tweezers." *Scientific American* (April, 1998).

Browne, M. W. "New Tool is Developed for Manipulating Genes of Organisms." *The New York Times* (August 4, 1992).

Goldsmith, D. "Human Artificial Chromosome Created." *Science* News 151 (1997): 204–205.

Hu, W. S., and Peshwa, M. "Mammalian Cells for Pharmaceutical Manufacturing." *ASM News* 59 91993): 65–73.

Kher, U. "A Man-made Chromosome." *Discover* (January, 1998).

Knight, P. V. "Baculovirus Vectors for Making Proteins in Insect Cells." *ASM News* 57 (1991): 567–571

Roush, W. "Counterfeit Chromosomes for Humans." *Science* 276 (1997): 38–39.

Zinkernagel, R. M. "Gene Technology and Democracy." *Science* 278 (1997): 1207.

Pharmaceutical Products of DNA Technology

 ## CHAPTER OUTLINE

LOOKING AHEAD

Some of the most useful applications of DNA technology occur in the pharmaceutical industry. This chapter explores the range of pharmaceutical products manufactured through DNA technology. On completing the chapter, you should be able to:

- understand how deficiencies of proteins, such as insulin, human growth hormone, and Factor VIII, contribute to ill health and how DNA technology can be used to produce these proteins.

- appreciate some of the biochemical problems encountered in manufacturing pharmaceutical products by DNA technology and recognize how biochemists circumvent these problems.

- discuss the synthesis and innovative uses of therapeutic drugs such as tissue plasminogen activator, interferon, and antisense molecules and summarize how they can be used to relieve disease.

- understand the basis of vaccine activity in the body and conceptualize the role of DNA technology in the production of vaccines.

INTRODUCTION

In the 1940s, the development and use of antibiotics marked a transition point in medicine. Before that time, doctors could do little for patients with cholera, typhoid fever, diphtheria, and other infectious diseases. With the introduction of antibiotics, however, they had powerful and decisive therapies at their disposal, and they could now alter the course of disease significantly. The antibiotics effected a radical change in medicine and charted a new course that has carried through to the present day.

Since the early 1980s, another radical change has been occurring in the treatment of diseases such as diabetes and hemophilia. **Diabetes** can result from the failure of the pancreas to produce insulin, whereas **hemophilia** is related to the liver's inability to produce an essential clotting factor. Both insulin and the clotting factor are proteins, and both can be replaced in the body from other sources. For example, insulin is obtained from animal pancreases and clotting factor from donated blood plasma. Unfortunately, obtaining the proteins is expensive, and there are possibilities for microbial contamination and allergic reactions in the patient.

Enter DNA technology. In the last fifteen years, scientists have used this innovative technology to develop safer modes of treatment for diabetes and hemophilia. For instance, it is now possible to obtain human insulin from genetically engineered bacteria (Figure 6.1); and it is feasible to isolate clotting factor from cultivated animal cells. In this chapter, we will see how both of these achievements came about.

Another medical improvement stems from enhancing the body's resistance to disease. Body cells, for example, are known to produce an antiviral substance called **interferon**. Physicians have found that supplying interferon to the body is an effective way of treating such viral diseases as herpes simplex, measles, chickenpox, and poliomyelitis. DNA technologists can now produce interferon in large quantities, and some optimistic biochemists foresee the day when interferon can be used for numerous viral diseases just as penicillin is used for many bacterial diseases.

The revolution in medicine also includes a number of so-called **designer drugs**. Physicians use these chemical products of DNA technology to inactivate a single molecule involved in the disease process. Certain drugs, called antisense molecules, react with messenger RNA molecules in cells, thereby interrupting the process of protein synthesis. We will examine how they work in the pages ahead.

For the prevention of disease, DNA technologists have brought forth an entirely new class of vaccines. The newer **vaccines** stimulate the immune system more efficiently than the older ones and have fewer side effects in the recipient. With these two advances, the new vaccines bode well for preventing disease in the years ahead.

hemophilia

an inherited disorder arising from deficiency of one or more blood-clotting proteins.

interferon

in″ter-fer′on

a cellular protein produced in response to viruses to yield protection against viral penetration of cells.

FIGURE 6.1

Crystals of purified bacteria-derived insulin. These crystals of human insulin were synthesized by engineered cells of the intestinal bacterium *Escherichia coli.*

Restoring and replacing missing proteins such as insulin and clotting factor encourage the body to perform its normal physiological processes. Augmenting the body's own defensive mechanism such as with interferon enhances its ability to resist infection naturally. And using an antisense drug to inactivate a specific molecule encourages physicians to focus on a certain disease-related substance. In all these cases, the pharmaceutical product works by a method that is rational and easy to understand. DNA technology thus brings to medicine a straightforward approach to the treatment of disease, one that is more precise than many hit-and-miss approaches of past eras. In those eras, the potions and concoctions of community healers may have been therapeutically successful, but no one really knew why.

HUMAN PROTEIN REPLACEMENTS

A human cell has about 100,000 genes, most of which encode an essential body protein. On occasion a gene is defective, with the result that its protein is incorrectly produced or, in some cases, not produced at all. A deficiency of the protein may go unnoticed, but sometimes the deficiency results in disease. The deficiency is generally inherited and thus the disease is genetically linked.

It is now possible to treat certain genetically linked diseases as long as the protein deficiency is identified. In this section we will discuss several instances in which the deficient protein has been pinpointed and where DNA technology is used to synthesize and restore the missing protein. In some cases, such as those discussed in Chapter 7, DNA technology can even help make the identification of a defective gene feasible.

INSULIN

An example of a genetically linked disease is **diabetes mellitus**. In the most widespread form of this disease, the beta cells of the pancreas fail to produce sufficient quantities of the hormone insulin, whose structure is shown in Figure 6.2. Insulin was originally identified by Frederick Banting and Howard Best in a series of classic experiments performed in 1921. It is distributed by the bloodstream to all cells of the body.

Insulin has considerable importance to the well-being of the body because it facilitates the cellular uptake of glucose for use in energy metabolism. Without insulin, the glucose remains in the bloodstream, and production of energy-rich adenosine

mellitus

mel-e′tus

Insulin

a pancreatic hormone composed of protein and responsible for facilitating the passage of glucose from the bloodstream into the cells.

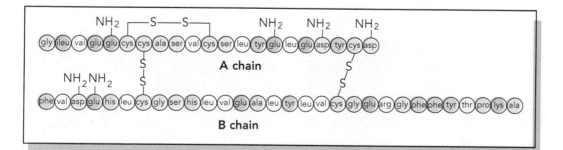

FIGURE 6.2

The insulin molecule. Insulin is a protein composed of two polypeptide chains: the A chain and the B chain. The A chain has 21 amino acids, each designated by its appropriate abbreviation; the B chain has 31 amino acids. The chains are bonded to one another at specific points by the side chains of selected amino acids. The bonds that form between the polypeptide chains are called disulfide (-S-S-) bonds.

triphosphate (ATP) is severely depressed. The patient with diabetes mellitus is very weak and tired because the cells are very low on ATP. The kidney removes excessive glucose from the bloodstream and excretes it in the urine ("mellitus" means sweet, a reference to the glucose in the urine). To keep the glucose diluted, the kidney also puts out an abnormally large about of water (the word "diabetes" is derived from the term for siphon, a reference to the large volume of water being siphoned from the body).

Diabetics are prone to eye damage and blindness, as well as kidney disorders, nerve damage, and circulatory diseases, including gangrene and stroke. They also must be careful to avoid diabetic coma, a loss of consciousness caused by pH imbalance arising from the body's breakdown of fats. Approximately 60 million people in the world are believed to be diabetic. They must take regular injections of insulin to avoid the serious consequences of the disease.

Insulin is a protein composed of fifty-one amino acids in two interconnected chains, an A chain of thirty amino acids and a B chain of twenty amino acids (Figure 6.2). Since the experiments of Banting and Best, insulin has been purified from the pancreases of pigs and cows obtained at slaughterhouses. Animal insulin is similar to human insulin, but there is a difference of one or three amino acids (for pig or cattle insulins, respectively). This difference in chemical structure causes some diabetic patients to have allergic responses to insulin. Therefore, human insulin is preferable to the animal form. Another problem is accessibility: seven to ten pounds of pancreas tissue (from 70 pigs) are required to obtain enough insulin to treat one patient for one year.

In the late 1970s biochemists saw the opportunity to apply the principles of DNA technology to the production of insulin. They recovered a segment of DNA from a gene library (Chapter 5) and verified that the segment codes for insulin. They then attached the structural gene for insulin to the bacterial plasmids of **Escherichia coli** depicted in Figure 6.3. Next to the structural gene they inserted a promoter site derived from the *lac*

glucose

a six-carbon carbohydrate used as a major energy source in the human body.

gene library

a collection of living cells that contains accessible genes for use in DNA technology experiments.

Escherichia

esh″er-ik′-e-ah

promoter site

a sequence of nucleotides that imitates the transcription of the genetic code in DNA to messenger RNA.

signal peptide

a small protein that facilitates the movement of a manufactured protein out of the cell into the extracellular environment.

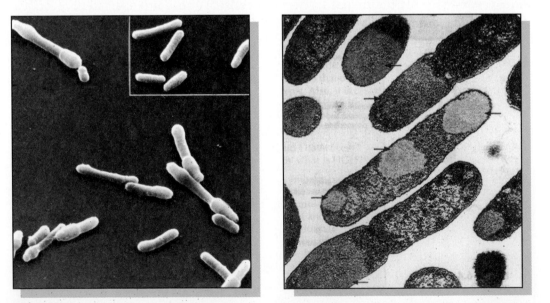

(a)   (b)

FIGURE 6.3

Insulin production in *Escherichia coli*. (a) A scanning electron micrograph of *E. coli* cells while producing the protein insulin. Prominent bulges can be seen in the cell walls, possibly reflecting a release of the gene product. The inset shows normal *E. coli* cells having no bulges. (b) A transmission electron micrograph of an ultrathin slice of *E. coli* cells during insulin production. Prominent inclusion bodies (arrows) are related to the intense protein production going on in these cells.

operon, as discussed in Chapter 3. Instead of promoting enzyme synthesis for lactose digestion, however, the promoter site initiated the production of human insulin. Because the insulin gene was a human gene, the insulin produced was human insulin.

In July 1980, seventeen volunteers received injections of genetically engineered insulin at Guy's Hospital in London, England. They were the first humans to receive human insulin produced by DNA technology. Indeed, they were the first to receive any pharmaceutical substance manufactured by DNA technology. The insulin worked as well as the traditional animal-derived insulin, and the experiments verified biochemists' hopes that DNA technology could be used to make substances with medical and commercial value.

Since those early years, insulin production has been streamlined further. Researchers have modified the bacterial plasmids by adding a short segment of DNA between the promoter site and the structural genes. The new segment directs the bacterial cell to attach a signal peptide consisting of twenty-four amino acids to one end of the insulin molecule. The signal peptide encourages the bacterium to secrete the insulin to the surrounding medium as it is produced, thus saving the trouble of retrieving insulin from the cell's interior. The peptide detaches from the insulin molecule as the latter passes through the cell membrane. The bacterium continues its metabolism undisturbed, as food is supplied at one end of the fermentation tank and insulin is removed at the other end.

Production has also been enhanced by synthesizing the two insulin chains, A and B, separately. In the human body, the pancreas cells first produce a large chemical molecule called **preproinsulin**. This molecule includes a signal sequence and three sections of amino acids. During natural insulin production, the signal sequence is removed, and **proinsulin** is formed. Then, a section of 33 amino acids is cut free at the cell's Golgi apparatus, and the remaining two sections (the A and B chains) are joined to form a molecule of insulin. Synthesizing preproinsulin is too difficult in bacteria, but DNA technologists found they could sidestep this process by programming separate bacteria with the genes for chain A and chain B, as Figure 6.4 demonstrates. Once the chains are synthesized, they are extracted from the culture media, purified, and joined chemically by disulfide linkages. Synthetic insulin is the result.

Golgi apparatus
a series of flattened membranes along which proteins are processed before synthesis is complete.

In 1986, the Eli Lilly Company received approval to market **Humulin**, the first genetically engineered insulin for general use. At the time, the successful use of DNA technology to produce insulin demonstrated without question that microorganisms could be persuaded to manufacture foreign proteins for use in humans. However, the following years brought some controversy because biotechnologists discovered a way to alter the chemical structure of porcine (pig) insulin to make it identical to human insulin. Thus, human insulin could now be manufactured by two methods: by DNA technology and by chemical alteration methods. It is unclear whether the altered-porcine insulin or the plasmid-derived insulin will eventually become the insulin of choice.

Humulin
hume'u-lin

■ HUMAN GROWTH HORMONE

Dwarfism is another metabolic disorder treated by restoring an underproduced protein in the body. Dwarfism occurs when the pituitary gland, a pea-shaped gland at the base of the brain, fails to secrete a sufficient amount of **human growth hormone (HGH)**. The hormone is a protein that stimulates overall body growth by increasing the cellular uptake of amino acids and encouraging protein synthesis. HGH also promotes the use of fat for use as body fuel.

**HGH
human growth hormone**
a hormone that promotes overall body growth..

When the pituitary gland produces insufficient HGH in childhood, growth is retarded. The child develops a chubby face, abundant "baby fat" at the waist, and a

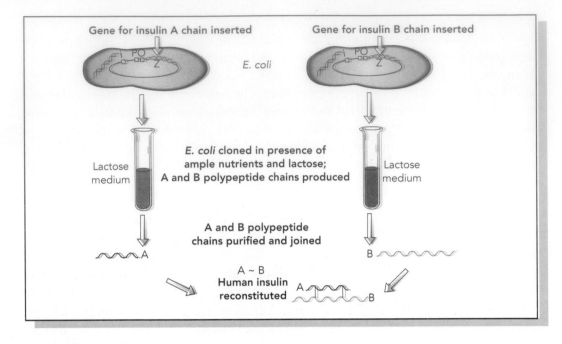

FIGURE 6.4

The mechanics of insulin production. The gene for chain A of insulin is inserted to the plasmid of one culture of *E. coli* cells together with an inducer gene (i), a promoter gene (P) and an operator (O) gene. Another culture of cells is genetically altered with the gene for the B chain. The operon used is the *lac* operon, so lactose is used in the culture medium as the inducer substance. The cells are cultivated, and the A and B chains are synthesized separately and isolated. Joining A and B chains forms the human insulin molecule.

height of about four feet. Though of normal intelligence, the adult has unusual body proportions and is a dwarf.

If dwarfism is diagnosed at an early age and while bone growth is still occurring, injections of HGH can be given. The traditional treatment is very expensive, however, because the pituitary glands from about eighty cadavers must be used to obtain enough HGH for a year's therapy. (Treatment for eight to ten years is common.) Children are also in danger of disease from infected brain tissue from cadavers. In 1985, for instance, the use of cadaver tissue was restricted in the United States and Great Britain because of the possibility of transferring **Creutzfeldt-Jacob (CJ) syndrome**. This disease of the brain is believed to be caused by a virus or a viruslike agent (a prion), and it is accompanied by tremors, convulsions, dementia, and wasting of the muscles. Indeed, by 1993, twenty-four cases of CJ syndrome had been identified in French recipients of HGH from cadaver tissue.

DNA technologists can now produce HGH exclusively through the mechanisms of genetic engineering (Questionline 6.1). The process is similar to that for insulin: the gene for HGH is spliced into a bacterial plasmid and the latter is then introduced to a bacterium. The gene product, HGH, appears in the extracellular environment (Figure 6.5).

To achieve suitable amounts of HGH, DNA technologists modified the basic production techniques to address a biochemical problem that arises. The natural HGH molecule consists of 191 amino acids. During HGH production in the body, an intermediate molecule having an extra twenty-six amino acids is encoded as a signal peptide. This amino acid chain is eventually cut free during secretion. Now, to synthesize the gene for HGH, the mRNA that encodes HGH is used as a model and a **complementary DNA (cDNA) molecule** is synthesized. When this cDNA is inserted into *E. coli*, the bacterium responds erratically because it does not "understand" the

Creutzfeldt

kroits´felt

Jacob

yak´ob

cDNA

the DNA molecule resulting from reverse transcriptase activity on RNA; the base code of cDNA complements the base code of the RNA molecule.

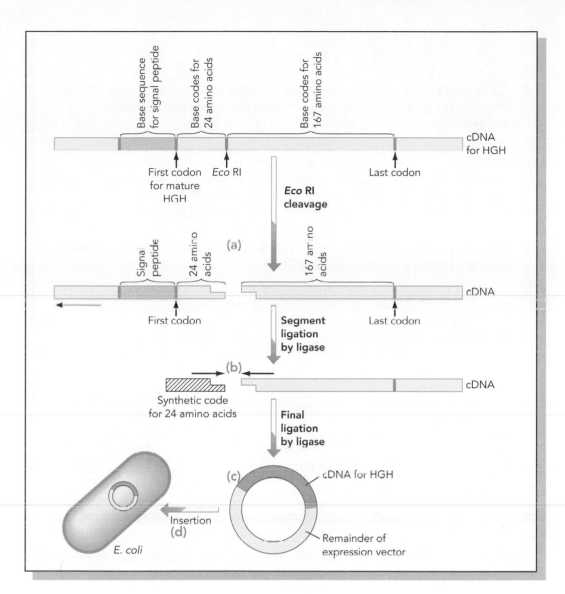

FIGURE 6.5

Gene production for HGH synthesis. The cDNA molecule encoding the human growth hormone (HGH) gene contains a signal sequence that cannot be interpreted in a bacterium nor can it be adequately removed by biochemical means. (a) To resolve this problem, the *Eco*RI restriction enzyme is used to remove the signal sequence genes plus genes for the first 24 amino acids in the mature HGH molecule. The remaining fragment has base codes for 167 of the total 191 amino acids in the HGH molecule. (b) Then the gene for the 24-amino acid sequence is synthesized separately and attached to the remaining cDNA molecule by ligase. (c) So constituted, the expression vector (plasmid) has no signal sequence to interrupt the protein synthesis process, and (d) the expression vector can be inserted to a bacterium such as *E. coli*.

significance of the signal peptide; it fails to remove the signal peptide from the intermediate HGH molecule. Nor can DNA technologists remove the signal peptide easily in the bacterium.

To resolve this problem, DNA technologists have worked with the cDNA molecule to cut away the signal peptide where it meets the HGH. Unfortunately, no known restriction enzyme can make this cut. The problem has been resolved by using the restriction enzyme *Eco*RI to remove from the cDNA molecule the base sequences for the signal peptide (twenty-six amino acids) plus an additional twenty-four amino acids (a total of fifty amino acids). Then the base sequence for the twenty-four amino acids

***Eco*RI**

a restriction enzyme from *Escherichia coli*.

QUESTIONLINE 6.1

1. **Q.** I have to take insulin to control my diabetes, and I was wondering whether Humulin, the synthetic insulin, is any different than the traditional animal-derived insulin.

   A. All forms of insulin are proteins that contain 51 amino acids. Insulin derived from animals has one or more amino acids that differ from human insulin, and so it could possibly be allergenic. Also, extracting and purifying insulin from animals is very expensive. Humulin, by contrast, is identical to human insulin and, therefore, the possibility of allergy is less. Moreover, it is less expensive to produce.

2. **Q.** My son is undersized for his age, and I have been told that he could benefit from HGH therapy. Exactly what is HGH?

   A. HGH stands for human growth hormone. It is a synthetic product of DNA technology that can make up the body's deficiency of the hormone. When used in undersized children, HGH stimulates body growth and helps a child reach his or her normal height and size. However, the hormone has to be used while bone formation is actively occurring, which means that adults would not benefit from HGH therapy.

3. **Q.** When will genetically engineered Factor VIII be commercially available for use against hemophilia?

   A. Factor VIII is a blood protein that contributes to the formation of a blood clot and, in doing so, prevents hemophilia. Synthetic Factor VIII, produced by DNA technology, is currently undergoing field trials and will probably be available for general use when you read this.

is synthesized and replaced on the cDNA molecule to formulate the complete HGH gene. Figure 6.5 displays this process. Without the signal peptide sequence, the gene can be inserted to *E. coli* cells for HGH production.

Genetically engineered HGH came into experimental use in 1985. Marketed as **Protropin** by the Genentech Company and **Humatrope** by Eli Lilly and Company, the synthetic hormone has tripled the growth rate of hormone-deficient babies in the first year and doubled it in the second and third years of therapy. Initially, the hormone was used only for children in the lowest three percent in height for their sex and age. Currently it is prescribed for children described as "very short." (Indeed, a 1996 study indicated that only four of ten children receiving HGH are actually hormone deficient.) To reach normal heights, daily intramuscular injections of the drug may be required for many years at a cost of about $20,000 per year. As of 1998, about 30,000 short children were taking hormone therapy each year in the United States.

It is conceivable that HGH therapy could also have a positive effect on children with normal hormone levels but who fall at the lowest end of the height curve. In 1990, the National Institutes of Health sponsored a ten-year study with 80 "short" teenagers, half of whom receive the hormone and half a placebo. Advocates of the drug's wider use cite studies that short stature can be a social handicap, but opponents question whether short stature should be treated as a medical disorder. The study is ongoing at this writing.

Another study published in 1997 questions whether HGH therapy helps a child

placebo
an inactive substance used in controlled studies to determine the medicinal effect of an experimental drug.

grow at a normal level after hormone therapy has ceased. For a period of 10 years, researchers at the University of Paris followed 3233 children having low growth hormone levels and calculated the target height anticipated as adults. Growth, on average, was about 2.5 inches less than expected, leading the scientists to conclude that HGH treatment had not restored normal growth patterns to the children beyond the initial spurt.

FACTOR VIII

In the United States, the most common inherited blood-clotting disorder is **hemophilia A.** This disorder affects about 1 in 10,000 males in the United States and is caused by a genetic defect for normal blood clotting. The defect is traced to the inability of the body to produce an important factor used in the blood-clotting process, the so-called **Factor VIII**.

Until the advent of DNA technology, Factor VIII was obtained from units of whole human blood. (An estimated 8000 pints of blood are needed to yield enough clotting factor for a single patient for a single year.) Unfortunately, contamination by unknown agents is possible, and this possibility was dramatically exhibited in the early 1980s by the passage of human immunodeficiency virus (HIV) to unsuspecting patients in batches of the clotting factor. During those years, thousands of hemophiliacs became infected with HIV and developed **acquired immune deficiency syndrome (AIDS)**.

Through the 1980s, DNA technologists worked to produce synthetic Factor VIII by using genetic engineering. They learned that the human gene for Factor VIII is located on the X chromosome and encompasses about 186,000 base pairs present in twenty-six exons among many stretches of introns (as shown in Figure 6.6.) Then they isolated the mRNA encoded by the DNA molecule and separated the mature mRNA (containing exons but no introns). This mRNA contains 9000 bases and encodes the final Factor VIII, a protein consisting of 2332 amino acids. Scientists synthesized a complementary (cDNA) molecule using this RNA as a template, then inserted it into hamster kidney cells where it encoded the Factor VIII protein. Mammalian cells were used because Factor VIII is very large and complex, with at least twenty-five sites where carbohydrate molecules are attached.

By 1992, two versions of Factor VIII were offered for licensure by the federal Food and Drug Administration (FDA). Within a year, both of the versions had been licensed: **Recombinate**, a product of Genetics Institute in Cambridge, Massachusetts, and

hemophilia

an inherited disorder arising from deficiency of one or more blood-clotting proteins.

AIDS

a viral disease characterized by loss of the immune system's T-lymphocytes and often accompanied by disease caused by opportunistic microorganisms.

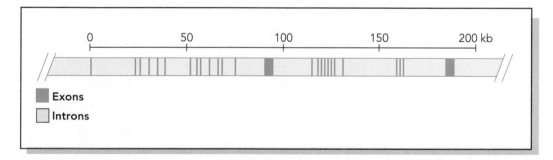

FIGURE 6.6

The complexity of the gene for Factor VIII. The gene for Factor VIII encompasses 186,000 base pairs (186 kb) organized into 26 exons. As the diagram shows, the exons are of varying length. Between the exons, are numerous introns having as few as 200 and as many as 32,000 base pairs. The final Factor VIII protein encoded by this large gene has 2332 amino acids (compared with 51 amino acids for insulin).

Kogenate, produced by Miles Laboratories. Symptoms of immunological intolerance have been observed in some recipients.

HUMAN THERAPIES

The promise of biotechnology has also extended into the realm of disease therapy as researchers and physicians seek to develop new drugs and approaches to treating disease using the fruits of DNA technology. Two examples of innovative drugs are tissue plasminogen activator (TPA) and interferon. Both drugs, once in very short supply, are now produced in large enough quantities for therapeutic use. In addition, an entirely new class of compounds called antisense molecules has been developed to hone in on the specific molecules associated with disease. In this section we will explore the production and uses of these drugs

TISSUE PLASMINOGEN ACTIVATOR

For many years, coronary thrombosis has been treated with streptokinase and other enzymes that destroy blood clots in the blood vessels. A recent addition to the list of therapies is a protein called **tissue plasminogen activator (TPA)**. TPA is a naturally occurring protease expressed in mammalian cells. It attaches itself to blood clots and stimulates other blood-borne components to break down the clot without reducing the blood's clotting power elsewhere in the body. By comparison, another clot-dissolving enzyme called urokinase decreases the clotting ability of blood throughout the body, with the consequent risk of internal bleeding. TPA can also be administered intravenously, whereas other clot dissolvers (i.e., urokinase or streptokinase) must be administered directly into the blocked vessel; TPA appears to work more quickly than other dissolvers.

TPA was the first drug to be produced by a mammalian cell culture. In the early 1980s, DNA technologists synthesized the complementary DNA molecule for TPA. Then they biochemically attached it to a synthetic plasmid containing a TPA signal sequence and promoter and termination sites. The plasmid was introduced to mammalian cells, and cells secreting high levels of TPA were obtained from the mixture, as Figure 6.7 demonstrates. Cultivated in large fermenters, the cells supplied sufficient TPA for pharmaceutical use.

In 1987, the biotechnology company Genentech received a license from the FDA to market TPA under the trade name **Activase**. Administered as a thrombolytic agent, TPA exhibits its protease activity by degrading an inactive precursor enzyme named plasminogen, changing it into plasmin. Plasmin is also a protease. It degrades fibrin, the protein of which blood clots are formed. Patients that demonstrate early signs of heart attack or stroke are treated with TPA. Some physicians maintain it is not an improvement over streptokinase or urokinase. Nevertheless, the reduced side effects of TPA compared with other clot-dissolving drugs is generally acknowledged.

INTERFERON

Interferon is an antiviral substance first identified in 1957 by Alick Isaacs and Jean Lindemann. Interferon is not a single substance but a group of more than twenty substances separated structurally and functionally into three subgroups: alpha, beta, and gamma interferons. All interferons are protein molecules, and many are also **glycoproteins**: they have carbohydrate molecules attached to the amino acids at various points along the chain. They are produced by the cells of various mammals on stimulation by viruses. The interferons trigger a nonspecific reaction that protects cells against numerous different viruses, including the ones that stimulated their production.

---

plasminogen

plas-min′o-jen

protease

an enzyme that digests protein.

urokinase

u″ro-ki′nase

streptokinase

strep″to-ki′nase

termination site

a series of nucleotides that signals the end of transcription of DNA to messenger RNA.

thrombolytic

able to break up a blood clot.

interferon

in″ter-fer′on

a cellular protein produced in response to viruses to yield protection against viral penetration of cells.

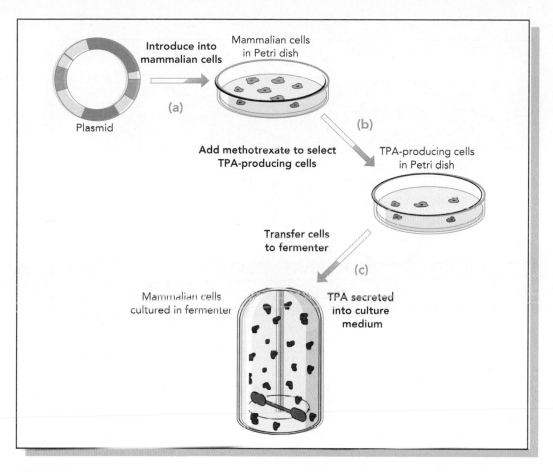

FIGURE 6.7

The production of tissue plasminogen activator (TPA) in mammalian cells. (a) The synthetic plasmid is introduced to mammalian cells. (b) A chemical called methotrexate is added to select out the cells carrying the plasmids and producing high levels of TPA. (c) The cells are transferred to an industrial tank called a fermenter, and as TPA is produced, it is secreted by the cells into the surrounding medium.

Interferons are produced by cells after a virus releases its nucleic acid into the cell's cytoplasm. The Interferons are secreted by the cell, and they bind to receptor sites on nearby cells. Here they trigger the production of protective proteins as illustrated in Figure 6.8. The proteins inhibit viral replication by methods not fully understood, although evidence exists that at least one protective protein binds to mRNA encoded by the virus. Unfortunately, human interferons are the only ones that operate in human cells; mouse, dog, or other animal interferons are ineffective.

The history of interferon study has been somewhat checkered. Biochemists were intensely interested in the activity of interferons in 1960s, but they could not obtain enough to work with (more than 90,000 pints of blood were required to produce a single gram of interferons); the research became too difficult to pursue. However, a major breakthrough occurred in 1980 when Swiss and Japanese biotechnologists deciphered the genetic code for alpha interferon and spliced the genetic code DNA fragment into *E. coli* plasmids. The bacteria-derived interferons reduced the symptoms of hepatitis, diminished the spread of herpes zoster (shingles), and shrank certain tumors.

In 1984, a Swiss biotechnology firm began marketing **alpha interferon** as Intron. And in 1986, the U. S. FDA approved the use of alpha interferon against a form of leukemia; approval for use against genital warts followed in 1988. At this writing, alpha interferon is used to treat Kaposi's sarcoma (a form of blood and skin cancer occurring

leukemia

a cancer of the blood characterized by abnormally high numbers of leukocytes.

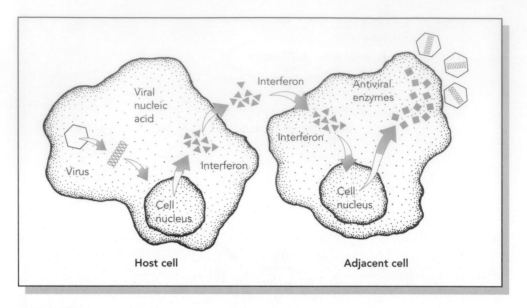

Host cell   **Adjacent cell**

FIGURE 6.8

The mechanism of action of interferon. When a virus enters a host cell, its nucleic acid stimulates the cellular DNA to begin the process of interferon production. The interferon passes out of the cell and enters an adjacent cell. Here it stimulates the cellular DNA to encode a series of antiviral proteins. These proteins appear to prevent viral replication and, thus, they exert a protective effect on the cell.

multiple myeloma

a malignant cancer of plasma cells arising in the bone marrow and accompanied by skeletal bone destruction.

in AIDS patients), as well as malignant melanoma, multiple myeloma, and certain kidney cancers. Another interferon, **interferon beta-1b**, was licensed in 1993 for use against a form of multiple sclerosis.

The outlook for interferons is either bright or gloomy, depending on one's perspective. On the plus side, the possibility remains viable that interferons could be clinically useful for a broad spectrum of viral diseases and cancers (Questionline 6.2). Indeed, researchers have shown that interferons could be useful as a nasal spray to interrupt common colds. On the negative side, interferons are costly to produce because bacteria cannot synthesize glycoproteins such as interferons. This drawback may be offset by inducing bacteria to produce a hybrid interferon, that is, an interferon produced from the genetic codes of various interferons spliced together. This method bypasses the carbohydrate attachment step and appears to produce a more reactive interferon.

■ ANTISENSE MOLECULES

Early in this century, the noted biochemist and Nobel laureate **Paul Ehrlich** introduced to medicine the concept of the "magic bullet." Ehrlich imagined a seemingly impossible drug that would destroy a microorganism without causing dangerous side effects. Eventually, Ehrlich synthesized arsphenamine, a magic bullet for syphilis (as portrayed in the movie *Dr. Ehrlich's Magic Bullet*), and he laid one of the cornerstones for modern chemotherapy.

syphilis

a sexually-transmitted bacterial disease caused by a spirochete and occurring in multiple stages, often resulting in organ damage, paralysis, and death.

Today's DNA researchers are seeking a different sort of magic bullet. They attempt to develop a molecule directed at a target within an infected cell rather than at the cell itself. Such a target is usually not essential for normal cell function, nor is it present in healthy cells. Rather, it is a chemical molecule produced when the cell is infected. Presumably, the "magic bullet" would react with this target molecule and restore the cell to normal health.

QUESTIONLINE 6.2

1. **Q.** Can tissue plasminogen activator (TPA) be used to prevent heart attacks?

   A. TPA is not intended for use as a heart attack preventive but rather as therapy for those having heart attacks. It is a protein product of DNA technology that attaches to blood clots and stimulates blood components to break down the clot. The protein works locally rather than throughout the body where it could cause hemorrhaging.

2. **Q.** I've heard interferon being referred to as "penicillin for viral disease." Is the label justified?

   A. Penicillin destroys bacteria by preventing their synthesis of the cell walls. Viruses have no cell walls, so penicillin is useless. However, interferon is known to prohibit the penetration of viruses to host cells, a necessary first step to viral replication. On this basis, interferon could be a valuable antiviral product in future years. DNA technology has successfully provided sufficient interferon for laboratory testing, and the drug has already been approved for certain uses.

3. **Q.** Can you make sense of "antisense molecules" for me?

   A. During the synthesis of protein, the DNA of a cell encodes a messenger RNA molecule that provides the genetic message for the amino acid chain in a protein. The RNA molecule is said to have "sense." An antisense molecule, by contrast, is an RNA fragment that unites with and neutralizes the messenger RNA molecule. By reacting with mRNA, it acts as a highly specific chemical projectile that indirectly inhibits synthesis of a particular protein in a cell.

Consider, for example, what happens when a person has **HIV infection** or its later stages, **AIDS**. In this situation, the HIV inhabits the infected cell's nucleus as a DNA molecule. Here it encodes the production of new viruses by serving as a template for the production of messenger RNA molecules (much as a normal cell's genes encode mRNA). In biochemical jargon, the mRNA molecules have a biochemical message that "makes sense"; the message can be used to synthesize the enzymes for the production of new AIDS viruses.

To interrupt the production of HIV particles, biochemists have devised a synthetic mRNA molecule to combine with and neutralize the sense mRNA molecules, as Figure 6.9 displays. The synthetic mRNA has a sequence of nitrogenous bases complementary to the sense molecule. Thus, the synthetic mRNA molecule is called an **antisense molecule**. Used therapeutically, it enters the infected cell and combines specifically with its complementary mRNA molecule, much like bringing the left and right hands together. When the mRNA molecules are inactivated, the synthesis of viruses comes to an end and the infection subsides. Figure 6.9 shows how the antisense molecule works.

Antisense technology was postulated in the 1980s by **Paul Zamecnik** (Figure 6.10), who was well known at that time for his groundbreaking work with transfer RNA molecules (Chapter 3). The first use of an antisense molecule against HIV occurred in 1992 when tests in laboratory-infected cells resulted in a significant reduction of the HIV population. Since then, scientists at Hybridon Inc.(founded by Zamecnik) have continued to develop a drug called GEM91 for use against HIV. By 1997, experimental results were showing that the drug not only blocks HIV replication but also prevents

nitrogenous
ni-troj´en-us

antisense molecule
a molecule that unites with and neutralizes an mRNA molecule that carries the genetic message from the DNA to the cytoplasm.

Zamecnik
zam´chek

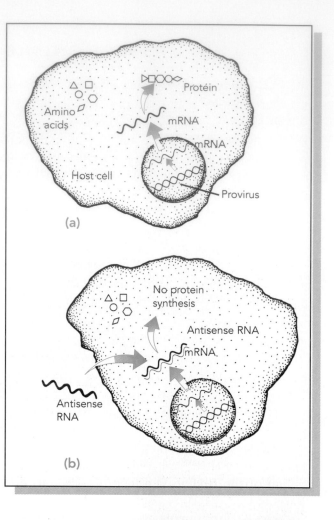

FIGURE 6.9

Antisense molecules as they are used to interrupt viral replication in an HIV-infected cell. (a)In an infected cell, the human immunodeficiency virus (HIV) inhabits the cell in the form of a DNA molecule called the provirus. Using mRNA, the provirus encodes proteins for the synthesis of new viruses, and the infection continues. (b)An antisense molecule has a base sequence complementary to that of mRNA. When it enters a cell, the antisense molecule unites with the mRNA (the sense molecule) and inactivates it. Without mRNA, protein synthesis comes to a halt, and new viral particles are not produced. Viral infection soon comes to an end.

FIGURE 6.10

Paul C. Zamecnik, winner of the 1996 Lasker Award for his achievements in DNA research and technology. Zamecnik was among the first to investigate the role of transfer RNA in protein synthesis; he later pioneered antisense technology.

binding of the virus to human body cells. (This accessory value can be a problem because it shows that antisense molecules attach to unintended targets.)

Another practical application of antisense technology was announced by British investigators in 1988. Seeking to produce better **tomatoes**, investigators successfully used antisense RNA to block the activity of the gene that contributes to the softening and spoiling of ripe tomatoes (Chapter 10). With softening inhibited, the tomatoes could ripen on the vine and withstand the rigors of shipping to market.

Antisense molecules can also be used to interrupt the activity of cancer-inducing

oncogenes. In 1989, for example, researchers at Georgetown University successfully used antisense RNA to block the expression of an oncogene called *raf*. The gene is contained in laryngeal cancer cells cultivated in the laboratory. The DNA encoding the antisense RNA was inserted into the cells, and the cells themselves produced the antisense molecule. Oncologists are hoping to interrupt leukemia-inducing genes in bone marrow cells in the same way. In the most optimistic scenarios, DNA-altered cells are placed into the bone marrow where they function as normal cells.

Patients with **Crohn's disease** have also benefited from antisense technology. Crohn's disease develops when tissues of the intestine become inflamed. The inflammation depends partly on a protein that adheres to the blood vessel wall and helps inflammatory blood cells penetrate through. An antisense molecule prepared by biotechnologists at Isis Pharmaceuticals has been used to inhibit production of the protein. Data presented in 1997 indicated that after one month of treatment, the disease went into remission in half of fifteen patients tested.

Antisense research focuses on the fact that only one strand of DNA encodes mRNA while the other strand remains dormant (Chapter 3). The dormant strand is a **natural antisense molecule**. To put it to use, scientists start with a copy of the double-stranded DNA of the original gene and perform a type of role reversal to activate the dormant strand. They accomplish this reversal by splicing a promoter site near the sequence they want the cell to express. This switches the polarity of the DNA sugar-phospate backbone. The strand once read in the backward direction now has a promoter site and is read in the forward direction. Previously dormant, the antisense molecule now has sense. Meanwhile the strand that was once active acquires antisense and becomes dormant. The molecule encoded by the newly activated DNA strand is antisense RNA.

One of the first to use inserted DNA to encode antisense RNA was **Harold Weintraub** of the Fred Hutchinson Cancer Center in Seattle. In the mid-1980s, Weintraub investigated the activity of **thymidine kinase**, an enzyme used in DNA synthesis. Weintraub's group successfully inserted a fragment of DNA into DNA-synthesizing cells and prompted the cells to produce antisense RNA to neutralize the mRNA that encodes thymidine kinase. The synthesis of thymidine kinase came to a halt and DNA synthesis was abruptly interrupted. Weintraub was among the first to demonstrate that cells could be endowed with a biochemical "program" for a magic bullet.

Biochemists can also synthesize small segments of antisense DNA identical to known sequences of DNA. These short segments consist of single-stranded synthetic DNA. They never enter the nucleus or act as the cell's normal DNA. However, they can enter the cytoplasm where they accomplish the same goal as antisense RNA; that is, they recognize and bind to target sequences on mRNA molecules.

The small segments of DNA are known as **oligonucleotides**, or simply **"oligos."** Oligos, shown in Figure 6.11, have approximately fifteen base pairs and are small enough to enter cells. Here they join with their complementary mRNA sequences to form an mRNA-DNA hybrid. The enzyme ribonuclease then attacks and destroys the hybrid. However, nuclease enzymes could destroy the DNA segment, so scientists replace oxygen atoms in the nucleic acid backbone with sulfur atoms to create a more resilient oligonucleotide containing phosphorothioate, a phosphate-sulfur compound.

Researchers have experimentally used oligos against *Trypanosoma brucei* (shown in Figure 6.12). This protozoan causes African sleeping sickness. They have also explored treatment of the viruses that cause AIDS, herpes simplex, warts, leukemia, and influenza. And they have designed oligos that migrate into the cell's nucleus and

oncogenes
genes involved in the development of tumors and cancers.

Crohn's disease
an inflammation of tissues of the intestine.

polarity
the direction in which the base code is read on a DNA molecule

thymidine
thi'mid-en

kinase
ki'nas.

oligonucleotide
ol'i-go-new'kle-o-tide

oligos
small segments of DNA that can act as antisense molecules.

phosphorothioate
fos-phor'o-thi'o-ate

Trypanosoma
tri-pan''o-so'mah

brucei
bru'ce-i

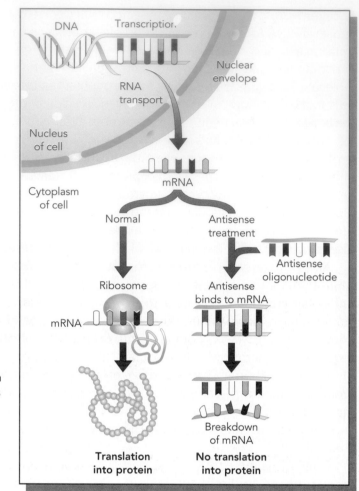

FIGURE 6. 11

The strategy underlying antisense technology with oligonucleotides. Under normal conditions, the cell transcribes a messenger RNA molecule, whose base is translated at the ribosome during protein synthesis. When an antisense oligonucleotide is used during therapy, the molecule enters the cell and binds specifically to the mRNA molecule and breaks it down. No protein is produced.

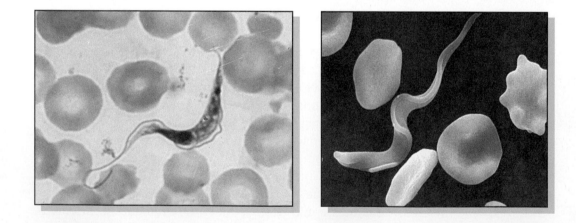

(a)   (b)

FIGURE 6.12

Two views of *Trypanosoma brucei*, the protozoan that causes trypanosomiasis ("sleeping sickness"). Antisense molecules have been used with some success to destroy this organism in the body and prevent its spread. The first view of *T. brucei* is with a light microscope; the second is with a scanning electron microscope. In both cases, the round objects are red blood cells.

squeeze themselves into specifically targeted portions of the spiral groove of the DNA's double helix. Once united with the DNA, an oligo forms a "triplex" and prevents the gene from passing its message to mRNA by inhibiting mRNA synthesis. In this way, a disease-causing gene can be stifled.

The optimism generated by antisense molecules is counterbalanced by certain concerns that must be addressed before the molecules become available for widespread practical use. For example, antisense molecules have been expensive to produce and are required in large doses to be effective. (A commercial manufacturer was selling the molecules for $2000 per gram in 1998.). Furthermore, the first generation of antisense molecules was unable to penetrate their target cells, even though they were coated with lipid-based coverings to get them past the lipid-based cell membranes. By 1997, however, scientists had shown that plain, uncloaked molecules are better suited for reaching their target cells in living animals, even though the reasons are ill-understood.

Through the years, the problem of unexpected behaviors also vexed scientists. They found that control antisense molecules (which do not complement the target RNA) were working almost as well as the therapeutic molecules. Apparently, the control molecules had the ability to mimic bacterial DNA, and they triggered an immune response that helped destroy the pathogen the antisense molecule was supposed to destroy.

In the early years of research, **side effects** also appeared to be a formidable barrier to success. Patients were experiencing increased blood clotting, high blood pressure, and decreased heart rates. Then the manufacturers discovered that the sulfur atoms in the phosphorothioate molecules caused negative electrical charges making the molecules stick to proteins both inside and outside cells. For example, the antisense molecules became entangled in growth factors and cell-anchoring proteins, effects that influence blood-related problems. This binding would also prevent the molecules from reaching the nucleus of target cells. To resolve the, difficulty biotechnologists developed a series of second-generation molecules with fewer negative charges, and thus, fewer side effects.

second-generation
referring to a class of substantially improved substances.

Another advance was made when scientists found why test animals were developing lethal **inflammations**. Apparently the antisense molecules were stimulating a reaction by components of the immune system, caused apparently by the large one-time injection. The inflammations subsided when the molecules were administered more slowly and in lower doses.

Beyond these problems are a number of other questions: can the antisense molecules resist the activity of cellular enzymes? Because the antisense molecules are directed at human mRNA, how can they be adequately tested in animals? How can researchers be certain that oligos will not integrate themselves to normal DNA? And, of interest to FDA reviewers, are the antisense molecules "biologicals" or "drugs?"

At least a dozen biotechnology companies are now investigating the use of antisense molecules, and in 1998, the first antisense preparation, **fomivirsen** (commercially known as Vitravene), was approved by the FDA for use in the eye to block replication of the **cytomegalovirus**. (Cytomegalovirus is an important cause of blindness-inducing retinal infection in AIDS patients.) Antisense molecules represent an attractive application of DNA technology and molecular genetics (albeit "backward genetics"). As the objections to antisense technology continue to fade, scientists look forward to a new generation of antisense molecules and a type of Renaissance in genetic medicine. "Finally," remarked one scientist, "we're make sense of antisense."

fomiversen
fom-i-ver'sin

cytomegalovirus
si"to-meg'a-lo-virus

■ OTHER INNOVATIVE PHARMACEUTICALS

As one of the cornerstones of the "new biology," DNA technology is central to the development of a wide range of innovative pharmaceuticals. One of the most successful products has been synthetic **erythropoietin**, a hormone normally produced by the kidneys. Erythropoietin stimulates stem cells in the bone marrow to mature into red blood cells. In 1989, Amgen Inc. received approval from the FDA to market a genetically engineered version of the hormone as Epogen (Table 6.1). The product is used to reduce the severe anemia that accompanies kidney disease. Amgen soon received FDA approval for a second drug, one that directs the stem cells to form a type of white blood cells called neutrophils. Neutrophils specialize in phagocytosis. Commercially know as Neupogen, the product helps the body resist bacterial infection by encouraging the proliferation of phagocytic neutrophils. In 1995, Amgen's profits from the two drugs totaled $1.8 billion.

A form of genetically engineered erythropoietin called **epoctin-alpha** received FDA approval in 1997 as an alternative to blood transfusions in anemic patients undergoing noncardiac, nonvascular surgery. The Ortho-Biotech Company markets the product as **Procrit**. Like the natural hormone, the drug stimulates red blood cell production. Used before surgery in anemic patients, the drug counters the effects of blood loss and lessens the need for preoperative transfusions. Fewer transfusions carry the concomitant benefits of reduced exposure to potential disease agents, decreased liability associated with mismatched blood, and fewer allergic responses to foreign substances in the blood. The major drawback is cost: the average cost for transfusing a unit of blood is $300; the average cost for epoctin-alpha therapy is $1500.

erythropoietin

e-rith"ro-poi´e-tin

a hormone produced by the kidneys to stimulate red blood cell production in the bone marrow.

epoctin

e-pok´tin

TABLE 6.1
Products of DNA Technology and Year of Approval in the United States.

PRODUCT	YEAR OF U. S. REGULATORY APPROVAL
Insulin	1982
Human growth hormone	1985
Alpha-interferon	1986
OKT-3	1986
Hepatitis B vaccine	1986
Tissue plasminogen activator (TPA)	1987
Erythropoietin-alfa	1989
Gamma-interferon	1990
Granulocyte colony-stimulating factor (G-CSF)	1991
Granulocyte-macrophage colony–stimulating factor (GM-CSF)	1991
Interleukin 2	1992
Factor VIII	1992
Beta-interferon	1993
DNase (Pulmozyme®)	1993
Glucocerebrosidase (Cerezyme®)	1994

Can genetically engineered molecules increase **memory**? Apparently they can, believe researchers at Cortex Pharmaceuticals. They have identified a series of molecules that seemingly improve the performance of volunteers in memory tests. The molecules appear to stimulate signal receptors of certain neurons. Then the receptors change structure, and in their new form they require a lower stimulus to generate an impulse. As a result, one remembers more efficiently. To be sure, memory is a very complex phenomenon, and improvement by means of DNA technology is still a far-distant dream for the most optimistic prognosticators.

Of more immediate concern are drugs to combat **cytokines**, a group of normally useful body substances that include interferons, colony-stimulating factors, and interleukins (substances that stimulate lymphocyte activities). Cytokines mediate numerous types of intracellular communications having to do with immunological responses, inflammatory reactions, and red blood cell production. However, during certain diseases such as rheumatoid arthritis, the body overproduces cytokines, disturbing the immune system and leading to pain, swelling, and stiffness of the joints. With genetic engineering techniques, researchers have developed **anticytokine antibodies** directed specifically against certain cytokines and have observed "striking improvement" in treated patients. The antibodies used against cytokines are called **monoclonal antibodies** because they are formed from a single clone of cells programmed to produce only one type of antibody.

Another form of monoclonal antibody was FDA approved in 1997 to help prevent rejection in kidney transplant patients. The antibody molecules attack T-lymphocytes (T cells) sent by the body's immune system to destroy cells in the transplanted organ. They bind specifically to target receptors on the T-lymphocyte's surface without affecting healthy cells. Moreover, the antibodies have been modified to be more like human antibodies than previous preparations, a factor that reduces the possibility of allergy.

And finally, biotechnologists have developed a bacterium known as the "**friendly Salmonella**." Long despised as a cause of severe food-borne infection, *Salmonella* cells have been genetically altered to fight cancer cells. Scientists have mutated the bacterium's DNA and disabled one of its RNA components so it is easily destroyed in the animal body—that is, unless cancer cells are present. Apparently, fast-growing tumor cells provide nucleotides and amino acids needed by disabled *Salmonella* cells. Thus, scientists at Yale University theorized that if harmless bacteria were placed near tumor cells, they might use up the nutrients needed by the tumor cells and the tumor might shrink. In experiments completed in 1997, new tumors grew in untreated animals to one gram weight in eighteen days; in mice treated with *Salmonella*, new tumors grew to the same weight in about thirty-five days, almost twice as long. Moreover, the untreated mice died in about twenty-five days; the treated mice survived for about fifty days.

By the end of 1999, the biotechnology industry was rapidly developing, with 1500 firms established for the expressed purpose of manufacturing biotechnology products. Close to 2000 therapeutic medicines were in various stages of development and testing, including drugs available for the first time against the common cold, Parkinson's disease, Huntington's disease, sickle cell anemia, and osteoporosis. Indeed, the Pharmaceutical Research Manufacturers of America estimated that the total investment in DNA technology was more than $16 billion. Table 6.2 delineates some forecasts made in 1996 for the years 2001 and 2006. The figures maintain the positive image that DNA technology has developed in the public and business consciousness.

cytokine

a cellular substance that encourages chemical communication during immune and other processes.

antibodies

protective protein molecules produced by the immune system when provoked by a foreign substance.

T-lymphocytes

white blood cells that function during the immune response by attacking and destroying parasites and by modulating the antibody response.

Salmonella

a genus of rod-shaped bacteria, one of which causes food-borne infection of the intestine.

TABLE 6.2				
Ten-Year Sales Forecast of the Value of DNA Technology Products in the United States.				
SECTOR	BASE YEAR 1996	FORECAST YEARS		AVERAGE ANNUAL GROWTH RATE (%) 1996-2006
		2001	2006	
Human therapeutics	7,555[a]	13,935	25,545	13
Human diagnostics	1,760	2,705	4,050	9
Agriculture	285	740	1,740	20
Nonmedical diagnostics	225	330	465	8
Totals	10,100	18,400	32,400	12

[a] Millions of 1996 dollars. Source: Consulting Resources Corp.

VACCINES

Tradition tells us that Chinese physicians were immunizing against **smallpox** as early as the eleventh century. They ground up dried scabs from the skin of smallpox patients and blew the powder into the noses of healthy individuals. People so-treated were less likely to contract smallpox when epidemics struck.

A less haphazard although equally effective way of immunizing individuals was developed in the late 1700s by the English physician **Edward Jenner**. Jenner noted that farmers and milkmaids who worked with cows developed a mild form of smallpox called cowpox, or vaccinia. He reasoned that if people were artificially given cowpox they might not develop the highly lethal smallpox. Jenner proceeded to inoculate volunteers with material from a cowpox lesion, and weeks later he gave volunteers an injection of material from a smallpox lesion (Figure 6.13). Smallpox failed to develop.

A hundred years passed before scientists understood the basis of smallpox immunity: cowpox viruses stimulate the body's immune system to produce protective protein molecules called antibodies; the cowpox antibodies neutralize cowpox viruses, as well as the more deadly smallpox viruses.

Modern vaccines work in the same way: they exploit the immune system's ability to recognize foreign organisms and respond with antibodies and other defenses that remain in the body and provide surveillance against future exposure to those organisms. But modern vaccines are more refined than Jenner's vaccine. Their components can consist of three types: dead bacteria or inactivated viruses, weakened (attenuated) bacteria or viruses, or viral fragments or bacterial molecules.

Of the three types of vaccine, the third is the safest because it contains no bacteria or viruses that can possibly infect the body. The bacterial molecules or viral fragments used in the vaccine are referred to as **subunits**. These subunits act as **antigens**: molecules that stimulate the immune system and elicit a specific response. Note that even though antigens are chemical molecules, they can provoke an immune response strong enough to protect the body against live organisms should they enter the body at a future date.

Streptococcal pneumonia is one disease against which a subunit vaccine has proven successful. Licensed in 1983, the vaccine consists of polysaccharides derived from the capsule of *Streptococcus pneumoniae*, the cause of streptococcal pneumonia. Unfortunately, not all subunit vaccines have been equally successful. For example, in the early and mid-1980s, the hepatitis B vaccine of that era was produced from viral fragments obtained from human blood. Contamination of the blood by HIV raised the

smallpox

a now-extinct, highly fatal viral disease of the skin and internal organs accompanied by bleeding pustules.

vaccinia

cowpox, a mild form of smallpox that affects cows.

attenuated

ah-ten´u-a˝ted

subunits

fragments of microbial proteins or polysaccharides that can be used as immunizing agents.

antigen

a foreign substance that provokes an immune response by B-lymphocytes, T-lymphocytes, or both.

pneumoniae

noo-mo´ne-a

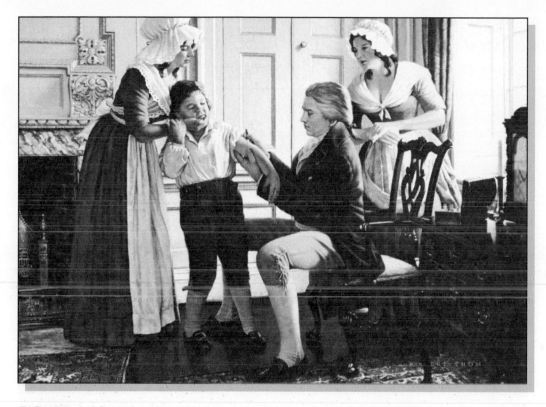

FIGURE 6.13

A painting by Robert Thom showing Edward Jenner performing the first smallpox vaccination. The young boy receiving the injection is named James Phipps. The woman to the right is Sarah Nelmes, the dairy maid from whom Jenner obtained the cowpox material for the vaccination (note she is holding a pad to her wrist, the site of the cowpox lesion).

possibility of HIV infection in recipients. This vaccine is no longer used. Instead a new vaccine manufactured by methods that use DNA technology is now in use, as we will see next.

■ HEPATITIS B

The new form of **hepatitis B vaccine** is produced through the biochemistry of DNA technology. A key element of the vaccine is a surface protein of the hepatitis B virus known as **HBsAg** (for hepatitis B surface antigen). DNA technologists have identified the gene that encodes HBsAg and have used yeast cells to synthesize copies of the antigen for use in the subunit vaccine. The yeast used to manufacture the vaccine is *Saccharomyces cerevisiae*, the harmless brewing and baking yeast.

To generate copies of the subunit, the gene for HBsAg is inserted to a plasmid from a **yeast cell**. The biochemist uses a strong promoter site from the yeast gene that encodes an enzyme called yeast alcohol dehydrogenase. A termination sequence from a yeast gene and an origin of replication from a bacterium are also inserted to the plasmid vector. In addition, the vector contains antibiotic resistance markers and a gene that permits yeast growth only in the absence of the amino acid leucine. The recombined yeast cells are selected out by cultivation in a medium without leucine. Then the cells are grown in high density in industrial size fermenters. One of the proteins they produce is HBsAg. This subunit is separated and used to immunize individuals against hepatitis B (Questionline 6.3).

hepatitis B

a viral disease of the liver caused by a fragile virus that passes from blood to blood or semen to blood to remain active.

Saccharomyces

sak´ah-ro-mi´-sez

cerevisiae

ser´e-vis´e-a

dehydrogenase

de-hi´dro-jen´´ase

leucine

lu´seen

QUESTIONLINE 6.3

1. **Q.** I've heard that the newer recombinant vaccine for hepatitis B is safer than the older vaccine. Is that true?

   A. The older hepatitis B vaccine was composed of viral fragments isolated from blood. As such, it was subject to contamination by blood-borne viruses and toxins. The newer vaccine contains viral fragments synthesized by DNA technology. Genes for the fragment are inserted to yeast cells, and they are stimulated to synthesize the fragments. Because there is no possibility for blood-borne contamination, the vaccine is considered safer.

2. **Q.** When the AIDS vaccine finally becomes available, what will it contain?

   A. Most current vaccines contain whole viruses biochemically crippled by chemical or physical methods. Most of the AIDS vaccines being developed contain no whole viruses, but rather a series of proteins or glycoproteins that stimulate an immune response. Produced by the methods of DNA technology, these proteins are safer to use than whole viruses because the proteins cannot possibly infect cells and cause AIDS. However, they provoke a less-efficient immune response than whole viruses, so this problem must be resolved.

3. **Q.** How can DNA technology be used to produce more effective vaccines in the future?

   A. Weakening or inactivating viruses for use in a vaccine is a hit-or-miss process that cannot easily be controlled. By comparison, DNA technology makes it possible for the manufacturer to select and synthesize those chemical components that are desirous for the vaccine. Moreover, the DNA technologist can bind the components to carrier molecules chosen in the laboratory. And an organism's genome can be selectively altered to destroy an organism's pathogenicity and formulate a vaccine.

allergic reaction

an immune system reaction where histamine and other mediators induce smooth muscles to contract.

In 1987, the hepatitis B vaccine was licensed and became the first synthetic vaccine for public use. Marketed as **Recombivax** by one company and **Engerix-B** by another, the vaccine is safe because it contains no whole viruses—either inactivated or attenuated. Nor is there a possibility of allergic reactions to living tissue components (as there would be in vaccines containing cultivated viruses) or any possibility of contamination by blood components or unexpected proteins.

Since 1987, the vaccine has become an accepted weapon to limit the spread of hepatitis B. It is used to immunize health care workers such as physicians, surgeons, nurses, dental hygienists, and medical laboratory technologists; and it is recommended for anyone coming in contact with blood or body secretions such as emergency medical technicians, morticians, firefighters, and police officers. Many pediatricians recommend it for newborns.

But there is still the problem of ensuring that the vaccine reaches all those who can be helped by it. The recombinant vaccine must be administered three times over a six-month period and many individuals do not receive the full series. Enter the "**hepatitis tomato**." Scientists at a Texas biotechnology company have successfully inserted hepatitis B genes into the cells of tomato plants by methods described in Chapter 10. The engineered plants produce hepatitis B antigens along with their own antigens. Company technologists look forward to the day when immunization to hepatitis B will simply involve having a tomato with lunch.

■ ACQUIRED IMMUNE DEFICIENCY SYNDROME

Success with the hepatitis B vaccine has encouraged scientists to use DNA technology to produce other vaccines. High on the list of priorities is a vaccine for **AIDS**. Several AIDS vaccines are currently being developed or field tested. Among the vaccines are a number composed of subunits, specifically the glycoproteins associated with the **human immunodeficiency virus**, or **HIV**.

Two glycoproteins, known as **gp120** and **gp41**, occur in the HIV envelope. The first, gp120, occurs in spikes projecting from the envelope; the second, gp41, lies beneath gp120. When HIV binds to T-lymphocytes in the body, the gp120 molecule unites with the cell's receptor site called the CD4 site. Then the gp41 molecule attaches to the host cell membrane and spurs entry of the virus to the cell's cytoplasm.

Vaccine manufacturers see in this process a possible avenue for inhibition. They have identified the genes for the gp120 and gp41 molecules and have successfully inserted the genes to *E. coli* cells. The bacterium has produced large amounts of gp120 and gp41 for use as a vaccine. When injected into volunteers, the glycoproteins elicit antibodies that bind to their respective glycoproteins, neutralize the gp120 and gp41 binding sites, and prevent binding of HIV to its host T-lymphocytes. Figure 6.14 shows how the process works.

gp120

glycoprotein 120, a molecule located in the envelope of HIV and used by HIV to attach to a host cell during viral replication.

CD4

a small protein at the surface of human T-lymphocytes; used as an attachment point by HIV during cell penetration.

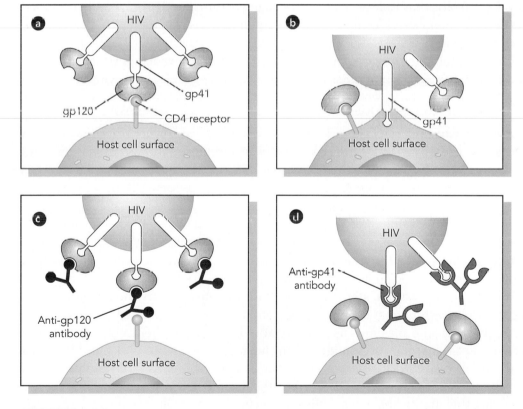

FIGURE 6.14

How a vaccine against HIV could work. (a) When HIV unites with its host cell, the gp120 molecule in the viral envelope binds to the CD4 receptor site on the host cell surface. (b) This initial binding uncovers the gp41 molecule, also in the viral envelope. When gp41 binds to the cell surface, it opens a passageway for entry of the virus. (c) A vaccine against HIV contains gp120 and gp41 molecules. The gp120 molecules stimulate the body's immune system to produce anti-gp120 antibodies. The antibodies prevent the union of gp120 and CD4. (d) The gp41 molecules stimulate the immune system as well, resulting in anti-gp41 antibodies. These antibodies also prevent virus-cell union. If union could be prevented, infection would not take place.

Before an AIDS vaccine becomes a reality, a number of problems must be resolved: For example, gp120 and gp41 molecules are poor stimulators of the immune system, so a suitable carrier molecule must be located. Testing is a problem because animal models are generally not available for vaccine experimentation. Moreover, HIV resides within host cells in a DNA form, and antibodies do not enter host cells. And HIV is known to mutate so that vaccine-elicited antibodies do not bind to the new virus. Nevertheless, some 20 million candidates for an AIDS vaccine live in the United States, and vaccine development remains a high priority.

■ OTHER TRADITIONAL VACCINES

herpes simplex

a viral disease of the skin and nervous system accompanied by lesions and tissue destruction.

genome

the sun total of all an organism's genes.

DNA technology has also entered the picture in vaccine development for such viral diseases as **influenza** and **herpes simplex**. For these diseases, a new form of vaccine is being researched. The vaccine consists of **vaccinia viruses** (the ones used by Jenner), which are spliced with genomes from several viruses. Vaccinia viruses are good candidates to be carriers because they are generally nonlethal and their biochemistry is well known. Moreover, the viruses do not remain latent in body cells (as other viruses such as herpesviruses often do), and they are relatively easy to manipulate because of their large size. In addition, the vaccinia virus can accommodate stretches of foreign DNA in its genome, and when it is released in the host cell cytoplasm, the genome expresses its own genes plus those it is carrying. The virus is relatively easy to cultivate, a factor that reduces the cost of manufacturing; and it can be freeze-dried and stored for many years, characteristics that permit transport without refrigeration or loss of potency.

The important advantage to using a **whole recombined vaccinia virus** is the high level of resistance it stimulates. A subunit vaccine stimulates B-lymphocytes to produce antibodies, but a whole virus vaccine stimulates T-lymphocytes as well. The T-lymphocytes attack and destroy virus-infected cells, thereby yielding a two-pronged attack on the virus.

The first use of vaccinia virus as a carrier was announced in 1986 by virologists at the New York State Public Health Laboratories. Since then, a **two-step procedure** has been used to engineer subunit genes to the vaccinia genome. The first step is to assemble a plasmid insertion vector. This plasmid contains the foreign subunit genes (for example, from a herpesvirus and a chickenpox virus), a vaccinia promoter, and other natural vaccinia genes that direct the plasmid to a specific region in the vaccinia genome. The recombinant plasmid is then amplified (multiplied) in *E. coli* cells, and the resulting plasmids are isolated and purified. These plasmids are the plasmid insertion vectors.

acyclovir

a-si´klo-vir

malaria

a mosquito-transmitted protozoal disease in which the parasites multiply within and destroy the body's red blood cells

Then comes the second step. Virologists take live animal cells and simultaneously infect the cells with normal vaccinia viruses and the plasmid insertion vectors. During viral replication, the plasmid insertion vectors incorporate themselves at the spot specified by the natural vaccinia genes on the vector, as Figure 6.15 illustrates. The incorporation takes place near the gene that encodes the enzyme **thymidine kinase (TK)**. The TK gene is lost in the process, and the recombined virus cannot produce TK. Destroying this ability has two significant effects: first, without the enzyme the virus is less infectious than the normal vaccinia; and second, cells containing the recombined virus can be selected out of the population when acyclovir is added. Acyclovir is an antiviral drug that kills cells infected with viruses expressing TK. The drug is harmless to cells infected with viruses not producing the enzyme.

By the early 1990s, vaccinia vaccines were available against hepatitis B, herpes simplex, influenza, and malaria, as well as against the animal diseases rabies and vesicular stomatitis. However, none of these vaccines has been licensed by the FDA at

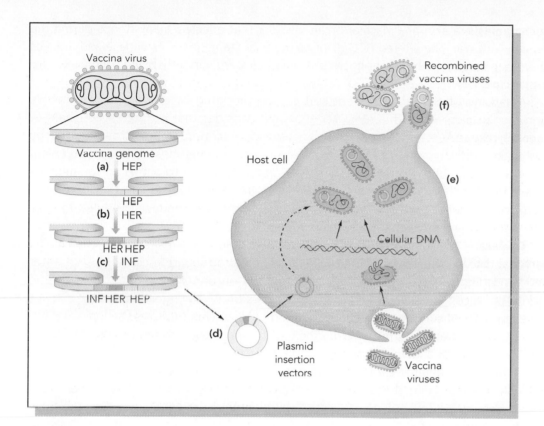

FIGURE 6.15

Production of vaccinia vectors for use in vaccines. (a) Fresh vaccinia (cowpox) viruses are treated to remove their DNA, and genes from hepatitis B viruses (HEP) are added. (b) Then, genes from herpes simplex viruses (HER) and (c) from influenza viruses (INF) are attached. The vaccinia DNA fragments are cloned in *E. coli* cells to increase their number and produce plasmid insertion vectors. (d) The plasmids are introduced to host cells simultaneously with fresh vaccinia viruses. The vaccinia viruses induce the cell's DNA to encode new vaccinia viruses. (e) During viral synthesis, the plasmids are taken up with normal vaccinia DNA to produce recombined vaccinia viruses. (f) These viruses are emitted from the cell and are suitable for use as vaccines.

this writing, partly because of concerns for safety. Critics argue that recombined virus might spread among humans and that life-threatening complications such as encephalitis could result. Proponents of the vaccines point out that careful screening of potential vaccines could minimize the two concerns. Another potential problem is that the vaccine could be ineffective in people already immunized with vaccinia to prevent smallpox. However, children are no longer given smallpox vaccinations so this concern may disappear in the future.

To bypass the concerns regarding vaccinia virus, extensive research has been done on **fowlpox virus** as a carrier. Closely related to vaccinia virus, the fowlpox virus replicates only in birds. Experiments reported by **Enzo Paoletti** in 1992 indicate that the fowlpox virus remains active long enough to deliver genes that encode the subunits of the protein capsid of rabies viruses.

Another possible carrier is a **bacterium.** DNA technologists are currently working on a vaccine that uses BCG to carry parasite genes. BCG stands for bacille Calmette Guérin. It is a benign animal strain of the tubercle bacillus that is used in many parts of the world to immunize individuals against tuberculosis (the United States is the notable exception). DNA technologists have genetically altered the bacillus by adding a gene derived from the protozoal parasite *Leishmania tropica*. This parasite causes

Leishmania

leesh-man'i-ah

leishmaniasis

leesh-man-i´a-sis

cholera

a bacterial disease in which bacterial toxins induce the loss of great volumes of water from the body.

leishmaniasis, a skin and visceral organ disorder that affected many troops during the Persian Gulf War. The altered BCG displays the parasite gene on its surface and induces an immune response by T-lymphocytes. Animal tests are currently being conducted with the vaccine.

A vaccine of a totally different sort can be prepared by altering an organism's genome. To appreciate this remarkable ability, consider that during earlier years of vaccine research, scientists prepared attenuated strains of infectious agents by prolonged cultivation. After weeks, months, or years (sometimes never), an organism would lose the ability to cause disease but retain its ability to act as an immunizing agent. However, no one was sure why. Today, DNA technologists can excise from an organism the genes that encode disease-inducing proteins. The result is a new form of attenuated organism useful in a vaccine.

Cholera is an example in which such a technique has proven successful. Cholera is a bacterial disease caused by **Vibrio cholerae**, a curved rod found in contaminated water and food (Figure 6.16 depicts the organism). On entering the human intestine, *V. cholerae* produces a toxic protein that induces cells lining the intestinal wall to release large amounts of sodium, bicarbonate, and other ions. Water follows the ions, and the patient soon experiences massive diarrhea, with loss of eight to ten quarts of water in a few hours. Death usually follows.

To immunize against cholera, it has been common practice to use dead *V. cholerae*. In 1992, researchers at the University of Maryland announced the development of a new vaccine. DNA technologists successfully identified the genes encoding the toxic protein, then removed the genes from *V. cholerae*. The cholera bacilli thus became attenuated and useful in a cholera vaccine.

But suppose one did not need to be exposed to the cholera bacilli at all? Would that be even safer? Yes, thought researchers at the University of California at Loma Linda, where a team of DNA technologists altered the toxic protein so it would not affect the intestinal cells but would still induce antibody production. And, as a delivery system for the new vaccine, they used the common household potato. In 1998, the California team reported success with a **gene-altered potato** that would induce immunity to cholera when eaten by mice. The researchers used the "medicinal spuds," as the potatoes were whimsically called, in both raw and cooked forms. FDA

FIGURE 6.16

A transmission electron micrograph of *Vibrio cholerae*, the agent of cholera. Note the distinctive comma-like shape and flagellum. Cholera bacilli as these have been prepared as vaccines by removing from the bacterial chromosome the genes believed responsible for the pathogenic potential.

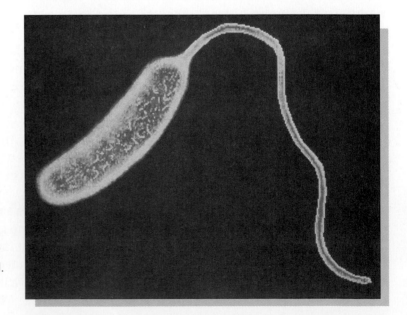

administrators wondered whether to call them a food or medicine.

The new vaccines for cholera are currently undergoing field trials, and the prospects for other innovative vaccines appear positive. Researchers, for example, are incorporating genes from infectious organisms into genetically engineered bananas, tomatoes, and corn, and they are using gene-altered goats to produce vaccines in their milk. These and other imaginative approaches to vaccine production are explored in Chapters 10 and 11.

DNA VACCINES

DNA vaccines came into being in 1990. Since that time they have been hailed as a novel approach to DNA technology. Entire scientific meetings are devoted to DNA vaccines, an Internet site has been created to disseminate germane information, and according to the respected journal *Science* "... biotech companies are tripping over each other to get into the business..." of producing DNA vaccines

DNA vaccines consist of plasmids engineered to contain a protein-encoding gene and all the necessary requisites. Unlike replicating viruses or live bacteria, they are not infectious or replicative, nor do they encode any proteins other than those specified by the plasmid genes. Thus, one of the outstanding features of the DNA vaccine is its safety.

But a DNA vaccine is not simple: A single plasmid might contain a promoter site, a convenient cloning site for the gene of interest, a poly-A tail used as a termination sequence, an origin of replication, and a selectable marker sequence such as an ampicillin-resistant gene (in addition, of course, to the protein-encoding gene or genes). The necessity for all these parts is discussed in Chapter 5.

One highlight of the DNA vaccine is that no special formulation is necessary. Animals can be immunized against disease merely by injecting plasmids suspended in saline (salt) solution. And **delivery of the plasmids** to target cells is relatively easy: as early as 1994, researchers identified five effective ways for delivering influenza virus plasmids to the animal body: by nasal spray; by injection into the muscle, vein, and under the skin; and by a so-called gene gun. The latter is a propulsion device that shoots DNA-coated gold beads into the skin (Figure 6.17). All techniques resulted in protective immune responses.

DNA vaccine

a DNA molecule that stimulates an immune response in the body.

poly-A tail

a series of adenine-containing nucleotides at the end of an RNA molecule.

saline solution

salt solution similar to that surrounding and within body cells.

(a)   (b)

FIGURE 6.17

Innovative methods for vaccine delivery. (a) Novel inhalation devices that deliver vaccines to the respiratory tract where antigens are absorbed to the bloodstream. (b) Antigen-containing gold microspheres seen between cells of the intestinal lining of a rat.

pathogen

an organism capable of causing disease in the body.

melanoma

cancer of the deep skin layers.

therapeutic

referring to treatment for disease.

spleen

a visceral organ containing cells of the immune system.

Furthermore, deploying plasmids as DNA vaccines has the added advantage of stimulating both antibody-mediated immunity (related to B-lymphocytes) and cell-mediated immunity (related to T-lymphocytes). This is because DNA vaccines encode proteins that are released from the cell and taken up by phagocytes for delivery to the B-lymphocytes for antibody production. At the same time, some of the proteins fix themselves to the target cell surface and stimulate a response by T-lymphocytes. Later when the pathogen appears, the antibodies and T-lymphocytes will still be available to neutralize the pathogen's proteins and destroy the pathogen itself (or the cell harboring the pathogen).

DNA vaccines also appear to elicit a **higher immune response** than scientists anticipated, despite the tiny amount of antigenic protein encoded. Researchers guess that this added benefit is due to an immune response to the other genes contained in the plasmid. The products encoded by these genes, like the main gene products, are interpreted as foreign and "...increase the wallop of a DNA vaccine," as one writer has said.

Even the production of DNA vaccines is advantageous. The technological methods for producing the plasmids are now in place, and biochemists can adapt their methods for various DNA vaccines. By using similar fermentation, purification, and validation techniques to produce a spectrum of DNA vaccines, manufacturers can reduce time and cost. Moreover, DNA vaccines are more stable at low, and high temperatures compared with conventional vaccines. They can be shipped easily and inexpensively and reach a greater population of vaccinees.

At this writing, DNA vaccines have been used experimentally to protect against influenza, *Salmonella typhi*, HIV infection, herpes simplex, hepatitis B, and several cancers including melanoma, colon cancer, renal cell carcinoma, and T-cell lymphoma (Table 6.3). Public health specialists envision that using DNA technology, vaccines can be developed quickly against emerging diseases (e.g., those caused by Ebola virus and drug-resistant bacteria), and they foresee that a number of different diseases can be controlled with a single vaccination by simply adding more genes to the plasmid.

There are **drawbacks** to consider, however. The DNA vaccines do little to address pathogenic mechanisms involving carbohydrates (as in streptococcal pneumonia, for instance). Delivery systems must be worked out, and ages determined for optimum immune response. The vaccines must also be evaluated for therapeutic uses in already infected patients and in those with chronic conditions.

Over the years, the research with DNA vaccines has been modified and expanded for various purposes. One approach has been to induce the body's B-lymphocytes to produce both antigens and antibodies. In 1997, researchers **Stephen Johnston** and **Maurizio Zanetti** at the University of California (Figure 6.18) described how they induced a mouse's B-lymphocytes to perform double duty: The researchers injected the B-lymphocytes with plasmids containing genes from malarial parasites and genes that encode antibodies against malarial parasites. Soon the spleen cells were producing antibody molecules bound to antigens from the malarial parasite. The antibodies reacted with the malarial parasite causing disease while the antigen portions stimulated other B-lymphocytes to produce still more antibodies. The "**antigenic antibody**" is an extremely novel idea in immunology.

Researchers at the University of Texas have been equally imaginative as they attempt to identify which plasmids are the key to defense against a pathogenic organism (in effect they hunt for the plasmid giving the strongest immune response). The scientists take a pathogenic organism's DNA and break it to fragments. Then they prepare perhaps 1000 different plasmids. Various plasmid groups are injected to animals

TABLE 6.3

A Summary of Pathogens, the Microbial Proteins Used in Vaccines, and the Type of Immunity Induced by the Vaccine.

PATHOGEN	PROTEIN(S)	INDUCTION OF[a]		
		ANTIBODIES	CTL	PROTECTION
Influenza virus	NP, HA, M1	+	+	+
HIV	Env, Gag, Rev	+	+	ND
Bovine herpesvirus	gp	+	ND	+
Hepatitis B virus	Surface and core antigens	+	+	+
Rabies virus	Gp, NP	+	+	+
Plasmodium sp.	CSP	+	+	+
Leishmania major	Gp63	+	ND	+
Mycobacterium tuberculosis	HSP65, Ag85	+	+	+
Hepatitis C virus	Nucleocapsid	+	+	ND
Herpes simplex virus	GB, gD, ICP27	+	+	+
Papillomavirus	L1	+	ND	+
Human T-cell	Env	+	+	+
Leukemia virus type 1				
Lymphocyte choriomeningitis virus	NP	+	+	+
Bacillus thuringiensis	endotoxin	+	ND	ND
Mycoplasma pulmonis	ND	+	ND	+
Salmonella typhi	OmpC porin	+	ND	ND

[a] Immune responses induced by some or all of the DNA vaccines encoding various antigens of a particular pathogen. ND, not determined; CTL, cytotoxic T-lymphocytes. Source: ASM News 62 (1996):479.

FIGURE 6.18

DNA vaccine researchers Stephen Johnston (left) and Maurizio Zanetti of the University of California. Research teams led by these scientists genetically altered B-lymphocytes to produce both antigens and antibodies at the same time. The result was antigenic antibodies.

as DNA vaccines, and the animal is given time to develop an immune response. Then the pathogen is injected. If the animal remains unaffected by one of the plasmid groups, the key element for protection is in that group. The group is subdivided into smaller groups and the test is repeated. Again, the uninfected animal points to the smaller plasmid group. Retest after retest follows, and the scientists eventually reach a single or few plasmids that are conferring protection. Those are the ones presumably useful for a DNA vaccine.

The fresh approaches to vaccine research bode well for future generations and represent an elegant application of the DNA technology. Scientists have had difficulty restraining their enthusiasm for the new vaccines. But that is not unusual because they have had difficulty restraining their enthusiasm for most prospects of DNA technology.

SUMMARY

Defective genes can lead to deficiencies of such proteins as insulin, human growth hormone (HGH), and Factor VIII, and, consequently, to such physiological problems as diabetes, dwarfism, and impaired blood clotting, respectively.

The natural proteins can now be replaced by proteins manufactured by the processes of DNA technology, and as such they represent a number of pharmaceutical products so produced. For insulin production, two protein chains are encoded by separate genes on recombined plasmids inserted to bacteria. A signal peptide is used to assist exportation of the chains. Then the chains are chemically joined. HGH production requires unique techniques to alter the protein because bacteria do not have the capabilities of producing mammalian proteins. Factor VIII, also a large protein, has been produced by mammalian cells, thus bypassing the limitations imposed by using bacteria.

In addition to replacing proteins, the products of DNA technology can be used for therapeutic purposes. For example, tissue plasminogen activator (TPA) is a protease that encourages a clot to dissolve. Encoded by an expression vector with several components, synthetic TPA is produced by recombined mammalian cells. Interferon, another therapeutic product, is produced in the bacterium *E. coli*. An antiviral protein, interferon has been approved for use against certain types of cancers and for genital warts. Antisense molecules represent a novel therapeutic device. They are molecules of RNA that react with and neutralize the mRNA molecules used for protein synthesis. DNA molecules called oligos can also be used as antisense molecules. Their high specificity for target mRNA molecules argues strongly for their use.

Vaccines produced through DNA technology represent another fortuitous application of this technological skill. The hepatitis B vaccine currently in use is composed of viral proteins manufactured by yeast cells recombined with viral genes. The vaccine is safe because it contains no viral particles. Vaccines for AIDS are in developmental stages using the same type of DNA technology. An important carrier of viral genes for vaccines is the vaccinia virus. This virus has several advantages that encourage its use, and it has been successfully engineered to carry other viral genes and to stimulate a strong immune response. Subtracting genes from an organism can also yield an immunizing agent, as has been accomplished for cholera.

DNA vaccines are among the exciting developments in vaccine technology. These preparations consist of plasmids that can be injected directly into cells without a vector organism. Both antibody-related and T-lymphocyte immunities are provoked by these vaccines. Investigators have even produced "antigenic antibodies" by plasmid injections into animals. The vaccines, therapeutic agents, and replacement proteins have generated a wave of optimism for the applications of DNA technology.

REVIEW

This chapter has focused on the pharmaceutical products derived from DNA technology and how those products are biochemically produced. To test your understanding of the chapter contents consider the following review questions:

1. Explain the importance of insulin to the body and outline the biochemical methods currently used to produce it commercially.

2. Which hormone can be used to treat dwarfism, and what modifications are used in the production methods for manufacturing this hormone?

3. What problems attend the use of Factor VIII derived from blood and why are mammalian cells instead of bacterial cells used to synthesize the factor?

4. Discuss the clinical use of TPA and indicate the components of the expression vector used in its synthetic production.

5. What is the theoretical basis for the activity of interferon and for which diseases is it currently used? What are some drawbacks to its use?

6. Using the human immunodeficiency virus (HIV) and acquired immune deficiency syndrome (AIDS) as models, describe how antisense molecules work. Explain the chemical basis for inserting the DNA for antisense molecules into the cellular genome.

7. What are the components of the currently used vaccine for hepatitis B, and how are those components produced through DNA technology?

8. Name the possible components of an AIDS vaccine and outline several problems that must be resolved before such a vaccine is available.

9. Describe how a vaccinia virus an be recombined for use as a carrier for subunits in a vaccine. What are advantages and disadvantages to its use?

10. Explain the experimental basis for a DNA vaccine and point out some advantages and disadvantages to this vaccine.

FOR ADDITIONAL READING

Agrawal, S. "Dollars and Antisense." *The Sciences* (May/June, 1997).

Caldwell, M. "The Dream Vaccine." *Discover* (September, 1997).

Cowley, G. "Vaccine Revolution." *Newsweek* (July 27, 1998).

Delessert, E. "Short Like Me." *Health* (November/December, 1993).

Haseltine, W. "Discovering Genes for New Medicines." *Scientific American* (March, 1997).

Roush, W. "Modified Microbe May Boost TB Vaccine." *Science* (1996): 271–447.

Roush, W. "Antisense Aims for a Renaissance." *Science* 276 (1997): 1192–1195.

Taubes, G. "Salvation in a Snippet of DNA." *Science* 278 (1997): 1711–1714.

DNA Analysis and Diagnosis

LOOKING AHEAD

One of the more remarkable uses of DNA technology is in the diagnostic laboratory to analyze samples and detect whether certain genes and DNA segments are present. How this technology is performed and how it is applied are the major thrusts of this chapter. On completing these pages, you should be able to:

■ understand the nature of a DNA probe and appreciate how it identifies a segment of target DNA.

■ summarize the biochemistry of the polymerase chain reaction and understand how it amplifies a sample of DNA.

■ explain how DNA probes and polymerase chain reaction technology can be used to detect infectious diseases such as AIDS, tuberculosis, and Lyme disease.

■ conceptualize the genetic basis for such diseases as cystic fibrosis, Duchenne's muscular dystrophy, and Huntington's disease and recognize how DNA analysis can make the diagnosis of genetic diseases possible.

■ give a synopsis of how certain forms of diabetes, cancer, and hearing and vision loss can be identified through DNA technology.

■ outline several instances in which DNA probes and polymerase chain reaction technology can be useful for monitoring microorganisms in the environment.

INTRODUCTION

Since the 1950s, scientists have discovered a broad variety of new techniques for medical analysis and testing. Where scientists once searched for disease-related antibodies in a patient, they can now identify the disease organism itself by methods not previously imagined. And they can focus on the organism itself rather than on evidence that it once was there.

Central to these techniques is the DNA molecule. It is now possible to reproduce DNA in a test tube, fragment it, determine its composition, change its structure, and map its genes. The principles learned from these breakthroughs have been applied in diagnosing infectious disease, screening individuals for cancer and genetic diseases, and ensuring the public health by identifying pathogens in the environment.

Many of these techniques have been made possible by a technology called the **polymerase chain reaction (PCR)**. Before 1985, hardly anyone had heard of PCR; by today's standards, a laboratory without a PCR machine is like an office without a photocopier. In a sense, the PCR machine even functions like a photocopier: It takes a fragment of DNA and automatically turns out millions of copies in a scant three hours.

A second essential element in the DNA analyses is a fragment of DNA called the DNA probe. Developed in the 1970s, the **DNA probe** (or **gene probe**) hunts for a complementary fragment of DNA within a morass of cellular material and signals when the fragment has been located. Tracking down a gene or set of genes can be a formidable task when one considers the size of the human genome: There are over three billion nitrogenous bases in the forty-six human chromosomes. A gene probe attempts to detect a stretch that is a few thousand bases in length, or about a millionth of the length of the genome.

In the pages ahead we will explore the fruits of PCR and DNA probe technologies as they apply to analyses and diagnoses. In many cases, these technologies are supplementing or replacing the traditional methods of analysis. Indeed, certain diagnoses cannot be made without the technologies involving DNA.

METHODS OF DNA ANALYSIS

In 1961, researchers **Sol Spiegelman** and **Edward Hall** discovered that single-stranded DNA forms hydrogen bonds with a complementary strand of RNA and yields a double-stranded DNA-RNA molecule. For the next twenty years, DNA technologists

polymerase chain reaction

po-lim′er-ase

a sequence of chemical reactions in which a fragment of a molecule of DNA is copied to yield millions of identical molecules.

changed the principle slightly and applied it to a DNA-DNA match. They researched the possibility of using a DNA strand to recognize a complementary DNA strand amid a mixture of other DNA strands, much as a key might seek out a lock. Eventually, they succeeded and brought forth the significant factor in DNA-based analyses: a molecule of DNA called a DNA probe.

■ DNA PROBES

A **DNA probe** is a relatively small (a few thousand bases in length), single-stranded molecule of DNA that recognizes and binds to a complementary segment of DNA on a large DNA molecule, as shown in Figure 7.1. Because DNA bases always pair A to T and G to C (Chapter 2), a DNA probe interacts very specifically with nucleic acid sequences in target molecules of DNA. Like a left hand seeking one specific right hand, the probe DNA mingles among a mixture of DNA strands until it locates one strand (or section of a strand) that is complementary. It then binds to that strand (or section).

The activity of a DNA probe is intimately linked to the biochemistry of DNA. A double-stranded DNA molecule tends to unwind and disassemble under certain laboratory conditions. These conditions include heating at a temperature over 90°C or subjecting the DNA to a pH higher than 10.5 or to organic compounds such as urea or formaldehyde. When so exposed, the hydrogen bonds between base pairs break and the complementary strands come apart. Such a process is called **denaturation.**

Denaturation can also be reversed. If the proper laboratory conditions of salt, temperature, and pH are established, the two single-stranded DNA molecules will reassemble to form the original duplex. This process is called **hybridization** (or renaturation) and is central to the use of the DNA probe. Under carefully controlled conditions, stable DNA duplexes will form only when complementary base pairing occurs perfectly along the entire length of the DNA strands.

DNA probe

a synthetic, small, single-stranded DNA molecule that will bind to a target, complementary DNA strand in a mixture of biological compounds.

denaturation

the separation of complementary strands in a DNA molecule as a result of chemical or physical treatment.

hybridization

the joining of complementary strands of a DNA molecule.

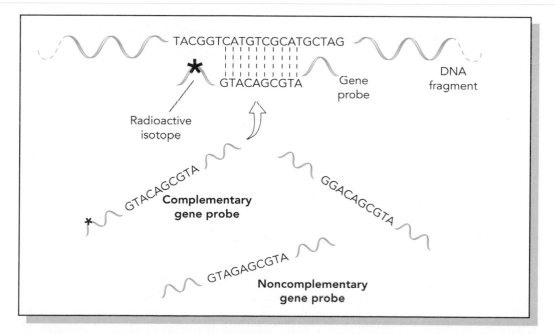

FIGURE 7.1

How a gene probe works. A gene probe is a single-stranded segment of DNA. When combined with a DNA molecule containing a complementary site, the gene probe seeks out the site and binds with it. If a radioactive molecule or atom is attached to the probe, the radioactivity accumulates at the binding site and signals that a reaction has taken place. Note in the diagram how the bases of only one probe complement the bases of the DNA fragment.

isotopes

forms of an atom or element with different mass but the same chemical properties; some isotopes emit radioactivity.

reverse transcriptase

an enzyme that will synthesize a DNA molecule using an RNA molecule as a template.

When used for testing purposes, the DNA probe is usually linked to a detector substance such as a **radioactive isotope.** When the probe binds to its complementary DNA, it takes along the radioactive isotope, and the accumulation of radioactivity signals that a union has occurred. Before the probe is added, however, the target DNA must be purified and split apart (or denatured). The labeled DNA probe will then hunt for its complementary segment among the single-stranded DNA molecules.

With current technology, a DNA probe can be developed to detect virtually any nucleic acid sequence. It is not even necessary to know the actual sequence of bases in the target DNA. For example, the DNA can be encouraged to form complementary messenger RNA (mRNA) molecules. The mRNA molecules can then be isolated and combined with the enzyme reverse transcriptase plus appropriate building block nucleotides and other materials. Reverse transcriptase will use the isolated mRNA as a template and synthesize a complementary molecule of DNA (called cDNA). This new DNA is identical to the portion of DNA (the exons) functioning in protein synthesis at one particular time. The new DNA molecule is now used as a probe to detect the nuclear DNA whenever an identification is required (Questionline 7.1).

A DNA probe can be as short as ten bases or greater than 10,000 bases in length. Clearly, the probe must be able to bind to (or "hybridize to") the target DNA, but it must also avoid hybridizing to other nucleic acid molecules that may be present. And once the probe-target molecule has formed, the combination must be stable.

During the early 1980s, the optimism generated by the use of DNA probes was counterbalanced by the problem of not having enough DNA to perform a reliable test. With insufficient target DNA, the probe's radioactive signal did not work well. Thus, it

---

QUESTIONLINE 7.1

1. **Q.** Occasionally I read about using DNA probes to detect various diseases. Exactly what is a DNA probe?

   A. A DNA probe is a small, single-stranded molecule of DNA. This molecule is constructed with specified nitrogenous bases so it will bind very exactly to a strand of DNA with complementary bases, much like a left hand brought together with a right hand while clapping. The DNA probe is a synthetic DNA strand manufactured to react with a naturally occurring DNA strand.

2. **Q.** How can a DNA probe be used for diagnostic purposes?

   A. Let's suppose you want to locate a virus such as human immunodeficiency virus (HIV). You take a patient's cells containing human DNA and viral DNA, and you break open the cells. Then you add the DNA probe specifically synthesized to react with HIV. The probe mingles among all the DNA strands and binds only with the DNA from HIV. A radioactive signal lets you know that a reaction has taken place, and you may assume that HIV is present.

3. **Q.** But suppose there was insufficient DNA to detect in the test. What then?

   A. That problem is solved by using the polymerase chain reaction, or PCR. The PCR is a process in which minuscule amounts of DNA can be amplified to enormous amounts in a matter of hours. The original DNA strand is used as a template to synthesize the strands, each of which then serves as a template for new strands, and so forth geometrically. The process is highly automated and results in sufficient DNA to give a reliable test when gene probes are used.

PCR

Polymerase Chain Reaction

uses a probe.

Probes

a few thousand bp. long

DNA denatures at 90°C or a pH higher than 10.5

or exposure to urea + formaldehyde

Denaturation breaks hydrogen bonds.

hybridization is putting DNA back together.

• Radioactive

Can be reversed with CDNA

PCR 3 steps

1 - denatured

2.

was important to amplify the target DNA. The technology for DNA amplification was developed in a procedure called the polymerase chain reaction, as we will see next.

■ THE POLYMERASE CHAIN REACTION

The **polymerase chain reaction** (**PCR**) permits the scientist to locate the molecular equivalent of a needle in a haystack. (Figure 7.2 shows the theory behind the process.) The PCR was developed at the Cetus Corporation in 1984 by **Kary Mullis** (Nobel laureate in Chemistry in 1993). It allows the DNA technologist to reproduce a single strand of DNA to a billion identical copies in a few hours. (A cancer cell, known for its high reproductive ability, would require a month to accomplish the same feat.) The DNA probe can then be used with a greater degree of accuracy and yield a more reliable result.

To multiply a DNA molecule with the PCR, four materials are required: the target DNA molecule (the one to be amplified); short strands of known "primer" DNA that will tag and identify the segment to be copied and provide a foundation for beginning the replication process; DNA polymerase, an enzyme that directs DNA replication in living cells (and for which the reaction is named); and a mixture of nucleotides, the nucleic acid building blocks from which new DNA will be formed.

DNA polymerase

an enzyme that synthesizes a DNA molecule using another DNA molecule as a template.

Once the materials have been assembled, the PCR involves three major steps, all performed over and over again in a highly automated PCR machine. In the first step, the target DNA is heated, thereby breaking the bonds that hold the two strands together and unwinding the molecule. Then, in the second step, the temperature is reduced and the primer molecules bind to and flank the target DNA, thus bracketing and identifying the area to be copied. The primers are the "start" and "stop" signals in the copying.

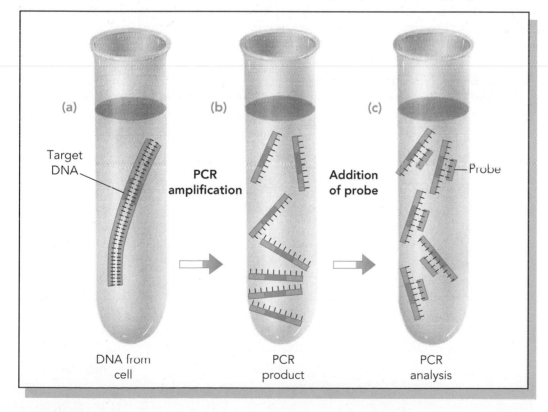

(a) Target DNA

PCR amplification

(b)

Addition of probe

(c) Probe

DNA from cell

PCR product

PCR analysis

FIGURE 7.2

An overview of the polymerase chain reaction (PCR) analysis. (a) DNA is obtained from a cell and placed in a test tube with appropriate other materials (Figure 7.3). (b) The enzyme DNA polymerase duplicates the target DNA millions of times. (c) With multiple copies of the DNA present, the DNA probe can easily locate its complementary binding site.

In the third step, DNA polymerase catalyzes the formation of single-stranded DNA molecules using the target DNA strands as templates. The synthesis begins at the spot marked by the primer DNA, and as the enzyme moves along each strand, it reads the sequence of chemical bases, using it as a template for assembling a complementary strand and a new nucleotide chain (the "chain reaction"). The DNA polymerase used is a special heat-tolerant enzyme called *Taq* **polymerase**. It is derived from the thermophilic bacterium *Thermus aquaticus*, first isolated from an oceanic thermal vent by **Thomas Brock** in the 1980s. The enzyme used must be heat-tolerant because a heat-fragile enzyme would have to be replaced after each heating step. At the conclusion of step three, four DNA strands exist where there were once two.

As the process continues, the PCR is repeated thirty to sixty times, each cycle beginning with a reheating of the mixture (step 1). A cycle takes about one to two minutes, and each new DNA segment serves as a template for many additional copies. Thus, the number of copies of DNA increases geometrically. Where there were originally two strands of DNA, there are now millions or billions. Instead of looking for a needle in a haystack, the DNA technologist has essentially made a huge stack of needles. Figure 7.3 summarizes the process.

But the PCR is not without **problems**. Chief among these is contamination. If any contaminating DNA is present, it is amplified along with the target DNA. To eliminate this possibility, great care must be taken in sample preparation. Such preparation tends to be labor intensive and costly. Moreover, the PCR is not quantitative—it can determine that a particular DNA segment is present but it cannot determine how much of it was originally present. And, because the process is so new, reproducible results may be difficult to obtain and a period of time for standardization will be required. Despite these drawbacks, the PCR used in conjunction with DNA probes holds great promise for the future.

■ SIGNAL AMPLIFICATION

The PCR operates on the principle of target amplification, that is, the target DNA is amplified before the DNA probe is added. An alternative principle is **signal amplification**. According to this principle, a minuscule amount of target DNA binds to the DNA probe and then the signal is amplified to indicate that a successful match has been made.

To achieve signal amplification, there are two general approaches. The first is to decrease background "noise" by eliminating excessive DNA. This is accomplished by separating the target-probe complex from the mixture of contaminating DNA. For the separation, strands of thymidine nucleotides (poly-T) are linked to magnetic particles. Complementary strands of adenosine nucleotides (poly-A) are then linked to the DNA probe. After the DNA probe has had a chance to bind to its target DNA, the poly-T particles are added. Now the adenine bases of the target-probe complex bind to the thymine bases of the particles. The particles can then be separated from the mixture, carrying along the target-probe complex. So concentrated and separated from the contaminating DNA, the signal (the radioactive DNA probe) is amplified.

The second approach is to amplify a detector that binds to the signal, that is, the probe DNA molecule. To achieve this goal, a second probe for the target DNA is used. This probe is composed of RNA having a unique tertiary structure. The probe works with **Q-beta replicase**, an enzyme that catalyzes RNA replication. The RNA probe binds to the target DNA along with the probe DNA. Once the three have bound together, the RNA-DNA-DNA mixture is separated from the mixture and Q-beta replicase is added. The enzyme replicates the RNA probe geometrically (thus amplifying the

Thermus aquaticus

a heat tolerant bacterium whose DNA polymerase is used in the PCR.

quantitative

able to determine exact quantity.

thymidine

a nucleotide containing deoxyribose, phosphate, and thymine.

adenosine

a nucleotide containing ribose or deoxyribose, phosphate, and adenine.

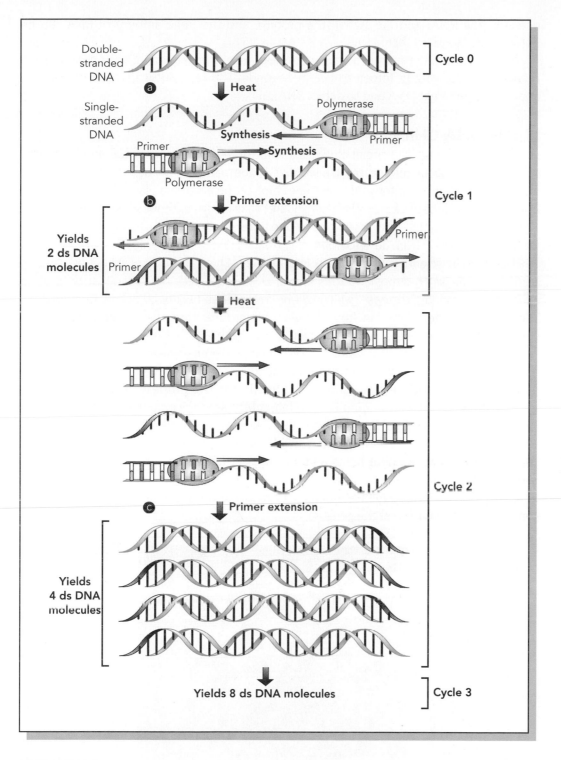

FIGURE 7.3

Details of the polymerase chain reaction (PCR). (a) Heat is used to separate the double-stranded (ds) DNA molecule. In cycle 1, the technologist adds a mixture of nucleotides, an enzyme called polymerase, and a segment of DNA primer. (b) The polymerase extends the primer segment with the available nucleotides and produces two ds DNA molecules. (c) The process is repeated, and at the end of cycle 2, four ds DNA molecules are present. Repeating the process in cycle 3 yields a total of eight ds DNA molecules. The strands increase in number geometrically in future cycles.

detector [the RNA] and the signal [the probe]), and the large amount of RNA can be detected by standard techniques.

The RNA probe system is advantageous because it is very rapid, each cycle of RNA replication occurring in fifteen to twenty seconds. The technique also can be used to quantitate the target DNA and to detect RNA.

■ THE DNA CHIP

Just as the integrated circuit chip has revolutionized the personal computer industry, so too the new **DNA microarray GeneChip® probe array,** may push the biotechnology industry a quantum leap forward. The GeneChip® prope array is a glass slide the size of a postage stamp (about two centimeters square) that contains thousands of DNA probes, which scan thousands of DNA samples at one time. The chip reduces the time necessary for a DNA detection from days to hours.

Designed by scientists at Affymetrix Inc., the GeneChip® probe array is built by allowing stencil-like masks to deposit various known DNA sequences (DNA probes) on a glass slide at sites called "features," which can be as small as twenty by twenty micrometers. The sequences then hang out from the glass slide. After thirty-two automated steps, a chip may contain more than 65,000 probes. The average length is about 20 bases.

To use the chip, an unknown DNA molecule is broken into fragments, and fluorescent marker molecules are attached to the fragments as signal devices. The fragments are pumped underneath the glass slide and forced over the probes; then the residue is washed away. Fragments finding a complementary sequence stick to the probes like zippers and deposit their fluorescent signals as Figure 7.4 shows. A laser beam now scans the slide, row by row, and excites any fluorescent molecules present. A computer records the pattern of bright lights and indicates which probes united with which fragments. Smaller probe sequences are constructed to form larger sequences, and soon a long map of a genetic sequence emerges. Comparing the sequence to a map of known genes reveals the nature of the gene.

DNA chip

a glass slide containing a series of DNA probes used for a genetic analysis.

Affymetrix

a-fim-et´rix

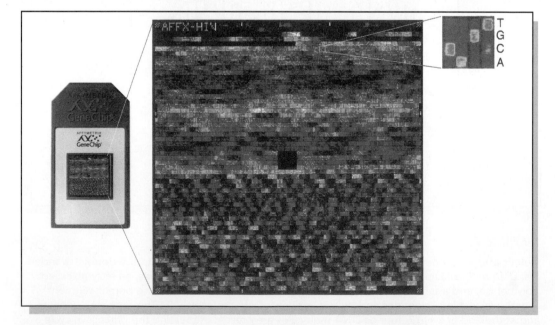

FIGURE 7.4

The DNA microarray GeneChip® probe array. This glass slide contains gene probes for thousands of different DNA sequences. When a sequence matches the probe(s), the latter glow, as seen along the top row.

A DNA microarray could also detect a **mutation** in a gene. For example, if a gene has a base sequence of …CCCAGGGG, it will bind at a feature where the probe contains the bases …GGGTCCCC. But if the gene contains a single base mutation, its sequence might be …CCCCGGGG, and it would bind to the feature where the probe has the sequence …GGGGCCCC. The signal emitted at that feature would indicate the presence of the mutation.

As of 1998, Affymetrix was producing GeneChip® probe arrays with more than 400,000 probes per chip. Although design is a lengthy process, production is automated and time-efficient, and the prospects for rapid gene probe analysis are positive. It is hoped that the DNA microarray may someday hold all the probes for the entire genome of an organism. By 1998, the GeneChip® probe array was being used to screen for mutations in the *p53* and *BRCA1* genes, both associated with cancer and both discussed later in this chapter.

■ RFLP ANALYSIS

It may be unnecessary to identify a particular gene if a suitable alternative exists. Such an alternative is the **restriction fragment length polymorphism,** or **RFLP** (pronounced "rif-lip"). A RFLP is a stretch of DNA serving as a marker for a specified gene. The RFLP has no apparent function in the cell, and various RFLPs are located randomly throughout a person's chromosomes as a type of genetic litter. They are useful as marker genes when the particular gene has not yet been cloned or is difficult to work with. Figure 7.5 shows how a RFLP is detected.

For a RFLP to be used as a marker, it must be less than 5 x 106 base pairs (or 5 centimorgans) away from the gene. If the RFLP and gene are within this distance, there is a high degree of certainty that they are linked biochemically and move together from parent to offspring during reproduction. DNA technologists are statistically certain, with ninety-five percent confidence, that if they can locate the RFLP in an individual, the gene they are seeking will also be present.

To understand the **nature of a RFLP**, we must note that a particular stretch of DNA can be broken into fragments of various size. The fragments are called **polymorphisms** (literally translated, "many forms"). Consider, for example, the following stretch of DNA extending from points A to F and where the double letters indicate points of restriction enzyme cleavage:

-A-BB-CC-DD-EE-F-

Let us now add a mixture of restriction enzymes (b, c, d, and e) that will cleave the DNA molecule at the double letters. The result will be five fragments of roughly equal size:

-A-B B-C C-D D-E E-F-
   1   2   3   4   5

But over the great expanse of time, changes have occurred in DNA molecules: spontaneous changes in base pairs that have no bearing in biochemistry or physiology but do influence enzyme activity. Thus, a change in base pairs may have occurred in the stretch of DNA to produce the following molecule:

-A-BB-CC-DX-EE-F-

If we now treat this molecule with the restriction enzyme mixture used before (b, c, d, and e), the following fragments will result:

-A-B B-C C-DX-E E-F
   1   2   3   4

Note that the result of this enzyme activity is only four fragments because enzyme "d" is inactive at the DX site owing to the change from the original DD to the present DX. Also, and of practical significance, one fragment (C-DX-E) is longer than the other three fragments. Thus the stretch of DNA exists in fragments of various sizes; the

mutation

a permanent change in the base sequence of a DNA molecule.

RFLP

a DNA segment of unknown functional significance that is isolated by restriction enzymes and exists in any of various lengths.

polymorphism

pol-e-morf′ism

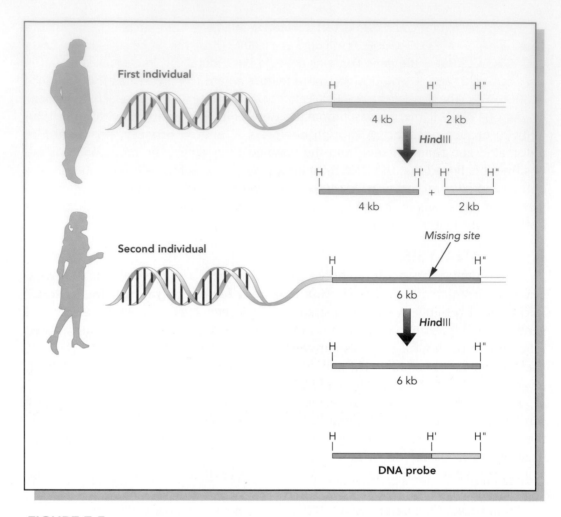

FIGURE 7.5

Detecting a RFLP. The first individual has DNA with three sites sensitive to *Hind*III, so that two fragments can be produced. The first fragment is 2 kilobases (kb) long; the second is 4 kb long. The second individual lacks one of the sensitive sites (H′) possibly because of an inherited mutation. Thus, when the *Hind*III restriction enzyme is added, it only cuts the DNA molecule at two sites and produces a 6 kb fragment. Note that the DNA probe for the DNA segment between sites H and H″ has complementary bases for the entire segment, so it will react with the 2 kb, the 4 kb, and the 6 kb segments. The probe can therefore be used to detect all three segments. The segments are RFLPs.

derived from "restriction" enzymes and the fact that the length of the fragments can be determined; hence FL "fragment length.")

To use RFLPs in a DNA analysis, the DNA technologist takes advantage of the fact that the base-pair length of a particular RFLP is known. A person's cells are obtained and the DNA is isolated and treated with the restriction enzyme mixture. Then the mixture is placed in an electrophoresis apparatus (discussed in Chapter 9), and the polymorphisms are separated according to size. The separations yield a pattern of bands similar to a supermarket bar code. When the pattern is analyzed, the band reflecting the particular RFLP may or may not be present, depending on the inheritance pattern of the individual. If the marker RFLP is present, it may be assumed with ninety-five percent confidence (nineteen chances in twenty) that the gene is also present (Questionline 7.2).

In practical terms, a particular chromosome and its genes can be followed from generation to generation by following the marker RFLP. When a disease-associated gene is

electrophoresis

e-lec″tro-for-e′ sis

QUESTIONLINE 7.2

1. **Q.** My daughter, who is a biochemist, sometimes talks about her work with rif-lips. What are rif-lips?

   A. A rif-lip is a way of pronouncing RFLP. RFLP stands for *restriction fragment length polymorphism*, a scientific term for a stretch of DNA along a chromosome. RFLPs can occur in varying lengths, and they can be used as identifying markers. They have no apparent function.

2. **Q.** If they have no function, why are they so important?

   A. RFLPs are important for two reasons: First they can be identified in an individual in a relatively easy way; and second, they are used as genetic markers for certain functional genes. This means that if the RFLP is present in an individual, then the gene is probably present, with 95% certainty.

3. **Q.** But why not just look for the gene without worrying about a marker?

   A. For certain genetic diseases, the responsible gene has not yet been pinpointed. However, DNA technologists can be reasonably certain that the disease is inherited because studies of related individuals show a pattern of inheritance; and in all the diseased individuals, the RFLP is present. So, until the disease gene is pinpointed and a test developed, scientists look for the RFLP and assume (with 95% certainty) that if you have the RFLP, you also have the gene.

analyzed, the RFLP pattern of an individual is compared with the RFLP patterns of normal relatives and relatives affected by the gene in question. Such comparisons make it possible to determine whether the individual has the marker RFLP and the disease gene. The information can then be used to make important decisions on how best to deal with the genetic disease in question, as we will see in the discussions to follow.

DIAGNOSING INFECTIOUS DISEASE

Diagnostic tests based on DNA probes and PCR represent a radical departure from past practices. Up through the current era, reagents for diagnostic tests took two general forms: biochemical reagents for detecting specific enzymes and antibodies for detecting disease-related proteins. DNA probes represent a totally new class of diagnostic reagents because they detect neither enzymes nor proteins, but instead, gene sequences. To effectively use DNA probes, large amounts of the DNA must be available. The PCR makes this possible. In recent years, a number of diagnostic tests for infectious diseases have come forth, as we will see next.

antibodies

protein molecules produced by the body's immune system during a time of infectious disease.

ACQUIRED IMMUNE DEFICIENCY SYNDROME

One place where DNA probes and PCR have made an impact is in detecting the **human immunodeficiency virus (HIV)**. This virus causes HIV infection and, in its later stages, **acquired immune deficiency syndrome (AIDS)**. Traditionally, the test used in detecting exposure to HIV has been an antibody-based test. The test works on the supposition that when a person is exposed to HIV, the body produces HIV antibodies that can be detected in the laboratory. Unfortunately, it may take several weeks for the body to produce sufficient antibodies to yield a positive test. Thus the person could test negative even though HIV is present in the body (a **"false-negative"** test). In the interim, AIDS could be transmitted.

T-lymphocytes

immune system cells that mediate cellular immunity and are infected by HIV.

With the introduction of DNA probes and PCR, it has become possible to develop a more direct test for the presence of HIV. During its infection cycle, HIV exists as a segment of DNA integrated to the chromosomes of its host cells. These cells belong to the immune system of the patient and are known as **T-lymphocytes**. To perform the diagnostic test, T-lymphocytes are obtained from the patient and disrupted to secure the DNA. The DNA is then amplified by PCR and up to a half-milligram of DNA can be obtained for testing purposes. (The unamplified weight of the DNA would be about a millionth of the final product.) Now an isotope-tagged DNA probe is added. The base sequence of the probe is complementary to the base sequence of the viral DNA. If viral DNA is present, the probe locates and binds to it. The binding is signaled by an accumulation of radioactivity. The results constitute a positive test.

Because the newer HIV diagnostic technique identifies viral DNA rather than viral antibodies, the physician can be more confident of a patient's health status. Drugs can be prescribed with more certainty, and measures to prevent HIV transmission can be instituted earlier. The technique also helps public health professionals track the AIDS epidemic more easily because they can know who is harboring the virus not just displaying antibodies.

transplacental

able to cross the pregnant woman's placenta and enter the developing fetus.

HIV diagnosis using a DNA probe also helps with a more assured detection of HIV infection in a **newborn**. The traditional way to test a newborn has been to ascertain whether it has HIV antibodies. The dilemma is that the antibodies may have come from the mother's blood by transplacental passage, and so may reflect the mother's HIV infection rather than the newborn's infection. By using the direct test, the DNA associated with HIV can be detected in the newborn's cells and a more definitive diagnosis can be made much earlier. The efficacy of this approach is enhanced by studies reported in 1992. Researchers found twenty-three infants within a day of birth to be positive for HIV by DNA probe testing and nineteen to be negative. When the infants were tested weeks later by the standard public health procedures, the results matched perfectly.

■ TUBERCULOSIS

Mycobacterium

my˝co-bac-ter˝i-um

Current diagnostic tests for **tuberculosis** can take several weeks to complete because the tubercle bacillus *Mycobacterium tuberculosis* multiplies very slowly. (A reproduction cycle takes place every 24 hours or so.) While waiting for a definitive diagnosis, physicians must make treatment decisions on the basis of limited information. This could lead to ineffective therapies and worsening illness; also the patient may transmit the disease to others as the wait goes on.

With help from the firefly, researchers have developed an innovative and imaginative diagnostic test for tuberculosis that could speed the time interval for detection considerably. Figure 7.6 summarizes the test. Only a few days may be required and the test could help determine whether that particular strain of *M. tuberculosis* is drug resistant. The new approach relies on the firefly enzyme **luciferase** to produce a flash of light in live *M. tuberculosis* cells.

luciferase

lu-cif′er-ase

bacteriophage

bak-ter′i-o-faj

The process works this way: a **bacteriophage** (a bacterial virus) specific for *M. tuberculosis* is genetically engineered to carry the gene for luciferase. (The phage is known as a **mycophage**, as well as a **luciferase reporter phage**.) A sample of phage is then mixed with a culture of unknown bacteria. If the culture contains *M. tuberculosis*, the phage penetrates the bacterium and inserts itself into the chromosome, carrying along the luciferase gene. The bacterium promptly begins producing luciferase. Now luciferin, a compound broken down by luciferase, is added to the culture together with the high-energy molecule adenosine triphosphate (ATP). If luciferase is present, the enzyme

luciferin

a compound that is broken down by luciferase in a light-emiting reaction.

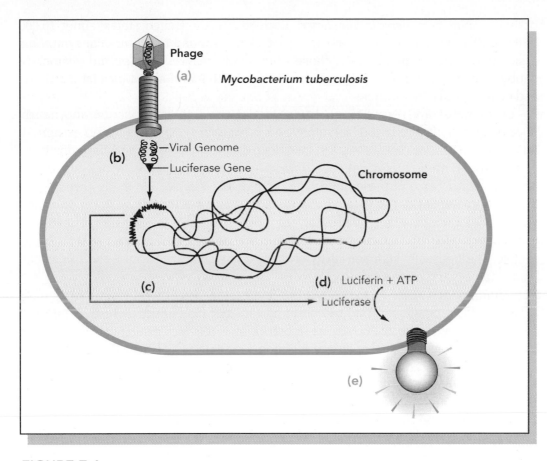

FIGURE 7.6

Detecting *Mycobacterium tuberculosis*, the agent of tuberculosis. (a) A bacteriophage (phage) specific for *M. tuberculosis* is added to a culture of bacteria. The phage carries the gene for luciferase production. If the bacterium is *M. tuberculosis*, the phage attaches to the cell wall, and the phage genome enters the bacterium. (b) The phage genome then inserts on the bacterial chromosome and brings along the gene for luciferase. (c) The bacterium begins to produce luciferase, and (d) when luciferin and ATP are added to the environment, the enzyme breaks down the luciferin. (e) The breakdown results in a flash of light, indicating that the bacterium is *M. tuberculosis*. If the bacterium were something other than *M. tuberculosis*, no phage attachment would take place, and no flash of light would be observed.

breaks down luciferin and the reaction results in a **flash of light**. A sensitive instrument (a luminometer) detects the light flash, and the culture is confirmed to contain *M. tuberculosis*. The report is made to the physician and the diagnosis is complete.

To determine drug susceptibility or resistance, the same procedure is used except a drug is added to the culture. If the bacteria are sensitive to the drug, the bacteria die and, literally, their "lights go out." If they are resistant, they continue to live and produce luciferase, and they give off light.

In the spring of 1993, scientists from New York's Albert Einstein College of Medicine and the University of Pittsburgh reported the test's development in *Science* magazine. As expected, when word of the successful test reached other journals and newspapers, headline writers had a field day. One cliche is particularly appropriate in this instance—the future of tuberculosis diagnosis "appears bright."

◼ LYME DISEASE

Lyme disease is another example of a bacterial disease that can benefit from DNA probe and PCR technology. **Lyme disease**, named for Lyme, Connecticut, is due to the

Lyme disease

a bacterial disease caused by a spirochete and characterized by a skin rash and involvement of the joints.

spirochete
spi´ro-ket

Borrelia
bo-rel´e-ah

burgdorferi
burg-dorf´er-e

tickborne spirochete ***Borrelia burgdorferi***. Each year in the United States, about 10,000 individuals in forty-five states have the fever, skin rash, and constitutional symptoms associated with the disease. For those who do not receive prompt and aggressive antibiotic therapy, the disease progresses to debilitating and sometimes fatal arthritis and neurological complications.

Spirochetes are very difficult to see under the light microscope and nearly impossible to cultivate in the laboratory, and *B. burgdorferi* (Figure 7.7) is no exception. Since 1975, when the disease was first identified, diagnoses have been made on the basis of symptoms, epidemiological considerations (e.g., presence of a tick bite), and antibody tests. None of these methods is completely reliable, and in the absence of a diagnosis, treatment can be delayed. However, researchers at the federal Rocky Mountain Laboratory in Montana have successfully used PCR to amplify the spirochetal DNA, and they have used DNA probes to detect as few as five spirochetes in a patient's blood sample. The unusually **high sensitivity** of the test is significant because the spirochetal load of the blood is often extremely low.

Adding to the value of the test is the finding that slightly different strains of *B. burgdorferi* react positively. This can be important because antibody tests would not necessarily detect all possible strains. Moreover, the test can distinguish *B. burgdorferi* from other *Borrelia* species that induce symptoms reminiscent of Lyme disease. Also, by identifying spirochetes directly (rather than indirectly by their antibodies), researchers will now be able to study the developing pathological process of the disease. For example, it is important to know whether spirochetes are actually present in the late-stage disease when complications set in or whether the complications are due to spirochetal fragments remaining after the spirochetes have been eliminated.

■ HUMAN PAPILLOMA VIRUS AND OTHER DISEASES

papilloma
a wartlike growth.

The DNA probe test that identifies the **human papilloma virus (HPV)** has gained approval by the Food and Drug Administration (FDA). This virus is know to cause **genital warts** in humans. Marketed as the ViraPap detection kit, the test uses a DNA probe to pinpoint HPV in samples of tissue obtained from a woman's cervix. The DNA probe has a base sequence complementary to a base sequence of DNA sought in the tissue sample. Because certain forms of HPV are also linked to cervical cancer, the test has won broad acceptance by clinical physicians concerned with identifying the virus and reducing the possibility of cancer.

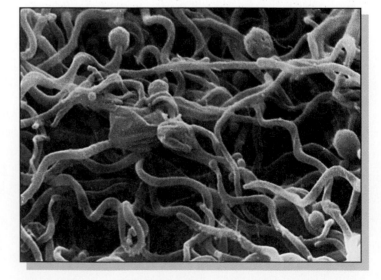

FIGURE 7.7

A scanning electron micrograph of *Borrelia burgdorferi*, the spirochete that causes Lyme disease. DNA technology techniques can now be used for rapid identification of this bacterium.

The development in recent years of many new antiviral drugs (such as acyclovir and ganciclovir) has increased the demand for new tests for viral diseases because treatment can begin early if the disease has been identified. (In previous decades there was little that could be done for the person suffering from viral disease.) The tests for HIV, Lyme disease, and HPV are forerunners of many other tests that we can expect to see in the future. DNA probe tests for hepatitis B, herpes simplex, and cytomegalovirus disease should be in widespread use in the years ahead.

A test based on DNA probe analysis has also been developed for **periodontal disease**. Periodontal disease is a degenerative infection of the gums that can lead to tooth loss. DNA probes developed can be used to detect three species of bacteria commonly associated with the disease. Efficient detection is the key to early diagnosis of the disease, improved treatment, and more successful prevention of tooth loss.

IDENTIFYING GENETIC DISEASE

DNA-based tests have also found great value in detecting the potential for genetic disease. Frequently a genetic disease cannot be diagnosed until its symptoms appear. With DNA tests, however, it is possible to locate the genetic basis for the disease. The individual can then be provided with counseling on what the future holds and advised on how to limit transmission of the gene. It may also be possible to institute gene therapy (Chapter 8).

Victor McKusick (Figure 7.8) of Johns Hopkins University has assumed the responsibility of cataloging all known genetic diseases in his voluminous book *Mendelian Inheritance in Man*. In its first edition (1966), his book listed 1500 phenotypes of diseases then known to be inherited. In its current 11th edition, the book has grown to 9000 phenotypes, most of which are linked to certain genes. Often regarded as a founder of the science of medical genetics, McKusick was honored in 1997 by the Lasker Foundation for a lifetime of achievement in science.

At this writing, DNA probes are being used to assist the diagnosis of such genetic diseases as sickle cell anemia, hemophilia A, Tay-Sachs disease, and phenylketonuria (PKU). In the pages ahead we will explore several other inherited diseases in which the probes have been useful in recent years.

CYSTIC FIBROSIS

Cystic fibrosis (CF) is a genetic disease diagnosed by a DNA probe. The individual with **cystic fibrosis** produces an abnormally thick, sticky mucus that clogs the

acyclovir
a-si′klo-vir

ganciclovir
gan-si′klo-vir

phenylketonuria
fen″il-ke-to-nur′e-ah

cystic fibrosis
a genetic disease in which abnormally thick mucus clogs the air passages and impairs respiration.

FIGURE 7.8
Victor McKusick, regarded by many scientists as the founder of the science of medical genetics.

respiratory passageways. (To help youngsters cough up the mucus, parents must slap them on the back repeatedly.) More than 30,000 cases of CF in children are reported annually in the United States.

Under normal conditions, a protein encoded by DNA regulates water absorption to keep fluid flowing freely along the surfaces of the respiratory passages. In the patient with CF, a defective gene (called the *cftr* **gene**) encodes a defective protein. The protein causes water to be pulled from the surfaces, which leads to the buildup of **thick mucus**, providing a breeding ground for infectious organisms. An estimated 12 million Americans carry the defective gene in duplicate. Figure 7.9 shows the genetic basis for inheritance of the disease.

In 1989, the gene for CF was identified on chromosome number 7 by **Francis Collins** at the University of Michigan and **Lap-Chee Tsui** at Toronto's Hospital for Sick Children. Soon after, a DNA probe was developed to help locate the gene in a sample of human DNA. The synthesis of the DNA probe was the important first step in allowing

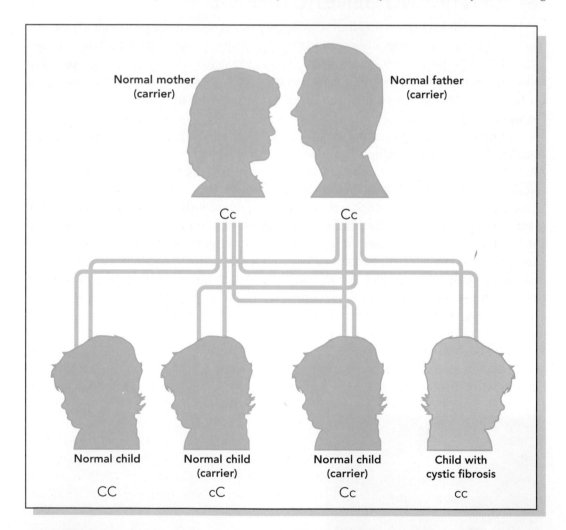

FIGURE 7.9

Inheritance of the genes for cystic fibrosis. Cystic fibrosis follows a recessive pattern, meaning that the disease develops when two recessive genes are present. In the diagram, both the mother and father are carriers of the recessive gene (c) for cystic fibrosis but are normal because each has a dominant gene (C) for the normal condition. Children inherit genes in pairs, one copy from each parent. Thus there are four possible gene outcomes in the children of this pair. Three children are normal because they have a dominant gene (two are carriers), but one child inherits both recessive genes and cystic fibrosis develops.

researchers to make copies of the normal and abnormal genes. In 1990, researchers forced normal CF genes into abnormal cells from patients with CF contained in test tubes. To their great satisfaction, the cells began acting like normal cells. For example, one characteristic of CF is that chloride is not effectively transported through channels of the cell membrane. In the laboratory experiment, the channels opened and the chloride ions passed freely in and out of the cells.

When the CF gene has been identified, counselors can let people know whether their offspring carries the CF genes in duplicate. In the months before birth, tests can be conducted on fetal tissue obtained by amniocentesis and by chorionic villus sampling. In **amniocentesis**, fetal cells are isolated from samples of the amniotic fluid that bathes the fetus. In **chorionic villus sampling**, a sample of the chorionic membrane is removed (this covering, like the amniotic membrane, encloses the fetus). Cells from the chorionic membrane provide the same information as cells from the amniotic fluid, and because the chorionic membrane forms before the amnion, detection of abnormal CF genes can be made earlier.

An advance of a different nature was announced in 1992. Instead of testing a fetus' cells for the CF gene, researchers reported the successful testing of an **eight-celled embryo**. In the research, embryos were conceived in the laboratory through *in vitro* ("test tube") fertilization. Because the parents were carriers of the CF gene, they had a one in four chance of conceiving a child with the disease. Researchers obtained cells from each of five embryos and identified them as having two, one, or no copies of the CF gene by means of DNA probes. Embryos containing one and no copies were implanted into the mother's uterus. Nine months later, a healthy baby was born.

◼ DUCHENNE'S MUSCULAR DYSTROPHY

A defective gene is also involved in a disease called **Duchenne's muscular dystrophy** (named for Guillaume Duchenne, the French neurologist who described the condition in 1868). In patients who have this disease, the muscles lack a muscle-strengthening protein called **dystrophin** (described as the biochemical equivalent of building plaster). Without dystrophin, the muscles become infiltrated with other tissues, and as the muscles enlarge, they begin to waste away. In 1986, researchers at Harvard University identified the gene that encodes dystrophin. The gene is one of the largest genes ever identified in the human genome.

Duchenne's muscular dystrophy is particularly insidious in children. Apparently normal babies develop with only the usual childhood problems until they are about three years old. Then they gradually become unsteady on their feet as they lose the strength and control of their muscles. Confined to a wheelchair by ten, they often die in their teens.

In modern medicine, a DNA probe is used to identify a telltale segment of DNA that lies close to the defective gene for muscular dystrophy. The segment is a **restriction fragment length polymorphism**, or **RFLP** (pronounced "rif-lip"). As noted earlier in this chapter, a RFLP is a marker segment of DNA that can occur in many different forms, with each form having a different length. If the RFLP is present, there is ninety-five percent certainty that the gene for Duchenne's muscular dystrophy is also present.

To use RFLP analysis for Duchenne's muscular dystrophy, DNA must be obtained from as many different relatives as possible, including parents, grandparents, aunts, uncles, and so on. The RFLP patterns from these relatives are then used to construct a pattern for the entire family. The individual's RFLP pattern is then compared with the patterns of affected and unaffected relatives to see which fits best. It is important to

amniocentesis

am″ne-o-cen-te′sis

chorionic

kor-e-on′ik

amnion

the membrane-enclosed sac of fluid that surrounds the developing fetus.

Duchenne's muscular dystrophy

doo-shen′

a genetic disease in which the muscles enlarge and waste away, resulting in unsteady gait and loss of muscle control.

dystrophin

dis′tro-fin

note that no universal test is available because RFLPs vary from family to family. A similar situation applies when RFLPs are used as markers to identify other genetic diseases (Questionline 7.3).

■ HUNTINGTON'S DISEASE

Huntington's disease

a genetic disease characterized by progressive deterioration of the nervous system and accompanied by thrashing movements.

chorea

ko-re'ah

glutamine

one of the twenty amino acids in protein; contains an extra amine group.

Well known as the disease that killed folk singer Woody Guthrie in 1967, **Huntington's disease** is an invariably fatal illness that destroys brain cells. The patient has progressive deterioration of the nervous system that eventually leads to constant thrashing and writhing movements, then insanity and death (the older name for the disease, Huntington's "chorea," is derived from the Greek *choreia* for dance, a reference to the jerky movements). The disease is named for George Huntington, the American physician who studied it in the early 1900s.

It is now known that Huntington's disease is due to a defective gene near the tip of human chromosome number 4. The gene is distinguished by an **excessive repetition of the base triplet CAG**, which encodes **glutamine.** In the normal version of the gene, the CAG triplet repeats itself between eleven and thirty-four times, but in the Huntington's patient, the gene occurs between forty-two and sixty-six times. (A similar pattern of triplet repeats is seen in fragile X syndrome, to be discussed next.) Moreover, the length of the repetition is related to the age of the patient: the younger the Huntington's patient, the more times the triplet is repeated. The abnormality, it appears, gives rise to a deformed or defective protein that encourages the death of cells in the basal ganglia, an area of the brain crucial to motor function.

The gene for Huntington's disease can now be identified by using a DNA probe to pinpoint an adjacent RFLP, first located in 1984 by **James Gusella**. Gusella and his

QUESTIONLINE 7.3

1. **Q.** I understand that cystic fibrosis is now considered a genetic disease. Why is the disease considered genetic?

   A. For decades, scientists were uncertain about the cause of cystic fibrosis. They now know that people with cystic fibrosis cannot produce a particular protein to keep fluid in the respiratory tract from becoming thick and mucoid. Because proteins are encoded by genes, it follows that a defective gene prevents protein production and encourages cystic fibrosis to develop.

2. **Q.** Are there other genetic diseases of this type?

   A. Yes. In Duchenne's muscular dystrophy, the cells cannot produce a protein that strengthens the muscles. Without the protein, muscles become weak and begin to waste away. Patients are usually confined to wheelchairs, and rarely do they live beyond their teenage years. Other genetic diseases considered in this chapter include Huntington's chorea, retinoblastoma, and quite possibly Alzheimer's disease.

3. **Q.** How can DNA technology help in detecting diseases such as these?

   A. DNA technology can help diagnose genetic diseases by detecting the defective gene or by locating a RFLP lying close to the defective gene. To make the detection possible, a DNA probe and PCR are used. Knowing whether a person is carrying a defective gene may help by suggesting a lifestyle change to prevent the symptoms from developing or interrupt gene passage to the next generation. A person carrying a defective gene may also be a candidate for gene therapy (Chapter 8).

colleagues, pictured in Figure 7.10, used a DNA probe designated G8 to pinpoint the suspected RFLP. But they could not be sure whether the RFLP that binds to G8 was the correct one. Fortunately they had a very large family to study—a family living near **Lake Maracaibo in Venezuela** and descended from a woman who had migrated to the area from Europe in the early 1800s and was afflicted with Huntington's disease. Working with clinical psychologist **Nancy Wexler**, Gusella and his colleagues were able to track the pedigree of this family through seven generations, and by use of the G8 probe, they found that the RFLP was present wherever the disease occurred. It was clear that a tight correlation exists between the RFLP and Huntington's disease.

For Nancy Wexler, the work in Venezuela has continued. Each year she returns to Lake Maracaibo and takes more data and blood samples to add lines and boxes to the growing pedigree chart of the family with Huntington's disease. More that 12,000 people in the group have now been tested, and the pedigree chart covers both sides of the corridor wall outside her office at Columbia University. The work has a personal interest for Nancy Wexler: She is a carrier of the gene for Huntington's disease.

Having found a marker RFLP for Huntington's disease, Gusella's group set to work on a ten-year project to identify the Huntington's gene. The researchers prepared mouse human hybrid cells in which each cell had only a few human chromosomes. Then they isolated the DNA from known chromosomes within these cells and combined it with DNA probe G8 to see whether binding would take place. Their results indicated that the DNA from cells containing human chromosome number 4 would bind to probe G8. Therefore, they concluded, chromosome number 4 carries the gene for Huntington's disease. In 1993, Gusella's group, together with an international consortium consisting of six research teams, announced that they had finally located the Huntington's gene near the tip of the chromosome.

At this writing, there is no approved test for the Huntington's gene, so the marker RFLP continues to be used as a flag to track the defective gene through successive generations. (A test for the gene itself is not yet possible for diagnostic purposes partly because the gene is so complex.) Knowing whether a person has the marker gene can be important because Huntington's disease usually shows no symptoms until middle age, long after the gene may have been passed to the next generation. Moreover, the trait is caused by a dominant gene, so only one gene needs to be present for expression. About 20,000 Americans are identified with the disease each year.

◼ FRAGILE X SYNDROME

The characteristic sign of **fragile X syndrome** is inherited mental retardation. An estimated one in 2000 girls and one in 1000 boys are believed affected each year. The syndrome is so named because the cell's X chromosome (the sex chromosome in

fragile X syndrome

a genetic disease caused by a defective gene near a weak spot on the X chromosome and accompanied by mental deterioration.

FIGURE 7.10

James Gusella and his colleagues at the Massachusetts General Hospital. Gusella's team identified the gene for Huntington's disease in 1993.

males and females) has an apparent weak spot at the tip where the chromosome snaps easily under experimental conditions.

In 1991, researchers located an abnormal gene adjacent to the weak spot. (The abnormal gene is now called the **FMR-1 gene**.). When they cloned the abnormal gene and examined its base sequence, scientists discovered a long stretch in which three nitrogenous bases (CGG), were repeated over and over. The normal gene has a small segment of repeats, but in the person with fragile X syndrome, there is an extraordinarily long segment of **trinucleotide repeats** (somewhat like an accordion being opened). Because the normal gene is active in the brain, the implication is that an abnormal gene might cause the retardation, although the mechanism is not currently understood.

Until 1991, detecting fragile X syndrome required cultivating an individual's white blood cells in the laboratory and microscopically examining them for broken X chromosomes. That year, however, a biotechnology company announced a DNA probe test for the defective gene. The announcement came barely a month after the gene's discovery. The probe tags the repeated nucleotides and identifies cells in which fragile X syndrome will probably develop.

How the repeating trinucleotide (CGG) brings about fragile X syndrome was partially resolved in 1997 by researchers at the University of Illinois. Biochemists found that the inclusion of repeating DNA trinucleotides upsets the transcription process, leading to a protein deficiency. The protein is necessary for normal nerve cell functioning at the synapse, the gap between adjacent nerve cells and nerve and muscle cells. As information processing declines, mental retardation may develop.

■ TRIPLE REPEAT DISEASE

triplet repeat disease

a genetic disease in which groups of three nucleotides repeat themselves many times.

spinocerebellar

spi″no-cer-e-bel′ar

Toward the end of the 1990s, it became apparent that fragile X syndrome and Huntington's disease were only two examples of a series of genetic disorders now known as **triplet repeat diseases**. Another example is **spinocerebellar ataxia**, in which forty to eighty CAG trinucleotide repeats occur on chromosome 6 and a neuromuscular disorder ensues. Degradation of the spinal cord develops in **Friedreich's ataxia**, in which the trinucleotide GAA repeats between 200 and 900 times in chromosome 9. Other triplet repeat diseases include **myotonic dystrophy** (CTG repeat) and **spinobulbar muscular atrophy**, also called **Kennedy's disease** (CAG repeat). In these diseases the number of repeats tends to increase with each generation and the disease becomes more severe, frequently revealed by decreasing age of onset. This phenomenon is called **genetic anticipation**.

Triple repeat diseases have captured the attention of geneticists in various fields. Classical geneticists, for instance, point out that the increasing number of repeats in succeeding generations undermines the principle that the genetic code remains constant; and they indicate that the rules of Mendelian genetics do not apply. Evolutionary biologists wonder why the repeats have been preserved in the human genome for so long and suggest they must have some value. And molecular biologists want to know what the repeats do in the normal individual.

The mystery deepens when one notes that triplet repeats are not seen in bacteria, fruit flies, or mammalian cells other than from humans. Researchers point out that the repeats give new insights to the development of human disease and are strong arguments for deciphering the entire human genome (Chapter 12) as a way of discovering still other mechanisms of human disease. They also show DNA to be a dynamic and, at times, unstable molecule, characteristics few biochemists thought it had.

RETINOBLASTOMA

Retinoblastoma is a rare cancer of the eye affecting about 200 children in the United States annually. Although it often requires removal of the eyeball, it can be cured if detected early by less radical treatments such as radiation therapy, laser surgery, and freezing.

In the 1970s, researchers observed that children of retinoblastoma patients also developed the disease, and it soon became clear that predisposition to the disease is an inheritable trait. Then, in the early 1980s, investigators used DNA probes to determine that a segment of DNA (a mutated DNA segment) was missing at the **q14 region** of chromosome number 13, as Figure 7.11 illustrates. They subsequently related the disease to the defective gene on the chromosome. Finally, in 1986, researchers at the Massachusetts Eye and Ear Infirmary in Boston isolated the gene believed to govern retinoblastoma. Apparently, this gene in duplicate (i.e., the homozygous condition) induces cancer not by its presence (as do other cancer genes found to date) but by its absence or **inability to function** (because of the missing or mutated segment). In this sense, the normal gene is an anticancer gene.

The finding of a retinoblastoma gene was a major breakthrough in science because

retinoblastoma

re˝tin-o-blas-to´mah

a genetically linked form of cancer of the eye.

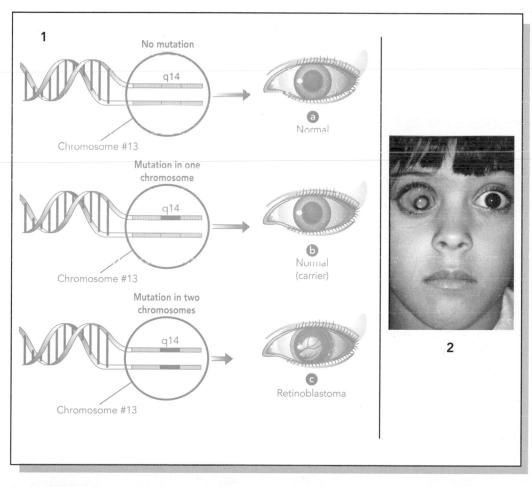

FIGURE 7.11

Retinoblastoma. 1. Development of retinoblastoma. (a) In the normal individual, the q14 segment of chromosome 13 is functional. (b) When one chromosome is mutated at the q14 region, the individual is still normal but is a carrier of the retinoblastoma gene. (c) When both chromosomes are mutated, the retinoblastoma develops. 2. A patient displaying the eye tumor characteristic of inherited retinoblastoma.

it generated interest in recessive cancer genes. Recessive cancer genes are permanently switched-off genes. Moreover, they must be present in dual copies (i.e., homozygous) if the cancer is to occur, and a mutation or random copying error could yield the second gene necessary for the cancer to begin. The hunt for other recessive cancer genes continues, even as you read this.

■ ALZHEIMER'S DISEASE

Alzheimer's disease is a degenerative brain disorder afflicting about 2.5 million Americans annually. It is the fourth leading cause of death among elderly Americans. Named for Alois Alzheimer, the German neurologist of the early 1900s, the disease is accompanied by organic **loss of intellectual** function (dementia), as well as memory loss and general incapacitation. Sufferers often cannot speak, walk, or tend to their most basic needs.

For years, scientists were at a loss to explain Alzheimer's disease. Then in 1987, researchers at several institutions identified a specific gene inducing the brain tissue abnormality, which characterizes the malady. Almost simultaneously, another research team announced that it was using a DNA probe to locate a genetic marker for the disease on human chromosome 21. Although these findings do not suggest that all cases of Alzheimer's disease are genetically linked, nevertheless they indicate that at least one form, **familial Alzheimer's disease (FAD)**, may be inheritable.

The gene for the FAD abnormality appears responsible for manufacturing a protein called **amyloid**. Amyloid is a major component of clumps (or plaques) of dead and dying nerve fibers that clog the brains of patients with Alzheimer's disease. (Figure 7.12 shows the plaques in the original patient.) Whether amyloid is the key to understanding the disease is uncertain at this writing. It is possible, for example, that the amyloid gene together with other environmental factors may produce the disorder. Many neurologists believe that a virus may be involved.

■ AMYOTROPHIC LATERAL SCLEROSIS

Amyotrophic lateral sclerosis (ALS) is a degenerative disorder of the motor neurons of the brain and spinal cord. ALS affects about 5000 Americans each year and is commonly known as **Lou Gehrig's disease** for the New York Yankees first baseman

Alzheimer's disease

alz-hi′mer

a genetic disease and a degenerative brain disorder, particularly in the elderly.

amyloid

am′e-loid

amyotrophic

a-mi″o-trof′ik

sclerosis

skle-ro′sis

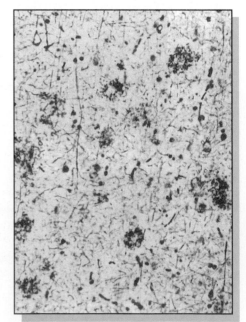

FIGURE 7.12

A light micrograph of a slide from the brain tissue of a woman identified only as Auguste D. In the early 1900s, the German neurologist Alois Alzheimer studied the brain tissue of Auguste D. and identified the plaques and tangles shown in this slide. On that basis, Alzheimer described the disease now named for him. Scientists questioned whether the patient actually had Alzheimer's disease until 1998, when they found some of Alzheimer's original slides and identified the typical symptoms shown here. The finding confirmed Alzheimer's original diagnosis.

who was its victim (Figure 7.13). With the destruction of motor neurons comes muscle weakness and "creeping paralysis" and death. Symptoms usually occur between the ages of 35 and 70, and death generally results in three to five years.

In about five to ten percent of cases, ALS passes from one generation to the next and therefore displays an inherited pattern (the remaining cases are said to be "sporadic," but the symptoms are so similar that a common chemical problem may be involved). It was this inherited pattern that attracted molecular geneticists to ALS in the 1980s, and their work culminated in the 1993 discovery of a gene for the inherited form called **familial ALS**.

Under normal circumstances, the normal gene, known as **sod1**, encodes an enzyme called **superoxide dismutase**. This enzyme helps cells eliminate toxic and highly reactive oxygen-containing molecules called **free radicals**. (Free radicals have been implicated in Parkinson's disease, Alzheimer's disease, and the aging processes.) With a defective superoxide dismutase (caused by a defective gene), the level of free radicals rises and toxic damage to the neurons ensues.

dismutase
dis-mu′tase

The research group making the 1993 discovery consisted of 31 scientists from four countries. It was led by **Robert Brown** of Massachusetts General Hospital and **Robert Horvitz** of MIT. The researchers first narrowed the gene for superoxide dismutase to chromosome number 21, then found that the gene was mutated in patients from thirteen families having members with ALS. Mutations in eleven different places in the gene were found, and each mutation led to a switching of a single amino acid at a key part of the enzyme. DNA from normal individuals has no such mutations.

neurons
cells of the nervous system.

On the basis of the results, it appears that ALS has a relation to defects in superoxide dismutase activity, and, by extension, to the accumulation of free radicals. A number of available drugs, including vitamins C and E, are known to help eliminate free radicals, and they could conceivably be used to reduce the effects of ALS. Moreover, a report published in 1996 indicates that the defective enzyme cannot control a transporter protein that normally clears glutamate from the nervous system. Large quantities of glutamate can be lethal to nerve cells.

FIGURE 7.13

Lou Gehrig, the renowned New York Yankees first baseman who died of amyotrophic lateral sclerosis (ALS) in 1941. With identification of the ALS gene in 1993, it is now possible to understand the disease better and to develop therapies to control its effects.

■ DIABETES

Individuals with **diabetes** cannot make use of glucose in their metabolism and have a difficult time obtaining the chemical energy it contains. Under normal conditions, the pancreas produces the hormone insulin, which promotes glucose metabolism. But without insulin, either of two types of diabetes can occur. In **Type I diabetes** (juvenile-onset diabetes), the insulin-producing cells of the pancreas function improperly, and **insulin** is not produced in sufficient amounts to sustain energy metabolism. This condition can be alleviated by injecting genetically engineered insulin described in Chapter 6. In Type II diabetes (maturity-onset diabetes), the body loses the ability to use insulin even though pancreatic cells synthesize sufficient quantities of the hormone.

Type II diabetes is the less severe form of diabetes, but it affects more than 10 million Americans annually. It can be treated by weight and diet control and by using oral drugs to make the cells more sensitive to insulin activity. Although the reasons for onset of the disease are not certain, a prevailing theory states that the body's cells cannot react to insulin because they lack the necessary **receptor sites** at their cell surfaces. Because insulin normally spurs the entry of glucose to the body cells, the lack of receptor sites would negate this function.

An uncommon form of Type II diabetes is **maturity-onset diabetes of the young (MODY)**. Whereas Type II diabetes usually develops after age 25 to 30, MODY develops in adolescents and teenagers. In 1992, researchers from the University of Chicago led by **Graeme Bell** reported that MODY could develop from a genetic defect related to an enzyme called **glucokinase**. The researchers followed up a series of linkage studies by French scientists, who identified a specific gene on chromosome 7 as the gene responsible for glucokinase production. The enzyme glucokinase is required by pancreatic cells to detect glucose concentrations in the blood. Bell's group pointed out that a gene defect causes MODY patients to produce insufficient glucokinase, and without glucokinase, the secretion of insulin by the pancreas is disrupted. (This is a novel approach to the development of diabetes because it does not depend on receptor sites or insulin insufficiency.) In subsequent studies, the researchers identified the same gene defect in a family with the more common form of Type II diabetes.

To perform their research, Bell's group used the PCR to amplify the coding parts of the glucokinase gene in DNA samples obtained from normal and diabetic patients. Then they scanned the genetic material with DNA probes, hunting for mutations. The researchers found that a single nitrogenous base was different in the diabetic DNAs. This defect is reminiscent of that in sickle cell anemia (Chapter 3). And conceivably it could be reversed—an exciting possibility for gene therapy (Chapter 8). Research reported in 1993 by an international team of investigators confirmed the work but expanded the number of possible mutations to almost two dozen.

Type I diabetes also appears to be gene-associated. In 1994, a team from Oxford University scanned the entire human genome by novel gene-mapping techniques and found eighteen different chromosome regions linked to the disease. Narrowing their search, they found regions on the long arms of chromosome 11 and chromosome 6 that appear to be central to the development of the disease. A third gene was found on the long arm of chromosome 18.

■ CANCER

Just as a genetic component is unfolding for certain forms of diabetes, it is becoming clear that the development of **cancer** is more than a result of random events and environmental hazards; it is a often product of a genetic predisposition. For example, oncologists estimate that approximately ten percent of individuals who have **melanoma** develop carry an inborn susceptibility to that cancer. Identification of the specific genes involved has obvious implications for detecting the susceptible individual and developing new treatments. Well-organized efforts by several research groups have pinpointed several melanoma-susceptibility genes on human chromosomes 1 and 9.

melanoma
a form of skin cancer.

A cancer-related gene of a different sort has also been investigated. This gene, named *p53* (because the protein it encodes has a molecular weight of 53 kilodaltons), is apparently a tumor-suppressing gene. First cloned in 1986, *p53* provides a common thread to all tumors and cancers because the gene seems to be guardian of the cell's DNA (it was *Science* magazine's Molecule of the Year in 1993). When damage occurs in the cellular DNA (e.g., the damage that leads to tumor formation), *p53* apparently encodes a protein that binds to the DNA alongside the damage and inhibits replication. In doing so, *p53* permits time for the cell to repair its DNA, thereby preventing the propagation of faulty, cancer-causing genes to progeny cells. Thus, *p53* acts as a tumor-suppressing gene to prevent cancer from establishing itself.

p53
a gene believed to suppress cancers by encoding DNA-repair proteins in cells.

The role of a mutated *p53* gene in cancer has been strengthened by the recovery of an **altered version** of *p53* in tumor cells from patients with colon, breast, brain, bone, lung, skin, bladder, and cervical cancers. The uses of DNA probes and PCR technology were prominent in these identifications. In 1992, researchers at Johns Hopkins University isolated four proteins encoded by the *p53* gene and postulated that the suppressor function depends on more than a single protein. If any of these four proteins are not produced by a mutated *p53* gene, DNA replication cannot be inhibited and tumor formation will proceed. To make matters worse, the tumor cells will all inherit mutated *p53* gene and continue forming damaged DNA.

The question of the mutated *p53*'s origin has also interested researchers. The prevailing wisdom is that environmental toxins can change a normal *p53* gene to its mutated form, but a biochemical misstep in DNA replication may also be the origin. Furthermore, the mutation may be inherited once it has arisen, thereby adding to the concept that certain cancers can occur in families.

toxin
a biochemical compound that disrupts a metabolic process.

Among the biggest pieces of news of 1994 was the location of a gene linked to **hereditary breast cancer**. The gene was named *BRCA1*. Its discovery was followed fifteen months later by identification of a second breast cancer gene, *BRCA2*. Researchers estimate that up to eighty percent of women who inherit mutated forms of either gene will have breast cancer develop in their lifetime, usually at a relatively early age, and women with a mutation in *BRCA1* have a high risk of ovarian cancer developing as well.

BRCA1
a gene believed responsible for certain forms of inherited breast cancer.

Many researchers believe that the proteins encoded by both normal genes operate in the DNA-repair pathway. In their mutated form, the genes cannot encode important repair enzymes, and cells accumulate mutations that foster cancer development. Other researchers present evidence that the two proteins (the first having 1863 amino acids, and the second, 3418 amino acids) act as gene regulators and activate transcription.

Only five to ten percent of cases of breast cancer are genetically inherited (about 15,000 cases annually in the United States.) Nevertheless, the findings have sparked hopes of developing diagnostic tests. And indeed, they did: by 1996 tests for *BRCA1* and *BRCA2* were commercially available. But soon the problems manifested themselves: (1) *BRCA1* was found to have hundreds of variations in its base sequence (spanning 100,000 bases), and screening for all variations became very expensive; (2) finding evidence of the genes creates dilemmas because there are no routine ways to prevent breast and ovarian cancers, and treatment methods vary considerably; and (3) the anxiety level in view of the dilemmas is extraordinarily high in women found positive.

Some forms of **colon cancer** also seem to be inherited. Indeed, medical researchers were long interested in why certain types of colon cancer seem to run in families. Was it blind chance? A shared exposure to a cancer-causing trigger in the diet? A rogue gene that increased their susceptibility?

An answer was provided in 1993 by researchers from Johns Hopkins University and the University of Helsinki, who found a gene causing an inherited form of colon cancer called **hereditary nonpolyposis colon cancer**, or HNPCC (also called **Lynch syndrome**). Although not the only colon cancer with a genetic root, it is believed to be the most common because the gene is carried by an estimated one in 200 individuals worldwide (a frequency ten times greater than that for cystic fibrosis). In the human body, the normal functioning gene encodes a protein used by cells to **repair DNA** when it is damaged. The protein removes nucleotides from the DNA chain when they pair up with incorrect bases. To complete the repair (called a "mismatch repair"), the protein fills the space with the correct nucleotides. A mutation in the gene prevents the protein's production and leaves the damage in place.

To locate the gene the researchers used a set of DNA probes designed to detect short repetitive DNA sequences through the chromosomal material of a cell. These short sequences are known as **microsatellites**. Because they vary greatly in length from one individual to another (as do the RFLPs discussed in this chapter), microsatellites are highly informative for linked genes such as the colon cancer gene. After 345 failures, the scientists found a microsatellite marker in family members who had colon cancer, but it was absent in those who did not. The marker and gene were on chromosome 2.

The pattern of cancer development in patients with HNPCC (Lynch syndrome) has been described as a "**caretaker defect**" because it involves a caretaker gene whose protein protects the cell by repairing its DNA. A second pattern associated with colon cancer is the "landscaper defect," first proposed in 1998. An example of the defect is seen when polyps develop in youthful patients having **juvenile polyposis syndrome (JPS)**. The polyps do not develop directly from the colon cells, but from the surrounding epithelial cells in the intestinal landscape. A third pattern, called the "**gatekeeper defect**," involves the *p53* gene. This tumor-suppressing gene normally inhibits replication of damaged DNA (as described above), but when mutated, it "opens the gate" to a tumor of the colon.

Colon cancers are among those that carry a favorable prognosis for successful treatment. Thus, a strong case can be made for presymptomatic DNA testing in the general population. Those with the defective form of the gene can be monitored closely for the first signs of cancer, and with early detection, the chances of cure rise dramatically.

■ OBESITY

In some individuals **obesity** is an intractable medical problem that increases the risk of diabetes, high blood pressure, and other life-threatening conditions. For many sufferers, the excess weight is an inherited problem with a molecular basis.

The nature of that problem remained obscure until 1994 when **Jeffrey Friedman** and his colleagues at New York's Rockefeller University provided a clue. After an exhaustive series of experiments covering eight years, they identified and cloned the gene that, when mutated, causes severe inherited obesity in mice. Then they found a very similar gene in human cells.

Friedman's group began by using breeding experiments to locate the gene

---

nonpolyposis

non˝pol-ip-o´sis

---

microsatellites

short repetitive stretches of DNA that serve as markers for detecting certain genes.

(designated *ob*, for obese) on the mouse chromosome 6. They identified the **ob gene** as a DNA segment containing 650,000 bases. Then they screened normal mice to see whether any genes from that area were operating in fat tissue, and they found a gene encoding a protein (containing 167 amino acids) in the fat and nowhere else. Furthermore, a strain of genetically obese mice had a mutated form of the gene, and no gene-encoded protein. The group concluded that the *ob* gene encodes a protein that keeps the animals' weight under control. Figure 7.14 shows the results of their experiment.

The activity of the weight-control protein suggests that it is a hormone. Researchers theorize that it contains a signal sequence most likely acting in the hypothalamus at the base of the brain. Neurologists have recognized this area as a control site for food intake and energy expenditure, both factors associated with obesity.

Despite the finding, biochemists were careful to construe that obesity can be resolved by focusing on a single gene. Researchers have found at least six other mutations associated with obesity in mice. One may encode the receptor where the gene product of *ob* reacts. Another gene (designated *tub*, for "tubby") works only after sexual maturity has been attained. Still another (designated *fat*) is similar to the gene that encodes insulin-making enzymes).

Nevertheless, the discovery of the *ob* gene and its product have excited the imaginations of medical geneticists, who see a possible treatment for inherited obesity. They envision the day when a hormone protein might be used to control weight as insulin is used to control diabetes. Indeed, in 1995, the biotechnology company Amgen paid $20 million plus royalties to Rockefeller University for exclusive rights to develop and market the protein for commercial use. The company's synthetic "ob protein," produced the next year, caused significant weight reductions in experiments with mice.

■ HEARING AND VISION LOSS

The successes achieved in human genetics spurred scientists to investigate inherited **deafness**. In a remote region of Costa Rica, researchers discovered a family whose

FIGURE 7.14

The effect of obesity genes. The mouse on the left has received the *ob* gene and displays the symptoms of gross obesity. Its littermate on the left is of normal size.

actin

a protein that helps support the hair cells of the cochlea of the ear.

members begin to experience deafness at age ten and then lose their hearing completely by age 30. A research team led by **Pedro E. Leon** traced the trait back through eight generations of the family, using genetic markers known to be inherited with the deafness. By 1992 they narrowed the responsible gene to chromosome 5. Then in 1997 Leon's team identified the exact position of the gene. Apparently its protein product helps the assembly of a protein called **actin**, which helps support the hair cells of the cochlea in the inner ear. Without the support, the limp hair cells are not able to detect sound waves, and the individual is deaf. (Actin in also a key protein of muscle cells.)

The Costa Rican gene is but one of several genes now related to deafness. Geneticists are currently studying a community on the north shore of the island of Bali, where an unusually high forty-eight of the 2000 inhabitants are deaf. The village has created its own sign language, with signed names replacing spoken names. In 1998, a gene responsible for the deafness was localized on chromosome 17.

In many cases, deafness is part of a broad constellation of symptoms ("**syndromic deafness**") and a number of genes are involved. In 1997, for instance, a gene mutation was found that affects the flow of potassium ions into cells. Deafness is accompanied by irregular heart rhythms and frequent loss of consciousness. Most cases of hearing loss, however, are not accompanied by other symptoms ("nonsyndromic deafness").

Complicating the situation is the observation that a mutated gene can sometimes cause syndromic but other times cause nonsyndromic deafness. During the early 1990s, for example, a mutated gene was related to **Usher syndrome** in which individuals who are born deaf slowly lose their **sight** (syndromic). In 1998, however, researchers found deaf individuals with no loss of vision (nonsyndromic). Apparently the gene product can be found both in the hair cells of the cochlea and among various cell types in the eye but the anomaly of the mutated gene remains unresolved.

glaucoma

glau-co'mah

Progress has also been made in the study of **glaucoma**, a disease that blinds almost 12,000 Americans annually. The disease develops when pressure builds up in the eye and damages the optic nerve. In 1997, molecular geneticists from the University of Iowa announced identification of a gene influencing juvenile-onset glaucoma, a hereditary form of the disease occurring in the teenage years. DNA markers were used to focus in on chromosome 1, and the search was eventually narrowed to a region containing about a million base pairs. Other researchers are pursuing the possibility of a gene for adult-onset glaucoma on chromosome 3.

trabecular

tra-bek'u-lar

The protein product of the normal gene is apparently related to activity of **trabecular meshwork cells**, a collection of cells that help regulate fluid drainage from the eye as new fluid is produced. Researchers have found that excessive amounts of the protein may clog the space between the meshwork cells and block the normal outflow of fluid. However, the work is very preliminary at this writing, and researchers are reluctant to make the leap from laboratory evidence to real-life situations.

macular degeneration

destruction of a portion of the retina called the macula.

A different form of vision loss is **macular degeneration**, a condition affecting a portion of the **retina**, the macula, where fine details are visualized. Ophthalmologists estimate that 1.5 million Americans have this disease and another 10.5 million display early signs.

As of 1997, a form of macular degeneration called **Sturgardt's disease** was linked to a gene mutation. This gene, identified in 1996 on chromosome 1, encodes a protein that uses high-energy adenosine triphosphate (ATP) to transport molecules across the membranes of retinal cells. The protein is called ATP-binding cassette *transporter*—*r*etina protein, or ABCR protein. A mutated ABCR protein results when two mutated alleles exist for chromosome 1. Although evidence is lacking on the normal role of

ABCR protein, researchers postulate that it may ferry molecules for recycling at the light-sensitive ends of the rod cells of the retina. Without the protein the material could build up and interfere with rod cell and retinal functions. Alternately, the altered ABCR protein could be toxic.

■ OTHER HUMAN DISEASES

Researchers have also focused on the genetic bases for various diseases of the brain. Although their work is still in early stages, DNA technologists reported in 1997, the identification of a gene mutation leading to a form of **Parkinson's disease**. In patients, the brain cells die and a neurotransmitter called dopamine drops in concentration; tremors are a typical symptom of the disease. A gene-encoded protein called **alpha-synuclein** apparently is misshapen and its accumulation in so-called **Lewy bodies** in the brain tissue of patients may lead to nerve cell degeneration.

synuclein
sĭ-nu′cle-in

In 1997, scientists also related a gene to another brain disease called **torsion dystonia**. In patients, the muscles twist, jerk, and contract involuntarily. DNA technologists used gene probes to search for a mutated gene in families in whom the disease occurred, and they located a gene and its defective protein in sufferers. Because dystonia does not develop in adulthood, they questioned whether the mutated gene is blocked in later years. Locating the gene makes it possible to distinguish these patients from those with cerebral palsy.

torsion dystonia

a genetic brain disease in which the muscles contract involuntarily.

Roughly one in 3500 in the Askenazic Jewish community is susceptible to **Canavan's disease**, a fatal brain disease in which the lipid sheath surrounding **nerve cells** is destroyed. The brain's nerve impulses cannot be transmitted, and virtually all nervous system functions are destroyed. Researchers at Miami's Children Hospital have now identified a gene mutation that leads to an enzyme in which a single amino acid is displaced. The normal enzyme regulates a part of the metabolism of aspartic acid (an amino acid), but the abnormal enzyme cannot function and abnormal metabolites build up in the brain.

One important "transport defect disease" now associated with a gene mutation is **copper deficiency disorder**, also called **Menkes' disease** (for John Menkes, who described it in 1962). In 1993, molecular geneticists observed a rare X-chromosome breakage in a girl with the disease. Gene probes were used to identify a spot at the breakage point where a **transport protein** was encoded. Without the protein, copper cannot be moved through cells of the intestine, and degeneration of the nervous system and mental retardation soon begin. Patients have extraordinarily kinky hair, which is very brittle because of copper deficiency.

Whereas Menkes' disease is caused by a copper deficiency, another gene mutation leads to **hereditary hemochromatosis**, or **iron-overload disease**. Reporting in 1997, scientists in France and Australia found a gene on chromosome 6 having a single base difference, thereby leading to a protein that has the amino acid cysteine where it should have tyrosine. The abnormal protein apparently causes **excessive iron absorption** in the intestine. Unusual amounts of iron are then deposited in the heart, liver, pancreas, and other organs, and as iron oxidizes the tissues, they literally rust away. Heart, liver, and organ malfunctions follow. At one time, hemochromatosis was believed to be a rare disease, but the U. S. Centers for Disease Control and Prevention now believe the disease afflicts one in 200 people.

hemochromatosis
hem″o-chro-mah-to′sis

On a more mundane level is the location of a possible gene for **baldness**. In 1998 Columbia University geneticists reported their work on a rare, inherited form of baldness called **alopecia universalis**. People with this condition have complete loss of hair on the scalp and other parts of the body. The researchers studied a gene on

alopecia
al″o-pe′ci-ah

chromosome 12 and identified a protein product that regulates part of the hair's growth cycle. Reports of the gene caused the expected stir among follically impaired individuals (especially men), so the researchers were quick to point out their work pertained to a "rare, inherited" form of baldness. Still, this was a jumping-off point for in-depth studies of baldness.

■ GENE BANKING

DNA probes are available for numerous purposes as these pages have shown, but they are not yet available for detecting a multitude of other conditions. In anticipation that probes will be available one day, individuals have been invited to place their DNA in storage units called gene banks (Figure 7.15).

A **gene bank** is established by an institution willing to extract, freeze, and set aside a person's DNA. For an established fee, blood is taken from an individual and the DNA is obtained from the white blood cells for storage. When a probe for pinpointing a particular condition has been developed, the DNA can be thawed and tested for the abnormal genes. In 1996, a Texas health center invited the public to use its gene bank. For $175, customers swab their inner cheek area with commercially available kits, then send in the samples for 25 years of storage.

lethal gene

one that predisposes an individual to illness or death.

According to researchers, gene banking is suitable for people wishing to know whether they are carrying a **lethal gene** such as for certain forms of cancer. Although contracting the disease probably depends on a complex interaction between genes and environmental factors, a person who knows he or she has the genetic marker for breast cancer may be able to minimize the risk by altering the lifestyle. To conclude that the gene marker is present or absent, the DNA must be analyzed from relatives from at least two generations, including someone with the disease. Gene banking ensures that the DNA can be obtained while the relatives are still alive. To be sure, knowing that one has the marker for a certain disease can be a disquieting experience, but to many, ignorance is even more frightening.

FIGURE 7.15

Gene analysis. A laboratory technologist studies an individual's DNA pattern and attempts to determine whether a particular gene is present by reference to a family pedigree. If the gene is present, the individual may receive appropriate counseling concerning future health expectations.

ENVIRONMENTAL MONITORING

In recent years, DNA probes and PCR have found use for detecting microorganisms from the environment—specifically, from samples of soil, sediments, and water. Very often the genes used as target DNA are those genes that encode ribosomal RNA. Other times, the genes are those encoding microbial enzymes or making up the genome of a virus. The process is particularly appealing because microorganisms need not be cultivated in the laboratory before they are identified. Instead, their DNA becomes a marker for their presence.

DNA-based analysis has found value in monitoring the distribution of genetically engineered microorganisms during **deliberate-release experiments** (Chapter 10). It would be useful to know, for example, whether gene-altered bacteria sprayed on a strawberry patch have spread beyond the bounds of the patch. A DNA probe used with PCR permits detection of the experimental bacteria and distinguishes them from other similar species that may be present. Indeed, the recombinant DNA molecule is commonly the target of the probe.

A similar technique can be used to conduct **water quality tests** on the basis of the detection of **indicator bacteria** such as *Escherichia coli*, shown in Figure 7.16. Traditionally, *E. coli* had to be cultivated in the laboratory and tested biochemically before it could be identified. With DNA technology, however, a sample of water is filtered, and the bacteria trapped on the filter can be broken open to release their DNA for PCR and probe analysis. By detecting certain specific genes, laboratory technologists can identify *E. coli*. Not only does the process save time, it is extremely sensitive: for instance, a single *E. coli* cell can be detected in a 100 ml sample of water. Moreover, the genuine pathogens transmitted by water (rather than the indicator organism) can be detected by DNA analysis. The identification of *Salmonella*, *Shigella*, and *Vibrio* species will become more feasible in the future as probe analyses become more widely accepted.

The same principle holds true for identifying other microorganisms in water. **Viruses**, for example, can be filtered from water samples and revealed by DNA probes directed at their genomes. Currently, a series of immunological techniques is required. Waterborne

Escherichia coli

a common intestinal bacterium often used as an indicator of water pollution by fecal bacteria; also used in recombinant DNA experiments.

Shigella

a genus of bacteria whose members can cause bacterial dysentery.

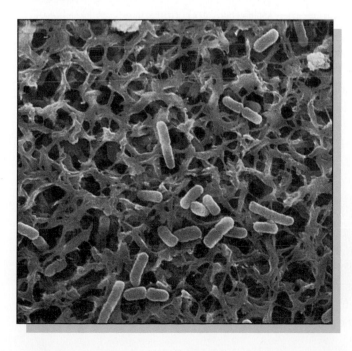

FIGURE 7.16

A scanning electron micrograph of *Escherichia coli* trapped on a bacteriological filter. Once trapped, these *E. coli* cells can be detected by rupturing the cells to release their DNA and performing a gene probe analysis. Considerable time is saved compared with performing a traditional isolation.

Giardia
je-ar'de-ah

lamblia
lam'ble-ah

Legionella
le-gion-el'ah

pneumophila
nu-mof'i-lah

protozoal pathogens such as *Giardia lamblia* could also be detected by hunting for their DNA rather than searching for them under a microscope. And water can be tested for **legionellae** by DNA probes and PCR. Legionellae such as *Legionella pneumophilia* are the causes of Legionnaires' disease, a serious form of pneumonia transmitted by aerosols from cooling towers and other water-containing devices. Already in place is the process to secure mRNA from legionellae, form complementary DNA (cDNA) by use of reverse transcriptase, and amplify the cDNA by PCR before DNA probe testing.

Ensuring the safety of potable waters is among the high priorities of public health officials. Use of DNA probes and amplification methods represents a revolutionary and exciting era in water quality testing. Whereas the previous procedures required many days of waiting, the DNA-based procedures often are complete within a few short hours. Quickly determining whether a health risk exists permits the introduction of health measures when they can benefit most people. And this is but one of the myriad values of DNA technology.

SUMMARY

The new techniques for DNA analysis and testing are based in large measure on the use of DNA probes and the technology centered in the polymerase chain reaction (PCR). A DNA probe is a relatively small, single-stranded molecule of DNA that recognizes and binds to a complementary segment of DNA on a larger, target DNA molecule. Once a match has been made, a signal such as a radioactive isotope indicates that a reaction has occurred.

To work effectively, a sufficient amount of target DNA must be available. The PCR is used to provide this DNA. In a highly automated PCR machine, the target DNA is combined with primer DNA, the enzyme polymerase, and a mixture of nucleotides. In geometric fashion, the enzyme synthesizes copies of target DNA in a three-step procedure. Heat is used to break and reform the hydrogen bonds, and a billion copies of the target DNA can be produced in a few hours. Among the problems encountered are the presence of contaminating DNA and the need for standardization. Nevertheless, PCR technology used with DNA probes has made the detection of minuscule amounts of DNA possible.

The DNA chip has added to the efficiency of DNA analysis by placing DNA probes on a glass slide the size of a postage stamp. More than 400,000 probes can be placed on the slide, and an equal number of DNA sites can be tested simultaneously. Genes can also be located by finding marker base sequences called restriction length polymorphisms, or RFLPs.

DNA detection is extremely useful in the identification of infectious disease and genetic disease. For example, the DNA encoded by the AIDS virus can be detected to give a direct diagnosis of HIV infection, an improvement over the indirect test for AIDS antibodies. Early detection of AIDS and of Lyme disease improves the opportunity for treatment. For genetic diseases, DNA probes and PCR make it possible to detect the genes for cystic fibrosis not only in the adult and child but also in the fetus and in the embryo shortly after fertilization. The genes for Duchenne's muscular dystrophy and Huntington's disease can also be identified by detecting RFLPs. Fragile X syndrome and certain kinds of diabetes, cancer, obesity, and hearing and vision loss are other conditions in which DNA diagnoses are now possible.

Outside the diagnostic laboratory, DNA probes and PCR are used to monitor genetically engineered organisms in the environment and to conduct water quality tests by detecting the DNA of *Escherichia coli*. Viral and bacterial pathogens can also be identified with gene probes to help ensure public health and safety.

REVIEW

The pages of this chapter have outlined the biochemistry and uses of DNA probes and PCR technology in the identification of genes and DNA segments. To test your knowledge of the chapter's concepts, answer the following questions.

1. What is a DNA probe and how does it work to identify a target molecule of DNA?

2. Explain the biochemistry that underlies the polymerase chain reaction (PCR) and indicate certain drawbacks to the technology.

3. Summarize the nature and value of the DNA chip and the RFLP. Explain how both can be used to detect defective genes.

4. Why is it more valuable to detect HIV directly by probe analysis than to detect HIV indirectly by testing for HIV antibodies?

5. What is the genetic basis for cystic fibrosis and how can a DNA probe be used to identify this disease at various stages in a person's life?

6. Briefly summarize the salient characteristics of retinoblastoma, Huntington's disease, Duchenne's muscular dystrophy, Alzheimer's disease, and fragile X syndrome, and indicate how DNA diagnosis is valuable to their identification.

7. Outline the genetic basis for one form of diabetes, and show how cancer can arise from a defect in the so-called tumor-suppressing gene.

8. Describe several types of hearing and vision loss that have a basis in gene defects.

9. What is a gene bank, and what do researchers use it to accomplish?

10. Describe how DNA probes and PCR technology can be used to monitor microorganisms in the external environment and help ensure public health.

FOR ADDITIONAL READING

Bottger, E. C. "Approaches for Identification of Microorganisms." *ASM News* 62(5) (1996): 247–250.

Carey, B. "Chance of a Lifetime." *Health* (May/June, 1994).

Cook-Deegan, R. "Private Parts." *The Sciences* (March/April, 1994).

Cullota, E., and Koshlund, D. E. "p53 Sweeps Through Cancer Research." *Science* 262 (1993): 1958–1962.

Fackelman, K. A. "Beyond the Genome: The Ethics of DNA Testing." *Science News* 146 (1994): 298–299.

Gibbs, W. W. "New Chip Off the Old Block." *Scientific American* (September, 1996).

Khachaturian, Z. S. "Plundered Memories." *The Sciences* (1997): 20–32.

Revkin, A. "Hunting Down Huntington's." *Discover* (December, 1993).

Rosen, J. "Genetic Surprises." *Discover* (December, 1992).

Travis, J. "Genes of Silence." *Science News*, 153 (1998): 42–46.

Gene Therapy  8

CHAPTER OUTLINE

LOOKING AHEAD

Gene therapy is among the most innovative and imaginative uses of DNA technology, as the pages of this chapter will illustrate. On completing the chapter you should be able to:

- understand how "foreign" genes can be delivered to the cells of a patient through the intervention of retroviruses.

- explain the effects of deficiency of adenosine deaminase and conceptualize how gene therapy can resolve that deficiency.

- appreciate the mechanism by which gene therapy can be used to treat certain forms of cancer.

- identify the molecular basis for cystic fibrosis and describe the uses of gene therapy to treat this disease.

- explain several approaches to treating AIDS patients by using gene therapy.

- name and explain several instances in which gene therapy will be useful in future years.

- discuss the levels of review and oversight of gene therapy experiments to ensure the patient's safety and that of the public.

⬛ INTRODUCTION

At approximately noon on September 14, 1990, history was made when a billion gene-altered cells dripped down a plastic tube and into the vein of a 4-year-old girl named Ashanti. Ashanti had an inherited immune deficiency that left her unable to withstand microbial infection. Now she would be provided with an army of cells bearing the critical gene she lacked. With that infusion of cells, researchers began the first federally approved use of gene therapy in a patient; and by administering the cells, medical practitioners took the first tentative steps into a new arena of DNA technology.

Gene therapy is one of the most radical notions ever put forward in medicine (and an ideal topic for armchair speculation). It is a process in which scientists take cells from the patient, alter their chromosomes by adding genes, then replace the cells back into the patient. Another approach, shown in Figure 8.1, is to insert the genes into a carrier such as a virus and place the viruses into human tissues. The new genes then provide the genetic codes for proteins the patient is lacking.

The idea behind gene therapy is seductively simple and direct: If disease is caused by faulty genes, then why not just replace the bad genes with good genes? Or, why not just add a few genes to cells and help them cope with the disease? Although conceptually straightforward, the process of gene therapy involves biochemical problems of gene delivery, gene control, and duration of gene action, each of which is a technical challenge. How, for example, does one "insert" genes to a person's genome?

gene therapy

the introduction of normal genes to the cell of a patient to replace defective genes failing to encode an essential protein.

genome

the complete set of genes in a cell.

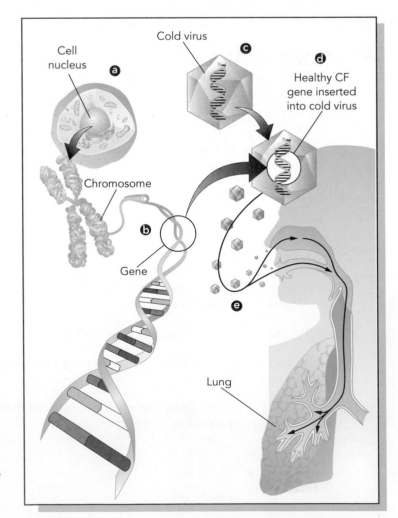

FIGURE 8.1

The general mechanism for gene therapy. (a) The approximate 100,000 genes in a cell encode the myriad proteins used by the body. (b) Each gene is a segment of the DNA molecule in a chromosome. (c) Researchers use a virus whose DNA naturally inserts into human DNA. (d) They splice a human gene into the viral DNA, and (e) place the modified virus into the human respiratory tract. The viruses invade the cells, carrying along the genes. The latter encode new proteins to provide relief from the disease.

QUESTIONLINE 8.1

1. **Q. I've heard considerable talk about gene therapy. Exactly what is gene therapy?**

   A. Gene therapy is a process for producing proteins in the body by inserting into cells the genes that encode the proteins. These proteins could be useful for correcting deficiencies that lead to genetic disease, for enhancing body resistance to disease, or for generally improving the quality of life. In one form of gene therapy, cells are removed from an individual and altered by inserting genes. The cells are then replaced, whereupon the new genes are expected to encode the desired proteins.

2. **Q. What kind of cells are used for gene therapy?**

   A. The cells used in gene therapy are nonreproductive cells, such as blood cells or skin cells. By using these cells, researchers ensure that the inserted genes are not carried over to the next generation.

3. **Q. How do scientists go about inserting genes into cells?**

   A. In one method for gene insertion, scientists attach the new genes to retroviruses and add the retroviruses to cells. Retroviruses have the ability to encode DNA using their RNA as a model and insert the DNA into human chromosomes. In doing so, they ferry the new genes into the cell.

And how does one deal with the numerous levels of gene regulation taking place in a cell?

The safety of human patients is also of paramount concern. This concern was somewhat muted by experiments performed in 1989. Six terminally ill cancer patients were infused with cells containing "marker genes" to determine whether the cells would survive in the patients and whether the patients would experience any serious side effects. Researchers found that the cells survived for a substantial time and did not adversely affect the patients.

After years of anticipation, the promises of gene therapy are being fulfilled. As we will discuss in these pages, two types of gene therapy are already in progress and several more types are planned for the future. It is possible, in fact, that the uses of gene therapy in medicine will outrace the ability of writers to report them (Questionline 8.1).

VECTORS IN GENE THERAPY

Gene therapy is performed with nonreproductive cells, generally known as **somatic cells**. These may include blood cells, skin cells, bone marrow cells, intestinal cells, or virtually any cells other than sperm or egg cells. The use of somatic cells ensures that there is no carryover of the inserted genes to the next generation. The exclusion of reproductive cells overcomes the objections of gene therapy opponents, who occasionally raise the specter of a "brave new world" in which scientists can create "superhumans" or attempt to eliminate "undesirable" genetic traits.

To deliver genes to somatic cells, researchers use some type of carrier molecule or particle called a **vector** (the Latin word *vehere* means to carry). A vector can be a plasmid such as one used in producing drugs (Chapter 5); or it may be a plasmid for carrying plant genes (Chapter 10); or it may be a cosmid (Chapter 5). For experiments in gene therapy, the vector most often used is a virus, specifically

somatic cells

nonreproductive cells in an organism.

vector

a molecule, particle, organism, or other carrier that transports molecules such as genes to a recipient.

retrovirus

a virus that penetrates a
host cell and uses its
RNA to encode a DNA
molecule that combines
with the DNA of the cell.

envelope

the membrane-like
structure of the surface
of many types of viruses.

murine leukemia

a cancer of the white
blood cells in mice.

**human artificial
chromosome**

a synthetic chromosome
that encodes human
protein in a cell and
replicates with other
chromosomes.

a type of virus called a **retrovirus**. A retrovirus contains RNA. After the retrovirus has entered its host cell, the RNA is used as a template to synthesize DNA. (This reverse flow of biological information accounts for the name "retrovirus" because "retro-" implies reversal.) The DNA then incorporates itself into the DNA of the host cell's genome, where it is for an indeterminate amount of time. In its DNA form, the retrovirus is called a **provirus**. Very often the provirus is harmless to the cell and no consequences arise from its presence. For some retroviruses, however, the consequences can be substantial; for example, a retrovirus called the **human immunodeficiency virus (HIV)** causes acquired immune deficiency syndrome (AIDS).

To use retroviruses as vectors, researchers use sophisticated biochemical methods to remove genes that might otherwise make the virus harmful. For example, a retrovirus can be crippled by artificially removing the genes that encode its envelope, thus rendering the virus incapable of combining with cells and beginning the replication process. Once a retrovirus is altered, scientists can insert human genes into the body for therapy. Because the retrovirus is an RNA virus, the human genes are inserted in their complementary RNA form, not the usual DNA form. It is also helpful to engineer the virus with certain genes so it unites with a specific target cell, rather than all body cells.

But billions of viral vectors are needed for a successful gene transplant. Therefore, the viruses must be replicated over and over again to obtain sufficient numbers for practical use, as Figure 8.2 illustrates. This multiplication is accomplished in the laboratory by infecting tissue cells simultaneously with two viruses: the **vector viruses** and a number of normal unmutated retroviruses called **helper viruses**. The helper viruses supply genes for viral replication. Repeated hundreds of times in fresh cells, the process will yield billions of vector and helper viruses. The vector viruses can then be separated from the helper viruses and purified for use in gene therapy in the body. (Successfully separating the vector and helper viruses remains a major concern because the helper viruses could conceivably pose a health threat to the patient.)

As the delivery system for human genes, the vector viruses are now ready for therapeutic use. Because they lack genes for synthesizing certain essential parts, the vector viruses cannot replicate in cells; thus, they have only a one-way passage into the cell. Once they deliver their genetic cargoes, they can never leave the cell.

In the retrovirus group, the most popular tool is a strain of **murine leukemia virus**, a virus that causes a form of leukemia in mice. The virus in this strain reacts with human cells (as well as with mouse cells) because its envelope protein docks with a cell surface protein that is very similar in mice and humans.

Should the most well-known retrovirus, HIV, be considered a vector? Perhaps so, suggest researchers. They point out that the murine leukemia virus can only infect cells that are dividing, and gene therapy is often directed at cells, such as brain cells, that do not divide or divide very infrequently. HIV, by comparison, invades nondividing cells and has the bonus of inserting therapeutic genes into the cells' chromosomes (many other vectors cannot perform this insertion).

But overcoming the public's fear of HIV is a substantial task. Nevertheless, researchers at the Salk Institutes in California are determined to try, and they have prepared a crippled HIV for use as a vector. The virus lacks all the genes for reproduction or forcing its way out of host cells but retains the genes for integrating its genome to the chromosomes. When bound to an enzyme-encoding gene and injected into an animal's brain, the vector invaded the cells and delivered the gene. The mouse leukemia virus could not deliver the gene. Despite the encouraging results, observers pointed out that a large divide remains between the laboratory and the clinic.

Perhaps a vector of a different sort, the **human artificial chromosome**, could resolve the safety issue (Chapter 5 discusses this vector). The first report of such a vector appeared in 1997 when researchers at Case Western Reserve University synthesized the key components of a human chromosome, forced them into a human cell, and found that the cell would assemble the parts. Moreover the cell cloaked the

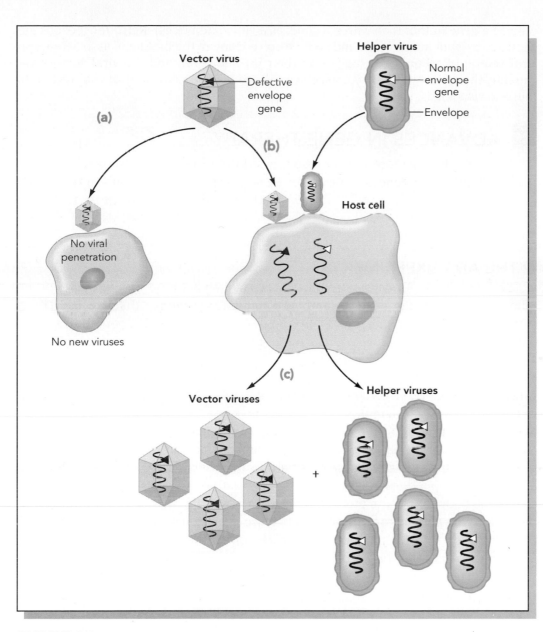

FIGURE 8.2

Production of vector viruses using helper viruses. **(a)** Vector viruses contain a gene defect such as a defective envelope gene. Without the envelope, the virus cannot penetrate host cells, and viral replication cannot occur. **(b)** Helper viruses are of a different type, but with a normal gene for envelope formation. With this gene, the helper viruses can penetrate cells and assist the penetration of the vector viruses. Now the genomes are released in the host cells, and **(c)** a large quantity of vector viruses and helper viruses are produced. Before the vector viruses can be used for gene therapy, the helper viruses must be removed.

synthetic chromosome with histone, a protein normally associated with a chromosome. Through the next 240 divisions, the chromosome duplicated itself and passed to daughter cells as if it belonged in the cell.

Other possible vectors include the liposome, gold bullet, and retrotransposon. A **liposome** is a microscopic sphere containing fat molecules. Coated with genes, the liposome is absorbed through the fat-based cell membrane into the cell cytoplasm, where the genes are released. The **gold bullet** is a microscopic gold sphere coated with genes and fired into a cell with a helium-pressurized gun. And, the **retrotransposon** is a small DNA segment from a cell that copies and inserts itself

liposome

a microscopic sphere of fat.

outside a gene (rather than within a gene, as a virus often does). Retrotransposons are known to exist in human cells and could resolve many of the problems associated with viral vectors. Research has lagged on their use, however, and the viral vectors are currently the delivery system of choice. We will see many applications of viral vectors in the following pages.

ADVANCES IN GENE THERAPY

Molecular biology took a giant leap forward in the 1990s with the application of DNA technology to gene therapy. After decades of speculation and anticipation, scientists realized that they might finally be able to insert functioning genes to replace defective ones in living cells and thereby treat genetic disease. Two experiments led the way, as we will see in the sections to follow.

THE ADA EXPERIMENTS

Each year in the United States about forty children are born with a disease of the immune system called **severe combined immunodeficiency disease (SCID)**. In almost half of the patients with SCID, the cells have a defective gene that cannot encode a particular enzyme. Without the normal gene, the enzyme is not produced (Figure 8.3 pictures one such patient).

adenosine deaminase

a-den´o-sin de-am´in-ase

an enzyme that breaks down a product of a cell's nucleic acid metabolism; lack of adenosine deaminase leads to destruction of T-lymphocytes.

The enzyme in question is called adenosine deaminase. **Adenosine deaminase (ADA)** helps break down certain by-products of a cell's nucleic acid metabolism. If ADA is not present in a T-lymphocyte, an enzyme called kinase converts one of the metabolic by-products to a toxin, and the toxin destroys the T-lymphocyte. T-lymphocytes are essential cells of the body's immune system. They not only participate in immune responses directly but they also influence indirectly, by controlling the activity of B-lymphocytes, the cells that produce antibodies. Thus, the ADA deficiency causes the body to lose the protection of both kinds of lymphocytes, and without that protection, the patient cannot mount a defense against infectious disease and soon dies. The gene for ADA production is located on chromosome number 20 and has 32,000 base pairs and twelve exons.

lymphocytes

white blood cells that form the foundations of the body's immune system.

In the summer of 1990, a research team from the National Institutes of Health (NIH) received approval to attempt gene therapy for the purpose of relieving ADA deficiency. The team, led by **R. Michael Blaese** and **W. French Anderson** (Figure 8.4), removed lymphocytes from the patient and exposed the cells to billions of retroviruses carrying the genes for ADA production. (This is the *ex vivo* approach to gene therapy compared with the *in vivo* approach where cells remain in the body.) After the genes were inserted into the chromosomes, the lymphocytes were returned to the

FIGURE 8.3

A photograph of a young boy known only as David. David had severe combined immunodeficiency (SCID) and was forced to live in a sterile plastic bubble because he could not mount an immune response to microorganisms. In 1983, David received a bone marrow transplant from his sister and left the bubble. Unfortunately he developed a type of blood cancer before an immune system could be established in his body, and he died shortly thereafter.

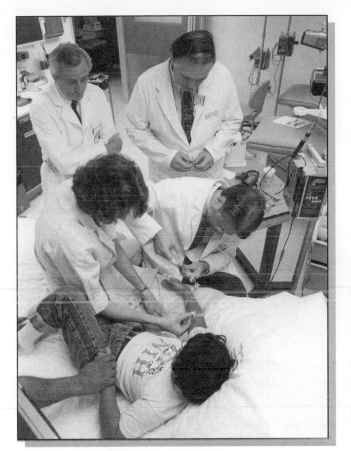

FIGURE 8.4

Gene therapists W. French Anderson (top left) and Michael Blaese (top right) supervising physicians who are treating a patient with genetically altered cells. Anderson and Blaese performed their pioneering research in gene therapy at the National Institutes of Health.

patient. After several weeks, the procedure was repeated because T-lymphocytes have a finite life expectancy of a few months in the body.

On September 14, 1990, the first gene-altered cells entered the bloodstream of a 4-year-old girl named Ashanti, the one mentioned in the opening paragraph of this chapter. She is shown in Figure 8.5. In the spring of 1991, Blaese reported that the modified lymphocytes were persisting in the girl's circulation. Moreover, the amount of ADA produced by the gene-altered lymphocytes appeared to be increasing with time, and the girl's ability to produce antibodies had increased substantially. There were even signs of tonsils, the storage sites for T-lymphocytes in the rear of the mouth. By that time, treatment of a second child, a 9-year-old girl named Cynthia, was also in progress. By September 1991 both girls were back in school, one to kindergarten, the other to fifth grade. One newspaper trumpeted in its headline "Back to School With Their New Genes."

As with any new treatment, some opposition was voiced to the ADA gene therapy. Chief among these are safety concerns. It should be noted, however, that before performing the experiments, the research team received approval from the NIH Recombinant DNA Advisory Committee (RAC); the Human Gene Therapy Subcommittee of RAC; the Director of the NIH; and the federal Food and Drug Administration (FDA).

Opposition was also voiced from those who point to alternative, less risky therapies. For instance, bone marrow transplants have been successfully used to infuse patients with normal cells carrying the ADA genes, but the donor and recipient bone marrows must be closely matched (usually a sibling is required.) Another possible problem is the rejection of the transplanted bone marrow. Proponents of gene therapy argue that rejection mechanisms could not possibly occur with gene therapy because the person's own cells are being infused.

RAC

Recombinant DNA Advisory Committee, a supervisory group within the U. S. National Institutes of Health that lends approval or disapproval to proposed experiments involving gene therapy.

FIGURE 8.5

The first two recipients of gene therapy. The girl on the left is named Cynthia; the one on the right is Ashanti. Both girls have lived a relatively normal life since receiving gene-altered cells to relieve ADA deficiency late in 1990 and early in 1991. This photograph appeared in *Time* magazine in mid-1993.

polyethylene

pol-e-eth´e-len

Questions also surfaced about whether the gene therapy was more useful than **enzyme replacement**. In early 1990, the FDA approved the use of purified ADA chemically bound to polyethylene glycol molecules. The molecule, called PEG-ADA, can be injected to patients with SCID to satisfy the ADA deficiency, but current estimates of costs are about $50,000 per year, a figure high enough to argue against drug use.

Another concern is that T-lymphocytes have a limited life expectancy and must be replaced periodically. A better approach might be to place the ADA gene into the blood's stem cells, the precursor cells that give rise to T-lymphocytes. Stem cells can be obtained from the umbilical cord blood at the time of a baby's delivery. In 1995, Blaese identified three fetuses with the ADA deficiency, obtained the cord stem cells when the babies were delivered, and successfully added ADA genes to those cells. Four days after birth, the infants received their cells back—and a permanent population of ADA-producing cells.

■ CANCER THERAPY WITH GENES

The second experiment with gene therapy was an attempt to insert into T-lymphocytes the genes for an anticancer agent, then return the cells to the patient (Questionline 8.2). The group conducting the experiment was led by **Steven Rosenberg** of the NIH in collaboration with **W. French Anderson** (also involved in the ADA experiment). In the procedure scientists used **tumor-infiltrating lymphocytes (TILs)**, special cells of the immune system. Under normal conditions, TILs are stimulated by cancer cells, and they enter tumors and destroy their cells. An important accessory weapon in this attack is an anticancer agent called **tumor necrosis factor (TNF)**. TNF is a protein product of macrophages, the amoeba-like cells that provide defense against cancer cells. The TNF protein can be produced by genetic engineering techniques (Chapter 5), and it can be used to enhance the cancer-fighting ability of tumor-infiltrating lymphocytes.

Rosenberg (Figure 8.6) and his colleagues treated patients having advanced **malignant melanoma**, a virulent form of skin cancer that normally fails to respond to

tumor-infiltrating lymphocytes (TILs)

certain cells of the immune system that attack and destroy tumor cells.

melanoma

mel-ah-no´mah

a tumor composed of cells that produce the pigment melanin; usually occurs in the skin layers.

QUESTIONLINE 8.2

1. **Q.** Newspaper reports talk about using gene therapy for patients who have ADA deficiency. What exactly is ADA?

   A. ADA stands for the enzyme adenine deaminase. When ADA is missing in the body, a series of metabolic events leads to the destruction of T-lymphocytes and a serious impairment of the immune system. Without the defenses provided by the immune system, most patients die from infectious diseases.

2. **Q.** How does gene therapy resolve the ADA deficiency?

   A. For gene therapy, researchers remove lymphocytes from the patient with ADA deficiency and expose the cells to retroviruses carrying the genes for ADA production. The viruses insert the ADA genes into the lymphocyte chromosomes, and the lymphocytes are returned to the patient's circulation. In tests conducted thus far, the cells remained active in the test patients and began producing ADA.

3. **Q.** What other diseases could be treated with gene therapy?

   A. Certain forms of cancer could respond to gene therapy because lymphocytes could be altered with genes that encode an anticancer agent such as tumor necrosis factor. There is also evidence that gene therapy could be successful in treating such diseases as AIDS, cystic fibrosis, familial hypercholesterolemia, and sickle cell anemia.

FIGURE 8.6

Steven Rosenberg, the NIH researcher who has performed seminal experiments to treat various types of cancer with gene-altered cells.

treatment. They removed a piece of tumor tissue from the patient (the *ex vivo* approach) and isolated a number of TILs from the patient's blood. They cultivated the TILs with cells from the tumor tissue to stimulate the TILs and enhance their selectability for the patient's melanoma cells. They included in the mixture a substance called **interleukin-2**, which is a naturally occurring lymphocyte protein, to encourage rapid multiplication of TILs. Then they mixed the stimulated TILs together with retroviral vectors carrying the genetic code for TNF. The viruses invaded the cytoplasm of the TILs and carried the TNF gene into the chromosomes. The genetically altered TILs

interleukin-2

in″ter-lu′kin

a protein produced by cells of the immune system that encourages the multiplication and activity of lymphocytes.

tumor necrosis factor (TNF)

a protein produced by human macrophages that encourages destruction of tumor cells.

ganciclovir

gan-ci'clo-vir

a drug that interferes with DNA replication by replacing thymine.

glioblastoma

gle''o-blast-o'mah

a cancer of the brain tissue primarily affecting the glial cells

p53

a tumor-suppressing gene that encodes a protein, which encourages repair of a damaged DNA molecule.

were cultivated for a time to produce a colony of **TNF-producing tumor-infiltrating lymphocytes**.

In the next step of the experiment, Blaese and his colleagues infused the TNF-producing TILs into the patient's circulation. With their enhanced ability to destroy cancer cells and their selectivity for melanoma cells, the TILs attacked the cancer cells and delivered their anticancer TNF. Healthy parts of the body were avoided because TILs were selectively stimulated to seek out melanoma cells.

As of 1993, Rosenberg's team successfully infused transformed cells into nine patients with malignant melanoma. There was question, however, about whether sufficient numbers of TILs were localizing at the tumor site and whether they were producing therapeutic quantities of TNF. To answer these and other criticisms, Rosenberg's group continued experimenting with new types of vectors to increase the ability of TILs to localize.

A less-publicized form of anticancer gene therapy was approved in 1993 for clinical trial. This form was carried out at the Iowa Methodist Medical Center in Iowa City. Researchers led by Michael Blaese and Kenneth Culver introduced into **brain tumor cells** the gene that encodes the enzyme **thymidine kinase**. Thymidine kinase is used by certain microorganisms in the synthesis of DNA during cell division. Its activity can be interrupted by the drug ganciclovir. It is hoped that the tumor cells will take up the genes, become dependent on thymidine kinase activity during cell division, and develop sensitivity to ganciclovir and die (the gene for thymidine kinase is often called a "suicide gene"). Ganciclovir could then be used to kill the tumor cells. Figure 8.7 depicts how the process works.

The vector used in these experiments is a retrovirus altered with genes for thymidine kinase production from herpes simplex viruses. Normal brain cells do not divide, but brain tumor cells divide excessively, and the retrovirus selectively inserts the thymidine kinase gene to brain tumor cells. (Normal brain cells do not produce much new DNA because they are not dividing.) To yield a steady supply of viruses, the researchers inject virus-infected skin cells from a mouse. Such infected cells give off viruses at a regular rate, thus ensuring that viruses are available whenever the tumor cells are multiplying. The mouse cells die after several days because of the body's normal rejection mechanism, but the tumor cells are infected and begin producing thymidine kinase. An injection of ganciclovir should then destroy the cells and the brain tumor.

As of 1998, fifty hospitals in the United States and abroad were using thymidine kinase therapy to treat **glioblastomas** and other brain tumors. Glioblastoma is suited to the therapy because the tumors grow very rapidly, sometimes doubling in size every two days. The virus-infected mouse cells are placed directly into the patients brain tissue during surgery (the *in vivo* approach to gene therapy). Two weeks later, ganciclovir therapy is begun. Gene therapy is useful for glioblastomas because the cells do not spread throughout the body as other tumor cells do, so treatment can be localized. Another advantage is that the body's immune defenses in the brain are weak, so the rejection to the mouse cells is not quite as dramatic as elsewhere.

Another approach to cancer therapy is to replace a damaged *p53* **gene**, the tumor-suppressing gene discussed in Chapter 7. The normal gene helps repair DNA damage in the cell, but when it is in the mutated form, the gene cannot perform its function and a tumor develops. Researchers are now investigating the possibility of using adenoviruses (to be discussed presently) to deliver correct copies of the *p53* gene to patients with liver cancer. At this writing, a limited number of tests have been performed, with encouraging results.

■ THERAPY FOR CYSTIC FIBROSIS

As the first experiments with gene therapy were taking place, scientists were formulating plans for future tests of this innovative approach to medicine.

Among the proposals was one to insert the correct gene to prevent **cystic fibrosis (CF)**. A major problem in patients with CF is the buildup of chloride ions within cells lining the body's organs and vessels. The ions normally pass out of cells through a channel-shaped, tunnel-like protein called **cystic transmembrane conductance regulator (CTCR) protein**. In persons with CF, the CTCR protein is not produced because of a gene defect, and the ions concentrate in the cells. There they draw water into the cells and leave a dehydrated, sticky mucus in the body's passageways, especially its airways, as shown in Figure 8.8. Conditions such as these are conducive to infection. Most CF patients die from respiratory tract and lung infections before reaching the age of thirty. (Chapter 7 contains additional insights on this disease.)

The discovery of the defective gene in CF patients was a notable achievement of 1989. Within a year, two research teams were proposing therapy to insert the correct gene. At the University of Iowa, scientists successfully used a vector virus to insert a normal version of the gene into laboratory-cultivated lung cells from a patient. The cells then produced the CTCR protein to transport ions and encourage release of excess chloride. The second team, working at the University of Michigan, had achieved

cystic fibrosis

a genetic disease in which sticky, dehydrated mucus accumulates in the airways and other ducts and makes breathing difficult.

CTCR protein

a protein that conducts chloride ions out of cells; locking in patients with cystic fibrosis.

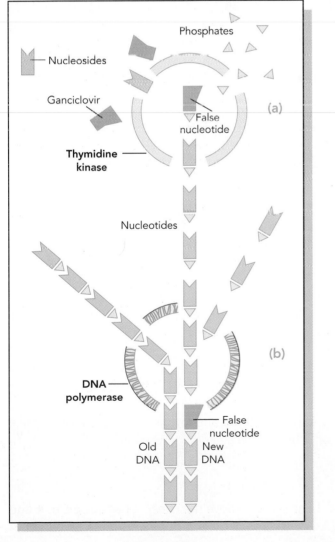

FIGURE 8.7

How ganciclovir interrupts cell division. (a) The enzyme thymidine kinase (top circle) acts by combining phosphate groups with nucleosides to form nucleotides. The drug ganciclovir bears a close structural resemblance to certain nucleosides, and the thymidine kinase mistakenly takes up the ganciclovir to form a false nucleotide. (b) During cell division, another enzyme called DNA polymerase (bottom circle) attaches the false nucleotide onto a developing DNA molecule during DNA synthesis. However, the false nucleotide lacks an attachment point for the next nucleotide, and the elongation of the DNA molecule comes to an abrupt halt. Because DNA synthesis ends, the cell cannot undergo division to two new cells. In gene therapy, inserting a gene for thymidine kinase to a cell would make the cell sensitive to ganciclovir activity. The ganciclovir could then inhibit cell division and kill the cell. For this reason, the thymidine kinase gene is called a "suicide gene."

FIGURE 8.8

Hairlike cilia extending from cells of the respiratory tract. Normally these cilia trap dust and microorganisms, but in cystic fibrosis patients, excess mucus makes them sticky and unable to function. In many instances, infection follows.

adenovirus

a common cold virus composed of a DNA genome and an icosahedral capsid; used as a vector in gene therapy.

similar success, but with pancreatic cells.

The optimism for treating CF with gene therapy was raised considerably by a 1991 report describing how a disabled **common cold viru**s transports normal genes into laboratory-reared target cells. Researchers at the National Heart, Lung, and Blood Institute demonstrated that genes coding for human enzymes could be spliced to a respiratory virus known as an **adenovirus**, shown in Figure 8.9. When rendered incapable of replication, the adenovirus is a suitable vector virus because it penetrates cells of the human respiratory tract and dispatches its genes to the nucleus. However, the adenovirus does not insert its genes into the chromosomes, thus avoiding the disturbance caused by retroviruses.

The year 1993 witnessed additional steps forward: a British team of molecular biologists reported the correction of the CF gene defect in mice. Also, the Recombinant DNA Advisory Committee (RAC) approved three gene therapy trials involving the CF gene bound to an adenoviral vector. The gene-vector combinations were administered by each of three research teams to ten selected patients (per team). Two teams sprayed gene-altered adenoviruses into the patients' nasal passages and lungs. The third team sprayed only into the patients' noses. Because adenoviruses do not integrate themselves into cells as retroviruses do, any positive results from the experiments would be transient. By 1996, successful insertions to both nasal passage and lung cells were reported but dramatic cures in patients were still elusive.

Another approach was reported in 1997 when molecular geneticists found a novel way to add a normal version of the CF gene to defective cells of developing **mouse fetuses**. About a week before birth, the investigators pierced the mother's amniotic sac and injected adenoviral vectors into the amniotic fluid. The mouse fetus breathed in the fluid, thereby exposing its lungs to the altered adenovirus. After birth, the mouse produced the CTCR protein, but only briefly, because the viruses integrated poorly to their lung cells.

◼ THERAPY FOR AIDS

Gene therapy may also be used one day to relieve the effects of **acquired immune deficiency syndrome (AIDS)**. Clinical tests have been approved for a novel gene-

FIGURE 8.9

An electron micrograph of adenoviruses (X 300,000), a type of respiratory virus used as a vector in gene therapy.

transfer therapy for persons infected with HIV. The test involves a mutant strain of HIV containing defective forms of the genes *rev* and *env*. This strain was produced by researchers at the University of California at San Diego. The new HIV strain cannot replicate because *rev* and *env* genes encode regulatory viral protein and envelope protein, both of which are essential for viral replication. To use the viruses as therapeutic devices, researchers remove T-lymphocytes from HIV-infected patients and insert the mutated viruses into the cells. Then they cultivate large numbers of the cells and inject them back into the patient. The **virus-containing T-lymphocytes** cannot produce viruses, but they do stimulate the body to produce cells called CD8 killer lymphocytes. These killer cells are specifically manufactured to interact with HIV-infected cells, and in laboratory tests they destroy those cells.

Another form of gene therapy involves attaching genes for HIV proteins to the **DNA of mouse viruses**, which are harmless for humans. The transformed viruses are then injected into individuals, who are infected with HIV but display no symptoms of disease. Researchers believe that the HIV genes will stimulate the normal body cells to produce HIV proteins. The proteins should stimulate the immune system to secrete anti-HIV antibodies. These antibodies may be useful in preventing HIV replication in the patient and forestalling the T-lymphocyte destruction that characterizes AIDS.

Still another possible form of anti-AIDS therapy involves a process that works in the laboratory but has been used in limited clinical trials to date. The process involves getting **HIV-infected cells to produce anti-HIV antibodies**. DNA technologists at the Boston's Dana Farber Cancer Institute have synthesized the gene for an antibody called F105. This antibody inactivates gp120, a glycoprotein found in the envelope of HIV and essential for HIV's binding to host cells. The scientists refined the gene so that the antibody molecule is produced at the cell's endoplasmic reticulum (ER), the membranous component where gp120 is also synthesized. And they added a gene segment so the antibody would anchor to the ER.

So constructed, the DNA technologists attached the gene to a plasmid and inserted it to HIV-infected cells from a laboratory animal, where it stimulated antibody production. Here was the unique situation where an infected cell produced antibodies against the very thing infecting it. The synthesis of gp120 was dramatically reduced,

rev and env

genes of HIV that encode regulatory and envelope proteins, respectively.

gp120

a glycoprotein located in the envelope of the human immunodeficiency virus (HIV); an essential substance for the binding of HIV to its host cell in the replication process.

and the synthesis of HIV particles slowed considerably. Moreover, the cells did not display any toxic effect.

The production of antibodies within the cell focused attention on a type of "intracellular immunization," first proposed in the 1980s. In contemporary jargon, the antibodies so produced have been called "**intrabodies**." Animal tests with infected monkeys have shown the method's efficacy, and researchers are targeting other components of HIV such as integrase, the enzyme used to integrate HIV into the host chromosome; gp160, another envelope glycoprotein of HIV; ribozymes, the RNA molecules capable of cleaving the RNA genome of HIV as it enters the cell; and *tat*, the gene whose protein product mobilizes the cell's enzymes for the replication of new viral particles.

■ OTHER ENDEAVORS

Another proposal for gene therapy is aimed at a rare inherited disorder called **familial hypercholesterolemia**. Patients with this disease are unable to remove cholesterol from their bloodstream. The source of the problem is a defective gene that normally encodes cholesterol receptors. **Cholesterol receptors** are protein molecules at the surface of liver cells that trap cholesterol in passing blood and remove it from the circulation. Without the receptors, the cholesterol is left in the blood to accumulate and clog the arteries and veins. High blood pressure and possible heart attack can result, even in the very young.

In 1993 scientists at the University of Michigan reported a successful attempt to insert a correct gene to replace the defective gene that leads to familial hypercholesterolemia. They secured a large sample of a woman's liver cells (about 3.2 billion cells, about 15% of the liver), then infected the cells with a retrovirus carrying the gene for normal protein receptors, and infused the gene-altered cells back into the patient. The woman's cholesterol level dropped significantly enough for the RAC to recommend extension of the experiments to additional patients. By 1998, the treatment was being used in clinical situations.

Nor does it end there. Researchers at the University of North Carolina have successfully inserted a gene to replace the gene involved in **sickle cell anemia** with a normal gene in experiments conducted with laboratory cells. The inserted gene encodes normal globin, the portion of the protein hemoglobin that is defective in patients with sickle cell anemia. Progress in treating sickle cell anemia is expected to be slow, however, because synthesizing the crucial protein demands precise regulation of gene activity, and controlling this regulation is currently beyond the abilities of biochemists.

Another candidate for gene therapy is **Lesch-Nyhan disease**. Annually, this rare disorder affects an estimated 2000 Americans in varying degrees. The disease is caused by a gene defect in which body cells lack the instructions for producing the enzyme **hypoxanthine-guanine phosphoribosyltransferase (HPRT)**. Without HPRT, the metabolism of guanine and hypoxanthine is interrupted and a buildup of uric acid occurs. The result is severe gout and kidney damage. Affected children exhibit cerebral palsy and mental retardation, as well as bizarre behaviors such as uncontrolled urges to spit, use profane language, gnaw at the lips and fingers, and bang the head against the wall.

Since 1984, scientists at the University of California at San Diego have been using retrovirus vectors to insert genes for HPRT production into isolated human cells. In their experiments the genes successfully raised the cellular levels of the enzyme. Experiments in the bone marrow cells of mice were equally successful. In humans, a major problem for consideration is the involvement of the brain—scientists are uncertain whether the enzyme can relieve the compulsive behavior and cerebral palsy, and they are not sure what the critical level of enzyme is in the body.

---

intrabodies

antibodies produced within a cell and acting within that cell.

hypercholesterolemia

hi″per-ko-les″ter-ol-e′me-ah

a genetic disease accompanied by a higher-than-normal level of cholesterol in the blood owing to the absence of cholesterol receptors.

sickle cell anemia

a genetic disease in which the hemoglobin is formed incorrectly and the red blood cells collapse and assume a sickle shape.

Lesch-Nyhan disease

lesh-ni′han

a genetic disease in which the absence of a certain enzyme interferes with guanine and hypoxanthine metabolism, resulting in a buildup of uric acid.

phosphoribosyltrans-ferase

fos-fo-ri″bo-sil-trans′fer-ase

Each year, more than 2 million Americans suffer the effects of blocked or clogged arteries in the legs, a condition called **peripheral artery disease**. Balloon angioplasty and bypass surgery can be used to treat the disease, but a significant number of patients do not respond. They face significant pain, skin ulcers, and possible limb loss.

During the 1970s, Harvard University researchers found that tumors coax nearby blood vessels to grow toward them by releasing a growth-enhancing protein called **vascular endothelial growth factor (VEGF)**. In the late 1980s, the gene encoding this growth factor was found. Investigators from Tufts University have now spliced the gene into plasmids and have injected the plasmids into patients with peripheral artery disease. In eight patients, the plasmids were placed onto the surface of a balloon coated with a gel, and the uninflated balloon was threaded into the femoral artery, where it was inflated to contact the smooth muscle of the arterial wall. The plasmids were taken into the cell and VEGF soon appeared. New vessel growth occurred within two weeks.

Gene therapy holds promise not only for helping patients with peripheral artery disease but also for those with coronary artery disease and angina. By 1998, fifteen heart patients were being treated with VEGF spliced to an adenoviral vector and injected into the heart tissue during bypass surgery. The technique, shown in Figure 8.10, has been named **therapeutic angiogenesis**. One advantage of the technique is that the genes do not have to keep working—once they have stimulated blood vessel growth, the vessels remain even if the gene is not activated.

Other candidates for gene therapy include Gaucher's disease and hemophilia B. **Gaucher's disease** is an inherited disease of the central nervous system accompanied by enlarged spleen and liver, erosion of the long bones, and yellowing of the skin in adults. In 1993, the RAC approved a clinical test to introduce into patients' cells the gene that encodes **glucocerebrosidase**. This enzyme is apparently missing or defective in patients. Enzyme production and activity may relieve the disease symptoms. Tests are ongoing at this writing.

balloon angioplasty

an″gi-o-plast'e

a procedure in which an inflatable device is inserted into an artery to enlarge it.

Gaucher's

gow-shaze'

glucocerebrosidase

glu″co-cer-re″bro-si′dase

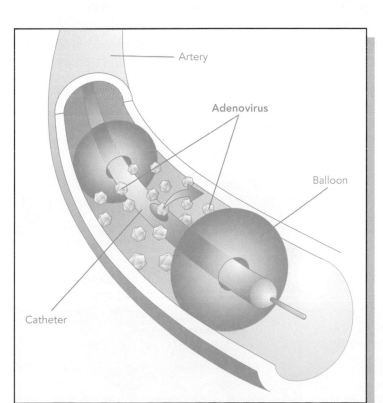

FIGURE 8.10

Gene therapy for clogged arteries. A catheter is inserted into the arteries, then two balloons are inflated to hold the artery open. Adenoviruses carrying genes emerge through an opening in the catheter and bring the genes that encode vascular endothelial growth factor to encourage arterial growth.

hemophilia B

a form of hemophilia in which the patient fails to produce an adequate blood clot because of the absence of an essential clotting factor.

hepatic portal vein

a major vein leading from the small intestine to the liver.

Patients with **hemophilia B** cannot produce a clotting protein called **factor IX**, and excessive bleeding takes place, as it does in hemophilia A (Chapter 6). In 1993, researchers from Baylor University reported the successful insertion of the genes for factor IX into the liver cells of dogs shown in Figure 8.11. A retrovirus was used to shuttle the genes into the cells; researchers found they could make gene insertions directly into live animals. To accomplish this, they surgically removed a portion of the dog's liver, thereby stimulating the remaining cells to divide rapidly (as if recovering from a disease or trauma). DNA technologists then injected the modified retroviruses into the dog's hepatic portal vein leading directly to the liver. The multiplying liver cells took up the viruses and the gene insertion was complete.

There is even hope for treating **baldness** with gene therapy, although the research is still rudimentary. To date, at least one mutated gene has been related to hereditary baldness (Chapter 7), and this finding has stirred thoughts of gene replacement. A necessary first step is determining whether genes can be introduced to cells of the hair follicle. In 1995, scientists showed this was possible when they reported the uptake and incorporation of enzyme-encoding genes in cells that give rise to the hair shaft. The genes were packaged in liposomes spread on the shaved backs of mice.

According to Victor McKusick's noted work *Mendelian Inheritance in Man*, there are thousands of diseases related to a single-gene defect (as Chapter 7 notes) and for which gene therapy could help. Genetic diseases are responsible for a significant number of miscarriages, as well as a large proportion of infant deaths and cases of mental retardation. In addition, they impose a severe economic burden on society, while extracting an enormous toll on human life.

SAFETY ASSURANCE

To calm excessive fears about gene therapy and to ensure proper societal safeguards, a multilayered review system is now in operation for scientific proposals. This system also provides an opportunity for public input into scientific endeavors. To proceed with their experiments, scientists will have to show that the inserted genes do not operate in tissues where they can cause harm (for example, it must be shown that genes of viral vectors will not activate human genes that should remain dormant). They

FIGURE 8.11

A successful gene therapy experiment. This dog, previously suffering from hemophilia, has received the gene for producing factor IX and no longer displays the symptoms of hemophilia.

must also demonstrate the safety of viral vectors in humans because a vector could conceivably insert at a chromosomal spot where it could cause a mutation by interrupting the genetic sequence. And they must show that helper viruses have been completely removed from the viral mixture because helper viruses are unaccustomed to human tissues and might cause disease if allowed to infect the tissues.

There is also the problem of inserting genes into the correct tissue cells. At this writing, insertions into bone marrow cells have the highest priority among scientists because they have been removing and transplanting bone marrow tissue for many years. The next major hurdle will be inserting genes into cells not easily accessed, such as brain cells. Indeed, a multitude of single-gene defects are linked to the cells of the brain.

Since 1984, a **multitiered review system** for gene therapy experiments has been evolving (Questionline 8.3). Federal overseers of DNA technology now require that any kind of research involving human subjects must be approved by the investigator's institutional review board. Human gene therapy proposals also need approval from the Food and Drug Administration (FDA) and, since 1997, comment (rather than approval) from the Recombinant DNA Advisory Committee (RAC) of the NIH. The RAC has established a special working group on human gene therapy proposals. Known as the **Human Gene Therapy Subcommittee**, the group consists of four laboratory scientists, three clinical scientists, three specialists in ethics, and two experts in public policy.

The Human Gene Therapy Subcommittee has prepared a document entitled **"Points to Consider"** for researchers planning to submit proposals for human gene therapy research. Researchers must supply detailed information on the disease they wish to treat, alternative therapies and their costs relative to gene therapy, characteristics of the genetic material for insertion and evidence for its purity, results of laboratory tests in cultivated cells and animals, possible side effects in patients and how they will be minimized, selection of subjects, and how patients' privacy will be protected. Certain ethical and social issues such as handling publicity and stemming public fears must also be addressed.

helper viruses

viruses that assist the replication of vector viruses but are then removed from the preparation.

QUESTIONLINE 8.3

1. **Q.** What type of safety concerns are there in gene therapy?

   A. To gain public confidence in gene therapy procedures, scientists will have to show that inserted genes are not toxic to the body; that the viral vectors are safe; and that any contaminants in the gene preparation have been removed.

2. **Q.** Who is in charge of reviewing gene therapy proposals?

   A. At present, there is a multilayered review system for gene therapy proposals. The institution must approve, and the federal National Institutes of Health must give its approval through advisory boards. The Food and Drug Administration is also involved because drugs are used in the procedures.

3. **Q.** What are the prospects for gene therapy in the future?

   A. The future prospects for gene therapy remain optimistic. Early successes and improved experimental protocols have spurred numerous researchers to develop novel proposals for the use of gene therapy. The experiences gained are also invaluable in treating patients, and gene therapy may become a standard treatment for genetic disease in the decades ahead.

In the minds of some scientists the requirement for multiple tiers of review and approval has slowed progress. Still, to retain public confidence, it has been necessary to undertake a painstaking effort to address many unsettled scientific problems and to anticipate some of the ethical questions that gene therapy could raise. About three and a half years were consumed obtaining the required approvals for the ADA experiment. It is clear that the operative recommendation for gene therapy has been "Proceed…but with caution."

■ THE FUTURE

Gene therapy may be the long-term answer to the problem of genetic disease. For successful therapy of this sort, the gene causing the disease must be isolated and reproduced; enough copies of the gene must be inserted into tissues that make the gene product; the genes must express themselves appropriately (that is, not too much or too little of the gene product); and the gene must be harmless to the patient. Researchers must also deal with the body's immune system as it attempts to reject foreign proteins encoded by the cell's new genes.

And there are the exaggerated expectations of the public, fueled in part by overzealous researchers seeking to acquire some of the hundreds of millions of dollars allocated by federal agencies for gene therapy studies. By 1999 about 1000 Americans were enrolled in clinical trials involving gene therapy, yet physicians lacked unambiguous proof that gene therapy had cured any of their patients. Indeed, various publications were questioning whether the research was worth the investment. A headline in *Time* magazine trumpeted: "Has Gene Therapy Stalled?" *Newsweek* magazine wistfully led its article with the headline "Promises, Promises." *The New York Times* said: "In the Rush Toward Gene Therapy, Some See a High Risk of Failure." And the respected magazine *Science* bemoaned in its headline "Gene Therapy's Growing Pains." The public seemed to be saying that more fundamental science was needed before the rush to application should begin.

Looking to the future, it will also be important that injectable vectors are widely available so that multidisciplinary teams (biochemists, physicians, surgeons, and specialized hospital personnel) are not needed to administer gene therapy. The concept of gene regulation will be a continuing issue because, although regulation is not critical in diseases now being treated, it will become a key factor in dealing with such diseases as diabetes. And the debate about gene insertions into somatic cells and reproductive cells will be revisited as protocols for insertions into reproductive cells are developed. Resolving these issues is a formidable task, which is why exaggerated hopes for gene therapy should always be tempered with realism.

Nevertheless, W. French Anderson, the NIH researcher at the forefront of gene therapy was quoted in 1995 as saying: "Twenty years from now, gene therapy will have revolutionized the practice of medicine. Virtually every disease will have [gene therapy] as one of its treatments." Stay tuned.

▧ SUMMARY

Gene therapy is a remarkable process in which cells are taken from the patient, altered by adding genes, then replaced back in the patient where the genes provide the genetic codes for proteins the patient is lacking. Because nonreproductive cells are used, there is no carryover of the inserted genes to the next generation. Certain of the vectors used for gene therapy are retroviruses, a group of RNA-containing viruses that insert their genes into chromosomes. Crippled so they pose no harm to human cells, the retroviruses are first combined with the human genes, then used to carry the genes into cells.

One place where gene therapy has been successful is in patients having deficiency of the enzyme adenosine deaminase (ADA). To resolve this deficiency, patients' lymphocytes are removed from the bone marrow, cultivated with vectors carrying genes for ADA production, then reinfused into the body. Once established, the gene-altered cells begin synthesizing the enzyme. Opponents to the gene therapy point to the availability of drugs for the deficiency and the possibility of bone marrow transplants. Proponents argue the high cost of drugs and the necessity of locating closely matched donors of marrow.

Gene therapy has also been performed for patients having melanoma, a virulent skin cancer. Tumor-infiltrating lymphocytes (lymphocytes that attack tumors) are isolated from the patient and treated with genes for an anticancer protein called tumor necrosis factor. The lymphocytes are then reinfused into the patient where they are expected to interact with cancer cells and produce their inhibitory protein. Other candidate diseases for gene therapy are cystic fibrosis, familial hypercholesterolemia, AIDS, Gaucher's disease, and Lesch-Nyhan disease.

About 2000 single-gene defects are believed to exist, and patients with these defects may benefit from gene therapy. Before therapy is instituted, however, multiple problems must be solved. To calm exaggerated fears, a multilayered review system exists to ensure the safety and efficacy of gene therapy proposals. Both the National Institutes of Health and the Food and Drug Administration are involved in this system, and multiple points of consideration must be addressed before approval is extended.

REVIEW

This chapter has reviewed the current and future prospects for gene therapy as a treatment for disease. To test your comprehension of the chapter's contents, answer the following questions.

1. What types of vectors are used for gene therapy experiments and how are they used?

2. Describe the first successful experiment involving gene therapy in a child with ADA deficiency.

3. Explain the process by which gene-altered tumor-infiltrating lymphocytes can be used to destroy cancer cells.

4. Summarize how insertion into cancer cells of genes for thymidine kinase can be useful for gene therapy.

5. What is an adenovirus, and how is this virus used in gene therapy?

6. Explain the protein deficiency that occurs in patients with cystic fibrosis and show how that deficiency can be resolved by gene therapy.

7. Explain the various approaches available for treating AIDS patients with gene therapy.

8. What is the basis for familial hypercholesterolemia, and what intervention can be made with gene therapy? How can gene therapy be used to prevent the development of AIDS?

9. Describe Gaucher's disease and Lesch-Nyhan disease and indicate how gene therapy can interrupt these conditions.

10. Summarize the multilevel review system that must be satisfied before a gene therapy experiment begins.

FOR ADDITIONAL READING

Anderson, W. F. "Gene Therapy." *Scientific American* (September, 1995).

Blaese, M. "Gene Therapy for Cancer." *Scientific American* (June, 1997)

Culver, K. W. "Splice of Life: Genetic Therapy Comes of Age." *The Sciences* (January/February, 1993).

Friedman, T. "Overcoming the Obstacles to Gene Therapy." *Scientific American* (June, 1997).

Glausiusz, J. "Gene Therapy: Special Delivery." *Discover* (January, 1996).

Goldberg, J. "Therapy of Hope." *Discover* (April, 1998).

Howard, K. "New Arteries for Old." *The Sciences* (July/August, 1995).

Jaroff, L. "Keys to the Kingdom." *Time* (special issue), (Fall, 1996).

Travis, J. "Inner Strength: Gene Therapy for AIDS." *Science News* 153 (1998): 174–178.

Medical Forensics and DNA Sleuthing 9

LOOKING AHEAD

The last decade has witnessed extraordinary advances in forensic medicine and molecular paleontology, both caused by the ability to isolate and analyze DNA from telltale specimens. On completing this chapter, you should be able to:

- understand the molecular basis for DNA identification and the biochemistry for its procedure.
- describe the theory and procedure for DNA fingerprinting and some of the controversy surrounding its use as an identification tool.
- discuss how DNA matching techniques can be used to decide paternity cases and to identify deceased individuals.
- understand how DNA analyses can assist microbial identifications and encourage understanding of a disease's pathology.
- explain the situations under which DNA can be isolated from fossilized remains and note the importance of these isolations.
- describe the studies being conducted on mitochondrial DNA and explain how the studies have been used to hypothesize the identity of the first humans.
- understand the basis for disconnecting Neanderthals from the lineage of modern humans and review the evidence for the identity of the first Americans.
- explain how DNA analysis opens a window on the characteristics of early humans and helps resolve modern problems.

INTRODUCTION

Every so often in scientific history a window opens and, suddenly, the theoretical becomes possible, then inevitable. Discoveries emerge in rapid succession and powerful new technologies drive researchers to unimagined heights.

One such window opened in biology with the successful use of DNA matching technologies (as explained in Chapter 7). In these procedures, synthetic pieces of DNA called gene probes stick like magnets to their chromosomal mirror images. Then a biochemical signal indicates that a match has taken place. The techniques permit a broad series of ultrasensitive genetic analyses, including the identification of numerous genetic diseases.

Another result of DNA matching techniques has been the process of **DNA fingerprinting**, which is used to link a suspect's DNA to the DNA recovered at a crime scene. Already used in thousands of criminal investigations, DNA fingerprinting helped exonerate or convict numerous suspects (Figure 9.1). We will examine its basis in this chapter.

The same DNA technology is helping to decide **court cases** where there is a question of paternity. Each year, more than 100,000 cases of disputed parentage are heard in courts in the United States. Before DNA matching techniques became available, the court's decision was based largely on the often imprecise determination of blood groups. Genetic analysis can match father to child with a much higher degree of accuracy, a degree estimated to approach ninety-nine percent.

Also taking advantage of DNA analysis techniques are the **forensic** pathologist and the biological sleuth. **Forensic pathologists** have discovered that DNA from the skeleton of a long-deceased human can be extracted and used to ascertain that individual's identity. **Biological sleuthing** has become the new wave of study in

gene probe

a segment of DNA that recognzes and bonds with a complementary segment of DNA in a DNA mixture.

forensic

having to do with laws and legal proceedings.

FIGURE 9.1

Technicians performing the decoding mechanisms of DNA fingerprinting. As a sophisticated application of DNA technology, DNA fingerprinting has considerably improved the science of forensic pathology.

anthropology and evolution, as molecular biologists seek to obtain and clone DNA from fossils (including from humans) that are millions of years old. Questions such as the possibility of Marfan syndrome in Abraham Lincoln and the origins of the first Americans may be answered through DNA matching techniques. Each of these topics will occupy our attention in this chapter.

DNA MATCHING TECHNIQUES

Barely two decades ago, DNA matching techniques were still within the realm of science fiction. Scientists believed it possible to identify a person by his or her DNA, but they did not think it would be possible until well into the twenty-first century. Today a form of DNA matching called **DNA fingerprinting** has become an extraordinarily powerful investigative tool. From a minuscule sample of biological evidence (i.e., tissue), forensic scientists can remove a set of genes. Should the genes match the suspect's genes, the prosecution has important evidence that the suspect was at the scene.

DNA matching techniques are also used in myriad identifications, including those associated with paternity cases, political situations, and medical diagnostic procedures. In the following pages, we will examine the biochemical basis for the matching technique, while examining their many uses.

DNA fingerprinting
the technique used to identify an individual based on the DNA content of the cells.

DNA FINGERPRINTING

DNA fingerprinting is a valuable law-enforcement technology, whose power comes from the technologist's ability to extract DNA from a single drop of blood, a few cells, or a tiny sample of semen at a crime scene and identify the owner with high certainty (Questionline 9.1). The technique is performed by extracting DNA from the tissue sample, digesting the DNA to specified fragments with restriction enzymes, and creating a recognizable pattern of the fragments. The pattern is then compared with the similarly prepared pattern of fragments from a known individual to determine whether the patterns match. If a match is made, the individual is identified because certain segments of DNA in the fragments are unique for a particular individual (except if one has an identical twin).

restriction enzyme
a digestive enzyme that biochemically cuts a DNA molecule at a restricted site.

The technology of DNA fingerprinting derives from observations reported in 1985 by the English geneticist **Alec J. Jeffreys** (Figure 9.2). Jeffreys noted that between the genes in chromosomes are certain segments of DNA with base sequences repeated many times over. For example, the DNA segment may read GCGCATG…GCGCATG…GCGCATG…and so on. In some individuals the base sequence

FIGURE 9.2

Alec Jeffreys, the English scientist who developed the gene probes used in DNA fingerprinting and made the technique useful as a device for matching DNA samples.

is repeated two times (GCGCATG…GCGCATG), but in other individuals, the sequence is repeated ten times (GCGCATG…8 repeats…GCGCATG), and in other individuals 200 times (GCGCATG…198 repeats…GCGCATG). The segments of DNA have no apparent function in the living organism and appear to be a type of "genetic gibberish."

For DNA fingerprinting, the nature of the bases in the sequence is not as important as the number of times the sequence is repeated. As Jeffreys found, the number of repeats determines the length of the sequence, and the length of the sequence is of prime importance. Put another way, the smaller fragments of DNA have fewer repeating sequences (for example, ten repeats), whereas the longer fragments have more repeating sequences (for example, 200 repeats).

The scientific name for the DNA fragments with repeating sequences is **variable-number tandem repeats** (**VNTRs**, pronounced "vinters") These are sometimes called **short tandem repeats**. VNTRs are somewhat similar to the **restriction fragment length polymorphisms** (**RFLPs**, pronounced "rif-lips"), which are discussed at length in Chapter 7. Both VNTRs and RFLPs are DNA fragments of different lengths. The difference is that RFLPs develop from random mutations at the sites of restriction enzyme activity on the DNA molecule, whereas VNTRs develop from the number of repeated base sequences between two points on a DNA molecule (Figure 9.3). A person's genome has many different VNTRs and RFLPs, and thus, the pattern of VNTRs and RFLPs for an individual is unique for that individual. (Figure 9.4 gives an example of VNTR analysis.) The pattern of VNTRs and RFLPs forms the basis for DNA fingerprinting.

To develop a DNA fingerprint, the biochemist begins with a tissue sample such as blood, semen, cells, hair, or any other suitable material likely to contain DNA. The DNA

VNTR

a segment of DNA that contains a specified number of naturally-occurring repeating units.

RFLP

a segment of DNA of a different length than another RFLP owing to random mutations.

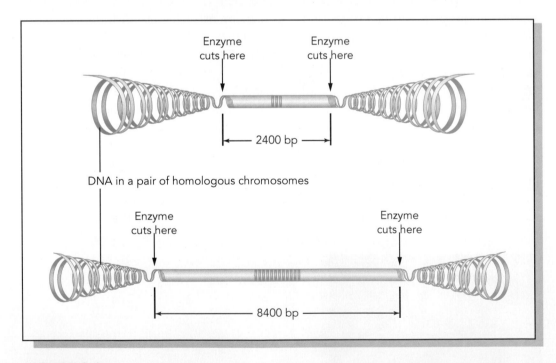

FIGURE 9.3

Formation of VNTRs. Two homologous chromosomes from an individual are shown. The arrows represent activity sites of a particular restriction enzyme. The bands represent short, repeating sequences in the DNA each with 800 base pairs. In the figure, the DNA of one chromosome has three repeats, whereas the DNA on the second chromosome has twelve. After enzyme activity, the first DNA fragment has 2400 base pairs, whereas the second has 8400 base pairs. The fragments move at different rates in an electrophoresis gel.

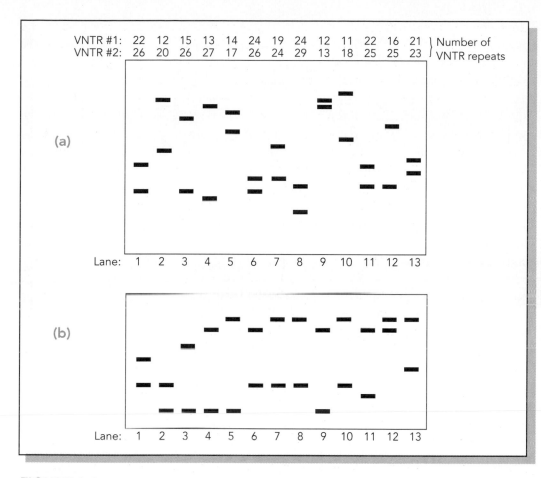

FIGURE 9.4

(a) Using variable number tandem repeats (VNTRs) to develop "fingerprints" of thirteen individuals. The diagram illustrates how many times VNTR #1 and VNTR #2 occur in the sixteen individuals. For example, in lane 1 (for the first individual), VNTR #1 occurs twenty-two times in this person's DNA, whereas VNTR #2 occurs twenty-six times. VNTR #1 also occurs twenty-two times in individual #11, whereas VNTR #2 occurs twenty-six times in individual #3. However in no other individual does VNTR #1 occur twenty-two times and VNTR #2 occur twenty-six times together. The VNTR pattern for individual #1 is therefore unique. The pattern illustrates what might be visualized after comparing only two VNTRs among thirteen individuals. By examining their VNTR fingerprints we can see that none are related to one another. (b) Now examine pattern b and determine which three of the thirteen individuals are related because they have the same two VNTRs.

is extracted from the tissue sample by standard laboratory procedures. Then restriction enzymes are used to digest the DNA into fragments at carefully selected points that flank the VNTR or the RFLP. The fragments to be analyzed are now separated from the mixture. Restriction enzymes are important tools in the digestion step because they scissor the DNA at the same spot regardless of the source of the DNA.

In the next step, the selected DNA fragments are separated according to size by a process known as **gel electrophoresis** (so named because it uses an "electric" current to "carry" molecules—*phorein* is Greek for carry). A gelatinous sheet of polysaccharide called **agarose** is used for this procedure. First a groove is cut at one end of the agarose and the fragments of DNA are placed in the groove. Now the agarose is electrified to establish a positive (+) electrical pole at the far end of the gel and a negative (-) pole at the DNA end, as Figure 9.5 displays. DNA fragments have a negative charge because of their phosphate groups, and they are attracted to the far

electrophoresis

e-lek″tro-fo-re′sis

a biochemical laboratory technique in which molecules move according to their size and chemical charges in an electrical field.

agarose

ah′gar-os

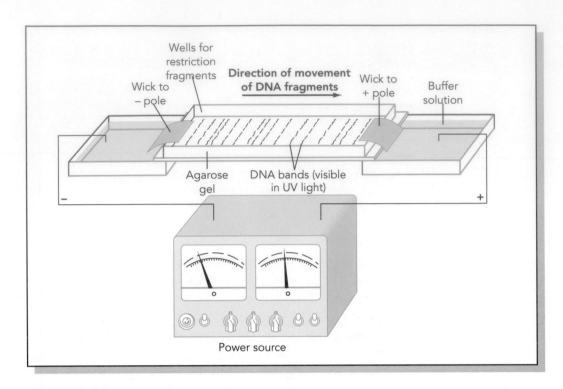

FIGURE 9.5

Electrophoresis. Electrophoresis uses an electric current to separate DNA fragments. The fragments have a negative charge and move toward the positive (+) pole. The smaller fragments encounter least resistance in the agarose gel and move the farthest in a set period of time. The process thus separates the fragments according to size, and bands show the location of the fragments.

QUESTIONLINE 9.1

1. **Q. What is a DNA fingerprint and how does it compare with a traditional fingerprint?**

   A. A traditional fingerprint is made by preparing an ink impression of the skin folds at the tip of a person's finger. Because the nature of these skin folds is genetically determined, a fingerprint is unique for an individual. A DNA fingerprint, by contrast, is an analysis of the nitrogenous base sequence in the DNA of an individual. Because the nature of the base sequence is genetically determined, the DNA is unique for an individual.

2. **Q. Is all DNA analyzed to obtain a DNA fingerprint?**

   A. No. Only certain segments of DNA are analyzed. These segments contain base sequences that are repeated a certain number of times in an individual. By determining how many times they are repeated, a pattern can be developed for a particular individual.

3. **Q. Does the pattern identify the individual?**

   A. Not really. The pattern is only used for comparison sake. If, for example, DNA is found at a crime scene, its patterns can be compared with the DNA pattern of a suspect. If the patterns match with high certainty, the suspect is placed at the crime scene. Traditional fingerprints are used the same way.

(positive) end because oppositely charged substances attract one another. The shorter DNA fragments encounter the least resistance in the agarose and migrate the fastest during a fixed amount of time. Because they move faster, they move farther through pores in the agarose. By contrast, the longer fragments meet more resistance, travel slower, and remain closer to the starting point during the same amount of time. Thus, the DNA fragments spread out in the agarose according to size, with the gel acting as a molecular sieve.

At this point, the location of the DNA fragments cannot be visually observed, so the next step is performed. The process is known as **Southern blotting**, for **Edward M. Southern** of Edinburgh University who developed it in the 1970s. The double helix of the DNA fragments is chemically separated into two strands, and a nylon membrane is placed over the gel. Then, absorbent material similar to a paper towel is placed over the membrane. The absorbent material attracts the single-stranded DNA fragments out of the agarose into the nylon membrane by capillary action. In the membrane the fragments stick and preserve their distinctive pattern.

Southern blotting
a laboratory technique in which molecules are blotted off one medium onto another for identification purposes.

The final step of the process was developed by Jeffreys. He synthesized **radioactive DNA probes** that mate perfectly with the single-stranded DNA fragments on the nylon membrane. Once the DNA probe has united with its matching DNA strand, the radioactivity signals the pattern of DNA fragments. When the nylon membrane is placed against X-ray film, the probes emit their radioactivity, and the positions of the fragments are revealed as charcoal-colored "sooty" bands on an off-white lane. Visual analysis and a computer scanning device yield a measurement of the size of each band. The image has been related to bar codes used to price and identify goods in the supermarket. Figure 9.6 shows the entire process.

DNA probe
a segment of DNA that recognizes and binds with a complementary segment in a DNA mixture.

A typical DNA fingerprint is obtained by using DNA probes in three identifiable chromosomal regions. (A word of caution: the technology for DNA fingerprinting is changing almost daily, and alterations in the technique may currently be in practice.) Because the number of repeats is highly variable (or "polymorphic") among humans, a database can be established by entering the frequencies of occurrence of particular DNA segments in certain ethnic groups in a population. Then a **statistical statement** giving the odds of an error in comparing two samples of DNA (the suspect's DNA and the crime scene DNA) can be offered; that is, investigators note how frequently a particular band occurs in their database compared with bands in other DNA fingerprints prepared with the same restriction enzymes and the same radioactive probes.

An example of how the **statistics** work is as follows: A particular band may occur in an ethnic group once in every 1000 people as determined by computer analysis. Then a second band is compared, perhaps with a frequency of one in a 100 people. The odds that the two bands will occur in a particular DNA sample are therefore one in a 100,000, or one hundred-thousandth (one thousandth multiplied by one hundredth). If a third band is found in the DNA sample and if this band appears in only one in 10,000 individuals, the odds that a person will have all three bands are one in a billion, or one billionth (one thousandth multiplied by one hundredth multiplied by one ten-thousandth).

With this type of information available, a laboratory investigator can report in court that the suspect's DNA and the crime scene DNA have been compared, and the chances that the DNA at a crime scene matches the suspect's DNA are 999,999,999 in 1,000,000,000. (Essentially, the laboratory has compared the DNA from two samples and found that the chances they do *not* match are one in a billion.) With such persuasive evidence, the suspect can be placed at the crime scene (but nothing further). Alternately, the suspect can be exonerated from a crime because the statistics

exonerated
found innocent.

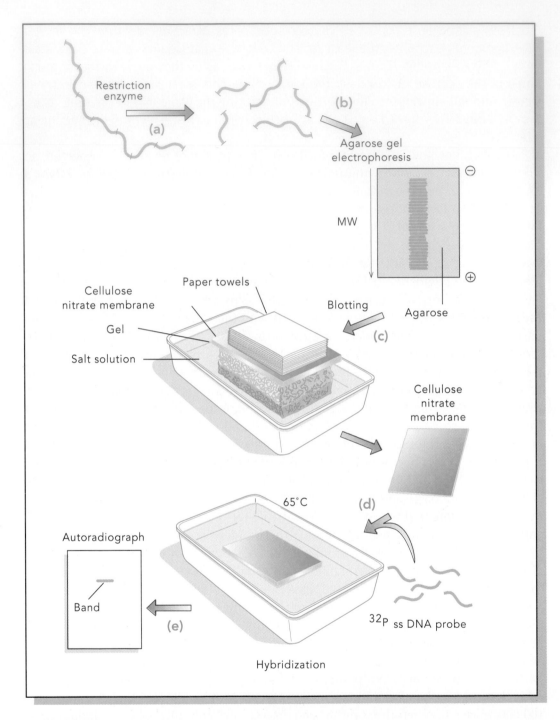

FIGURE 9.6

DNA separation by electrophoresis and Southern blotting. **(a)** A restriction enzyme is used to cut the DNA fragment into fragments. **(b)** The fragments are then subjected to electrophoresis on an agarose gel, a process that separates the fragments according to size. **(c)** Southern blotting follows. In this procedure, the gel is placed together with a nylon membrane of cellulose nitrate and several sheets of paper toweling. The fragments enter the nylon membrane. **(d)** To detect a particular DNA segment, a gene probe for that segment is combined with the membrane to see where a union will take place. The probe carries a radioactive isotope (^{32}P) and is single stranded (ss). **(e)** If the segment is present, the union will take place and radioactivity will be emitted at the site of union. A band of radioactivity will show up on the X-ray film.

show he or she was not at the crime scene. Several innocent individuals, especially in rape cases, have been set free on the basis of DNA evidence.

The statistics were taken a step further in 1997. That year the Federal Bureau of Investigation (FBI) stated that it would declare two samples identical if its laboratory found the chances of a match to be 259,999,999,999 in 260,000,000,000.

As with any new technology, the process of DNA fingerprinting has **drawbacks** that require attention before the technology receives unwavering acceptance. One inadequacy is the relatively small number of DNA records in the databases of laboratories. It may happen, for instance, that only several hundred individuals were used for the broad categorization of an ethnic group. These individuals may or may not be a representative sample of the group. Thus, the frequency of bands in the test sample may not reflect the real frequency of the bands in the general population.

Other **inadequacies** involve the actual performance of the test. For example, "pure" DNA from a crime scene is hardly ever available—an article of clothing, a weapon, or a utensil may have been contaminated with DNA from cells deposited on the item or from bacteria. It is also possible that the semen from a rape victim's vaginal tract may contain DNA from her own cells or from bacteria normally present in the reproductive system. Any such contaminants could be broken down by the restriction enzymes, causing the formation of erroneous DNA fragments.

Poor laboratory technique could also be a determining factor in the procedure's validity. For instance, the methods for DNA isolation and concentration are possible points where errors could be made, and the inability to follow directions precisely could alter the results. Moreover, any contaminating DNA in the solutions or apparatus could hinder the activity of the DNA probe on the target DNA. The matter of interpretation has also raised concern because the placement and intensity of the charcoal-colored bands are open to interpretation during visual and/or computer screening.

Despite these drawbacks, DNA fingerprinting is a very useful technique because recovering DNA at a crime scene is relatively easy. DNA is more resistant to environmental decay than proteins or other biochemical molecules normally sought. Moreover, conventional fingerprints are generally sparse compared with the DNA in blood, hair, semen, or other tissues. Indeed, DNA fingerprinting can detect as little as a billionth of a gram of DNA. (A single hair fiber has roughly ten billionths of a gram of DNA.) To aid the DNA detection, researchers use the **polymerase chain reaction** (often called the "PCR") to amplify certain genes in a sample. A full explanation of the PCR is given in Chapter 7.

polymerase chain reaction

the biochemical laboratory process in which a segment of DNA is multiplied millions of times over through the activity of the enzyme polymerase.

On the negative side, DNA fingerprinting has raised certain *civil rights issues* and questions. For example, what degree of probable cause must law enforcement officials demonstrate before a judge can order a DNA test? And, can blood collected for other purposes be used for DNA fingerprinting?

Another question is whether DNA results can be retained in a "DNA mug book?" Is retaining the results an invasion of privacy? Apparently not because most states now have laws establishing DNA databanks, where DNA profiles are stored from defendants convicted of murder, felony sex offenses, felony assault, incest, or escape. A national databank overseen by the FBI is also in operation, as of 1998, with DNA profiles of 250,000 state felons and DNA evidence taken from the scenes of 4500 unsolved crimes. And in 1999, New York City began investigating the possibility of banking DNA fingerprints for all children born in the city. A typical DNA fingerprint is dsplayed in Figure 9.7.

databank

an electronic, paper, or other storehouse of information.

Still another major issue is regulation. Which state or federal agency should set up standards for DNA testing? How are statistical standards established? What is the

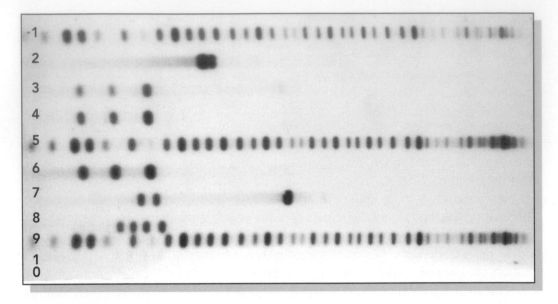

FIGURE 9.7

Use of DNA fingerprinting in a practical situation. Two suspects have been accused of attacking and raping a young woman, and DNA analyses have been performed on various samples from the suspects and the woman. Lane 1 contains the DNA from a sample of virus. It is used to denote the presence of fragments in the test samples. Lane 2 is DNA from the blood cells of suspect A. Lane 3 is DNA from a semen sample found on the woman's clothing. Lane 4 is DNA from the blood cells of suspect B. Lane 5 is another DNA control marker. Lane 6 is DNA obtained by swabbing the woman's vaginal canal. Lane 7 is DNA from the woman's blood cells. Lane 8 contains DNA used a control sample, as is the DNA in lane 9. Lane 10 is also a control sample. Partly on the basis of this evidence, suspect B was found guilty of the crime (note how his DNA fingerprint in lane 4 matches the DNA from the semen in lane 3 and the vaginal swab in lane 6).

role of private laboratories compared with governmental laboratories? And how are laboratories monitored?

As researchers study questions such as these, the promise of DNA fingerprinting continues to remain optimistic. As with all new technologies, however, the process needs fine-tuning and the arguments must be resolved before DNA fingerprinting can become a mainstay of forensic analysis.

■ GENETIC IDENTIFICATIONS

In forensic pathology, DNA fingerprinting is used to single out one suspect from many possible subjects and to provide odds that the DNA at a crime scene matches the suspect's DNA. In paternity testing, by contrast, DNA fingerprinting is used to identify a particular individual and relate that person to another individual.

paternity testing

testing used to determine whether a certain individual is the father of a particular child.

In most cases, the reason for **paternity testing** is to determine whether a man has fathered a particular child. Traditionally the blood type of the man was compared with that of the child. The results might exclude the man as the child's father but they could not definitely point to him as the father. For example, if the man had blood type AB and the child had blood type O, the man could not possibly be the father because he would contribute the gene for A or B blood antigens to any child he fathered; and neither gene is present in this child. By contrast, if his blood type was O, then he could possibly be the father—but many other men could be the father as well.

The advantage of DNA fingerprinting in this situation is that only two types of DNA must be compared: the man's and the child's. Either the DNAs will match, in which case he is declared the father, or they will not match, in which case he is excluded from paternity. In a match, half the child's genetic pattern will be identical to half of the father's pattern because half the genes have come from him. The gene patterns are therefore compared as in DNA fingerprinting to see whether the father and child have the same number of base sequence repeats. DNA fingerprinting of this type yields such persuasive evidence that trials are often considered unnecessary when the test results are presented (Questionline 9.2). An estimated 120,000 paternity suits are heard annually in the United States.

An unusual paternity case surfaced in 1997. A woman in Spain gave birth to a pair of fraternal (nonidentical) twins, and the husband, suspecting they were the result of an affair, demanded DNA fingerprinting to see whether he was the father. To his surprise, DNA analysis showed that he was the father of one child and another man was the father of the second. After questioning, the woman admitted she was having an affair. Apparently she had produced two egg cells that month, and each was fertilized by a different man's sperm.

During the late 1980s, research evidence indicated that DNA could be extracted from bones at archaeological sites. This finding is critical not only to the hunt for ancient DNA (as we shall see presently) but also to modern forensics because it means that genetic identification can be made of the recently deceased. During the **Persian Gulf War** of 1991, for example, DNA analyses helped identify victims of the shooting and bombing. (A special DNA identification laboratory was established for this purpose.) And in 1996, after the explosion of **Airline Flight #800** over Long Island, DNA testing was used to identify the remains of disaster victims. In that instance, VNTRs were used in the testing.

Genetic identifications have also brought to an end the concept of the **Unknown Soldier**. In 1998 scientists exhumed the remains of a serviceman entombed in

unknown soldier
a deceased war participant who symbolizes all those remaining unidentified by conventional means.

QUESTIONLINE 9.2

1. **Q.** How much DNA must be recovered at a crime scene to perform a DNA fingerprint?

   A. The amount of DNA needed for a reliable DNA fingerprint analysis is remarkably small. For example, a single hair fiber has more than enough DNA for starters. The DNA can be amplified by the polymerase chain reaction ("PCR"), and an abundant amount of DNA can be obtained for analysis.

2. **Q.** What are some of the problems associated with DNA fingerprinting?

   A. Like any other new technique, DNA fingerprinting has problems that must be resolved. It will take time, for example, to gather enough information for statistical determinations; also, the analysis requires pure DNA, and often there is extraneous DNA present (such as from bacteria); finally, laboratory technicians must develop sufficient skill in performing the test and interpreting the results so as to make their determinations reliable.

3. **Q.** Can DNA fingerprinting be used for anything other than identifying criminals?

   A. Most certainly it can. DNA fingerprinting can be used in paternity cases to link a child to its father or to exclude a man as the father. It can also be used to identify a deceased individual by matching DNA from its tissue to previously obtained DNA such as that in the cells of a blood sample.

Arlington National Cemetery and brought a bone sample to federal laboratories for analysis. Here they learned the identity of the soldier by comparing his DNA to that of a woman believed to be his mother. The soldier was an Air Force pilot shot down over Vietnam in 1972.

The idea of perpetuating the memory of an unknown soldier as a symbol of national remembrance arose after World War I when President Warren G. Harding dedicated the Tomb of the Unknowns on November 11, 1921. Sophisticated DNA analyses make it unlikely that the remains of any service personnel will go unidentified in the future. However, it is probable that unknowns from earlier wars will remain unidentified because relatives needed for matching blood samples have since passed away.

An identification of a different sort was made in 1992. Bones found in 1985 were believed to be the remains of **Joseph Mengele**, "doctor" at the Auschwitz concentration camp during World War II. (Mengele had disappeared after the war.) Family members maintained that the bones were Mengele's, and his mother and son agreed to give blood samples for DNA matching tests. Alec Jeffreys' laboratory conducted the DNA analysis, and he concluded that the DNA from family members matched that taken from the bones. The case was closed.

In 1994 scientists identified the long-lost remains of **Czar Nicholas II** of Russia Figure 9.8. The czar, Czarina Alexandra, four daughters (Olga, Maria, Tatyane, and Anastasia), a son (Alexis), and four others were executed by their Bolshevik captors on July 17, 1918. In 1991, a mass grave was found near Yekaterinburg, a Russian city where the murders presumably occurred. Then, in 1994, British DNA technologists extracted DNA from the bone cells, amplified it by the PCR, and compared its base composition with DNA obtained from blood cells of living family descendants (including Prince Philip of Great Britain, a grand-nephew of Czarina Alexandra). The base sequences of DNA were close enough to conclude that the remains were those of Czar Nicholas II and others. On July 17, 1998, eighty years after his death, the czar was given a state funeral in Russia.

But the mystery was not completely ended. Czar Nicholas had a daughter named **Anastasia**, and soon after the executions, a number of women came forward proclaiming themselves Anastasia and demanding the czar's fortune. The most famous "Anastasia" was a woman named Anna Anderson, who maintained she was the czar's daughter until her death in 1984. In 1995, DNA technologists obtained a sample of Anna Anderson's intestinal cells (preserved by a hospital after surgery) and compared its nuclear DNA with the czar's. There was no match. When they compared the mitochondrial DNAs (to

Yekaterinburg
ye-kat′er-in″burg

FIGURE 9.8

Czar Nicholas II and his family. DNA analyses were used to identify the remains of the czar and his family after their graves were discovered in 1991. Further analyses helped eliminate the possibility of impostors surviving the murders.

be discussed shortly), the samples once again failed to match. The evidence indicates that Anna Anderson was an impostor. Indeed, the prevailing wisdom is that Anastasia perished with her family (despite the story in the 1998 animated movie).

■ MICROBIAL IDENTIFICATIONS

Another important genetic identification was made in 1994. That year, a strange epidemic of viral disease broke out in the southwestern United States where four states come together (Arizona, Colorado, New Mexico, and Utah). The disease came to be known as the "**four corners disease**," and it spread quickly among Native Americans, causing severe hemorrhaging, pneumonia, and kidney disease and resulting in thirty-five deaths. Scientists performed a battery of tests, hunting for antibodies against myriad viruses. They came up with only one possibility—a remote virus known as the **hantavirus** (for the Hantaan river in Korea, where the virus has been found). DNA was obtained from cultures of known hantaviruses and cloned to produce primers. Then DNA was extracted from diseased tissue and connected to the primer DNA to produce a DNA probe specific for hantavirus. When the DNA probe was combined with patient's tissues, it bound to the viral DNA and pointed to the presence of hantavirus. A diagnostic tool had been developed even before the responsible virus was isolated or cultivated.

hantavirus

an RNA virus known to cause a serious pulmonary syndrome.

Scientists now realize that identifying bacterial DNA from human remains can shed light on the spread of infectious disease. There has been considerable debate, for example, on whether **tuberculosis** was brought to the New World by explorers of the 1500s or whether it predated them. The question was answered in 1994. That year, DNA technologists from the University of Minnesota extracted and identified DNA from *Mycobacterium tuberculosis*, the tubercle bacillus, from preserved lung tissue of a **Peruvian mummy**. The mummy traces back to about the year A.D. 1000. Tuberculosis, it appears, was already endemic in the Americas before the European settlers arrived.

tuberculosis

a transmissible disease of the lung tissues.

In 1994, the **leprosy** bacillus *Mycobacterium leprae* was also identified by its DNA. British researchers studied bones from an Israeli cemetery adjacent to the supposed location of a leper colony in biblical times. The scientists probed the metatarsal foot bone and extracted a few milligrams of dust. Using gene probes they identified the DNA of *M. leprae* in the dust and confirmed that the leper colony was at that site. Moreover, they were able to compare the DNAs of modern and ancient leprosy bacilli.

leprosy

a degenerative bacterial disease accompanied by sores of the skin and internal organs.

A microbial identification can also be used in a forensic case: In 1997, DNA analysis was used to match the strain of **human immunodeficiency virus** *(HIV)* infecting a woman to the HIV strain in a physician's patient. The woman accused the physician of injecting her with the patient's blood and using the virus as a murder weapon after she ended their relationship. The physician was convicted of the crime. This **phylogenetic analysis**, as it is called, was also used in 1992 to show that a Florida dentist was the inadvertent, but probable source of HIV in six of his patients.

phylogenetic

phy″lo-gen-et′ic

Another viral disease attracting interest is **influenza**. Epidemiologists have reported that the 1918 pandemic of influenza was the most virulent of all time, with up to 30 million people its victims worldwide. They were unsure, however, whether the virus was related to the **swine influenza virus** currently found in pigs. In 1997, **Jeffrey Taubenberger** and his colleagues extracted samples of viral RNA from the paraffin-preserved lung tissues of World War I soldiers whose tissues were preserved at the Armed Forces Institute of Pathology after they died (Figure 9.9). They recovered RNA fragments from five viral genes and deciphered their base sequences. Then they compared the sequences with the sequences of every other known swine influenza gene. Finally they found a match and confirmed the relationship to the current virus.

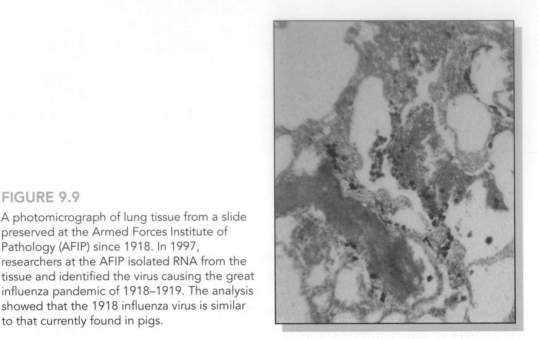

FIGURE 9.9

A photomicrograph of lung tissue from a slide preserved at the Armed Forces Institute of Pathology (AFIP) since 1918. In 1997, researchers at the AFIP isolated RNA from the tissue and identified the virus causing the great influenza pandemic of 1918–1919. The analysis showed that the 1918 influenza virus is similar to that currently found in pigs.

The research also confirmed a suspicion in existence for more than eighty years.

Whereas a pig was the "mixing vessel" for the 1918 flu pandemic, a bird was the apparent source of the **avian flu outbreak** occurring in Hong Kong in 1997. Scientists isolated a virus from a young boy who succumbed to influenza, then they matched its RNA to that of a flu virus in chickens. The ability of the virus to infect two different species is unusual, and the jump from chickens to humans is alarming because humans have no immunity to the virus. Moreover, transmission of the virus from human to chicken could devastate a country's poultry industry. To help stem a possible epidemic, millions of chickens were slaughtered.

Knowing the gene compositions of the avian and swine flu viruses proved helpful to pathologists trying to understand the mechanism of disease. The focus of their research was **hemagglutinin**, a protein at the surface of the influenza virus that assists penetration of host cells during replication. In the avian flu virus, scientists found that the gene for hemagglutinin was mutated. Apparently, the mutation altered the gene, and the gene encoded a defective hemagglutinin molecule, which made it particularly easy for the virus to penetrate human cells, thus increasing its virulence. Thinking that the same mutation may have been responsible for virulence in the 1918 swine virus, the researchers looked for a match between the two hemagglutinin genes. However, none was found. The virulence of the 1918 virus remained unexplained.

DNA WINDOWS TO THE PAST

Twenty years ago, few scientists believed that organic molecules could be retrieved from fossils—indeed, Michael Crichton probably could not have conceived his best-selling novel *Jurassic Park* at that time. Today, fossil molecules not only exist, but they are one of the most-discussed aspects of DNA technology and the subject of intensive research. Fossil molecules from ancient specimens give biologists a window to the past, while making it possible to conceive what life was like millions of years ago. Hunting for ancient DNA molecules has become as exciting for the DNA technologist as hunting for ancient bones has always been for the archaeologist.

hemagglutinin

hem″ag-glu′tin-in

an enzyme in the spike of the influenza virus envelope that facilitates penetration to a host cell.

Crichton

Kri′ton

Jurassic Park

a popular novel and movie in which dinosaurs are cloned from ancient DNA specimens in amber.

■ ANCIENT AND SPECIMEN DNA

The opportunity to recover DNA from archaeological material and museum specimens holds the prospect for answering multiple questions about the kinships of animals, the migrations of ancient peoples, and the taxonomic relationships and evolutionary rates of long-extinct species. Organized under the imposing name of **molecular paleontology**, the study of ancient and specimen DNA represents still another practical application of DNA technology.

The scientific community's attention to fossil molecules was attracted by the cloning of DNA from a zebralike mammal called a **quagga**. Like the dodo in Mauritius and the passenger pigeon in the United States, South Africa's quagga was a victim of mass extermination, its last free-living specimens shot by European settlers at the end of the 1800s (the final animal died in an Amsterdam zoo in 1883). All that remains of the quagga are twenty-three stuffed and mounted specimens in European museums.

In 1984, a research group led by **Alan Wilson** of the University of California extracted and copied a small amount of DNA from the hide of a quagga about 140 years old. His research team cloned the sample and sequenced its 229 bases, less than one millionth of the total genome. Because the technology of the PCR was not available at the time, the DNA had to be inserted into live bacteria and the latter cultivated to produce copies of the DNA studied. Although the task was laborious and the amount of DNA cloned was minimal, the successful experiment fired both the scientific and public imaginations. Zoologists were interested to learn that the quagga was probably a subspecies of the plains zebra rather than a separate zebra species or a relative of the horse.

Next came humans. In 1985, **Svante Pääbo**, then a biochemist at Sweden's Uppsala University, obtained DNA from the remains of an **Egyptian mummy** 2400 years old (Figure in 9.10). Pääbo's group amplified and sequenced 3400 bases of the mummy's DNA. In 1989, an Oxford University group led by **Bryan Sykes** and **Robert Hedges** drove back the genetic clock even further when they extracted DNA from a **human bone** 5500 years old. By this time, PCR technology was available, and the DNA cloning did not require the intervention of bacterial cells.

quagga
qwa′kah

Mauritius
Mor-e′shus

FIGURE 9.10

A photograph of an Egyptian mummy from which a DNA fragment was cloned in 1985.
The successful cloning was performed by Svante Pääbo and his research group.

In 1991, the age record was smashed by **Edward Golenberg** of Wayne State University. Golenberg's team extracted DNA from **magnolia leaves** obtained from the shale bed of an ancient lake in Idaho. The magnolia leaves were 18 million years old.

It only took one year for the clock to be pushed back 30 million years. In 1992, DNA was extracted from a **fossil bee** by California researchers and from a **fossil termite** by investigators at New York's American Museum of Natural History. (It should be noted that the results were questioned by scientists who claimed DNA contamination of the specimens.) In both cases, the insect specimens were preserved in pieces of **amber**, the fossilized material obtained from tacky, odoriferous "pitch" resin produced by plants. Amber-encased fossils have long been valued for their excellent three-dimensional detail, and they are equally suitable for molecular studies because the amber preserves organic tissue extremely well. The amber seals the tissue from the decay-promoting effects of air, microorganisms, and water, and it entombs the tissue to form a natural mummy. In 1994, Raul Cano and his colleagues at California Polytechnic State University isolated DNA from the spores of a *Bacillus* species of **bacteria** obtained from the intestinal contents of a bee encased in amber 25 million years old.

During 1993, Michael Crichton's *Jurassic Park* opened in theaters and, coincidentally, a research group led by paleontologist **Jack Horner** claimed a successful DNA extraction from a fossil of the 65-million-year-old **dinosaur** *Tyrannosaurus rex*. Blood cells had apparently fossilized within the animal's bones, and they served as a source of the DNA. A year later, a group headed by **Scott Woodward** of Brigham Young University claimed isolation of DNA from dinosaur bone fragments found in a coal mine and dated to 80 million years ago. The bone had not fossilized in the coal (coal is derived from peat, an excellent preservative), and its cellular structures were well preserved.

Was the DNA truly dinosaur DNA? Not in the minds of many critics. One reason is the possibility for microbial or other contamination, so extracted DNA sequences must be compared with those of hundreds of possible contaminants. (If the sequence turns out to be similar to those of a fungal gene, for example, then the sample has been contaminated.) And because there are no living dinosaurs, the extracted DNA must bear a comparison to a crocodile's DNA (because dinosaurs and crocodiles came from the same stock) and to a bird's DNA (because birds are descended from dinosaurs). Moreover, confirming the results is difficult because DNA-containing fossils are extremely scarce, and the amount and quality of the DNA varies in the samples. Because the results of the two research groups did not perfectly fit these criteria, scientists were soon disputing their claims.

By the mid-1990s, scientists were voicing their concern for reproducibility. They were having difficulty reproducing the work of Horner, Woodward, and others, and doubts were creeping in about the ability to extract DNA from fossils and animals encased in amber. Many scientists were asserting that the remarkable extractions were really a type of contamination. Other scientists stood by the extractions, as the lively debate continued.

One possible way of resolving whether ancient DNA is intact is to apply a less involved test developed by **Hendrik Poinar** of the University of Munich: The protein associated with the ancient sample is analyzed, and its **aspartic acid** (an amino acid) is studied. The conversion of aspartic acid's normal form to an abnormal form is known to correlate with the destruction of DNA, so if a high degree of conversion has occurred, it is likely that the DNA has been degraded and is not worth pursuing.

Another way of resolving the dilemma is to hunt for ancient DNA in new sources.

---

amber

the fossilized remains of a pitchlike flowing material that oozes from pine trees and their relatives and hardens to preserve materials within it.

spore

a highly resistant form of a bacterial cell formed by a limited number of bacterial species.

Tyrannosaurus rex

a huge carnivorous dinosaur of the Jurassic Period.

aspartic acid

one of the twenty amino acids found in protein molecules.

Indeed, a new chapter in ancient DNA study opened in 1998 when Hendrick Poinar and Svante Pääbo presented evidence on DNA extraction from the **fossilized dung** of an extinct giant sloth, as shown in Figure 9.11. The dung samples, know as **coprolites**, were treated with a novel chemical compound to release the trapped DNA. Cells sloughed off from the intestinal wall were the apparent source of the DNA because the latter matched to DNA extracted from modern sloths. Although less glamorous than amber, animal feces opens another possibility for obtaining ancient DNA.

coprolite

kop'ro-lite

Studying the DNA from ancient specimens can give important information to the evolutionary biologist. In 1992, the DNA was sequenced from the bone of an extinct saber-toothed "cat" obtained from the tar pits near Los Angeles. The animal was 14,000 years old. For decades, folklore had displayed the animal as a saber-tooth "tiger" until biologists of the 1970s presented evidence that it was more correctly a cat. The DNA evidence shows that the original depiction was correct. With its knifelike canine teeth, the animal pictured in Figure 9.12 is closely related to the great cats and should probably be reinstated as a "**saber-tooth tiger**."

The evolutionary kinship of an ancient bird, the **moa**, has also come under scrutiny. Until several hundred years ago, moas prospered on the islands of New Zealand. Then, with the arrival of settlers and predators, the numbers of moas declined and eventually, they became extinct. This flightless, ostrichlike bird ranging up to three meters in height had an uncertain kinship until DNA research reported in 1992. Investigators led by Svante Pääbo sequenced DNA from remains of four different moa species and concluded that moas are not forerunners of the kiwi (another flightless bird), nor do they have the same origin. Moas, it seems, originated from a separate line of birds.

Moa

mo'a

an ostrichlike bird that is now extinct.

Studies on fossil DNA offer the opportunity to relate evolutionary adaptations to gene sequences and to study possible reasons for the demise of such beasts as the **mammoths, mastodons**, and **giant ground sloths**. One reason for those demises may be the presence of pathogenic microorganisms. For instance, traces of viral DNA among the animal remains might show that an epidemic of disease was responsible for an animal's extinction. Studying the viral DNA could also help pathologists understand modern epidemics in contemporary species.

mastodon

a now-extinct elephant-like animal.

In recent years, DNA technology has given a new view to **mammalian evolution** as it occurred during ancient history. The traditional view was that mammalian ancestors were tiny, shrewlike creatures living in the shadows of the dinosaurs. The theory goes on that mammalian ancestors evolved to the modern mammals only after room developed on Earth following the mass extinction of dinosaurs 65 million years ago. But, in 1998, investigators from Pennsylvania State University presented evidence that

FIGURE 9.11

A photograph of fossilized dung from an ancient ground sloth. Researchers led by Hendrik Poinar recovered DNA from this sample after discovering that the DNA was bound and preserved within long-chain sugar molecules.

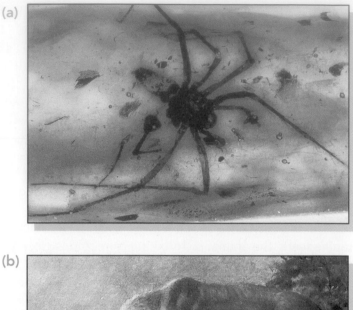

(a)

FIGURE 9.12

Ancient DNA studies.
(a) A piece of polished amber containing an extinct insect. Amber begins as the sticky resin produced by certain trees. When it entraps a biological object and hardens, the amber can preserve a specimen intact for millions of years. (b) The sabre-toothed "cat," now believed to be related closer to the great cats, such as the tiger, than to the smaller house cats.

(b)

the modern orders of mammals could have been present on Earth 100 million years ago, as Figure 9.13 indicates and that they coexisted with the dinosaurs. **Blair Hedges** and **Sudhir Kumar** compared DNA sequences from hundreds of mammalian and other vertebrate species and studied how much variation exists among members of a group. This information told them how extensively the DNA sequence changed per million years. By use of sequence differences as a type of "molecular clock," they then dated mammals to the predinosaur eras.

paleontologist

a scientist who studies specimens of ancient forms of life.

The data of Hedges and Kumar were met with skepticism, however. Paleontologists pointed out that remnants of predinosaur mammals do not exist in the fossil record. (One scientist exclaimed in exasperation; "you can imagine how maddening this stuff is to a paleontologist!"). Other skeptics questioned whether enough DNA sequences were compared. The controversy continues.

■ MITOCHONDRIA AND HUMAN ORIGINS

Although most lay people assume that DNA resides solely in the chromosomes, ample evidence—some of it a quarter-century old—indicates that DNA also exists outside that domain. In the 1970s, a number of research groups reported cell isolations of DNA from the cell membranes of mammalian cells. Previous to that report, DNA was known to exist in the mitochondria of chick cells, the first evidence having been presented in 1966.

Although the roles of membrane-bound DNA and mitochondrial DNA are not firmly established, scientists have used the two forms of DNA in their study of

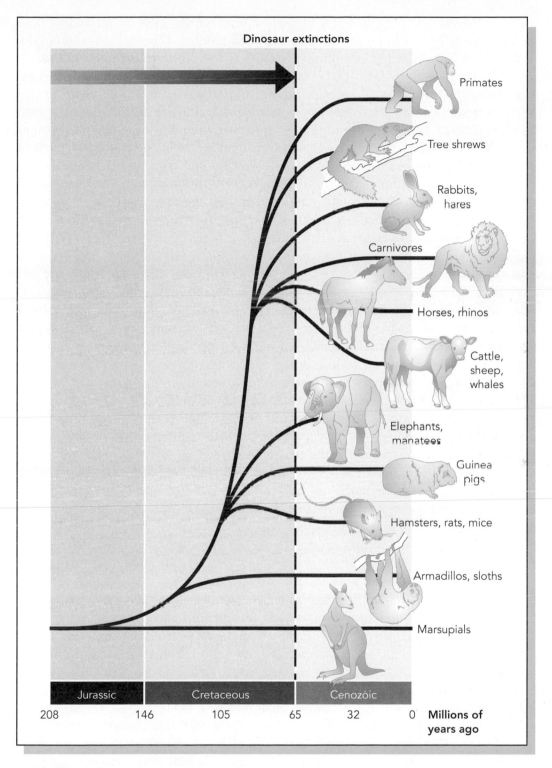

Dinosaur extinctions

Primates

Tree shrews

Rabbits, hares

Carnivores

Horses, rhinos

Cattle, sheep, whales

Elephants, manatees

Guinea pigs

Hamsters, rats, mice

Armadillos, sloths

Marsupials

Jurassic	Cretaceous	Cenozoic
208	146 105	65 32 0	**Millions of years ago**

FIGURE 9.13

A view of mammalian evolution showing the existence of many orders of mammals concurrent with the existence of the dinosaurs. Traditional theories have suggested that mammals evolved after the dinosaurs became extinct, but recent DNA studies indicate that numerous mammalian types could have been present at the time of the extinctions. Paleontologists have described the theory's findings as "maddening."

anthropology. In one notable series of studies, biochemists used mitochondrial DNA to trace the origins of humans. The object of their studies is the identification of the mythical first human female, or group of females, the so-called **mitochondrial Eve**.

mitochondria

microscopic bodies within a cell where carbohydrates are metabolized and their energy released for cell functions.

cristae

kris′tae

Mitochondria are oval-shaped ("sausage shaped") organelles of eukaryotic cells. The organelles are about a half-micrometer in diameter and roughly two to five micrometers in length. These dimensions approximate the size of many bacteria such as *Escherichia coli*. Mitochondria (literally "threadlike granules") are bound by outer membranes and have extensive inner, folded membranes called **cristae**, as Figure 9.14 displays. Along the cristae are the enzyme systems and cofactors that use energy released from carbohydrates to synthesize adenosine triphosphate (ATP). Because of this biochemistry, mitochondria are often referred to as the "powerhouses of the cell."

During the 1960s evidence accumulated that mitochondria possess DNA and a type of miniature genetic system existing alongside the major one in the nucleus and replicating on its own. During succeeding decades, scientists identified a series of genes in the mitochondrial DNA (Figure 9.14b). The studies of molecular biology and human evolution eventually merged in controversial assertion made in 1987. That year, **Allan Wilson** (of quagga fame), **Rebecca Cann**, and **Mark Stoneking** at the University of California published data in the respected British scientific journal *Nature* asserting that the mitochondrial DNA of all modern humans is derived from a common ancestor or population of ancestors, a mythical and symbolic woman who lived in Africa about 200,000 years ago. To understand their assertion, two important questions must be answered: Why a woman? and why Africa at that particular date?

mitochondrial eve

the mythical woman or group of women from which all human mitochondrial DNA was obtained.

The reason our mitochondrial ancestor is an "Eve" (rather than an Adam) is based on **reproductive physiology**. Both egg and sperm cells contain mitochondria, but the sperm contributes no mitochondria to the fertilized egg. Instead, only the sperm nucleus enters the egg cell to effect fertilization. Thus, all a person's mitochondria are derived from mitochondria in the egg cell. Our mitochondria are obtained solely from our mothers, and our mothers from their mothers, and so on back to the first human woman (a type of endless bucket brigade).

One of the most important aspects of sexual reproduction is a mixing of genes from the nuclei of male and female reproductive cells. Allied to this mixing is a rearranging of genes to new genetic combinations, a rearranging that leads to immense biological diversity among the human species and ensures that a brother will be similar to but not exactly like his sibling (except, of course, if they are identical twins). **Mitochondrial genes**, by contrast, remain intact, unshuffled, and unmixed. In doing so, they represent an unbroken line of genetic information back to the first woman, or population of women, the mythical Eve.

Using mitochondrial DNA for evolutionary analyses is biochemically **advantageous** because it is a relatively small molecule (15,000 to 18,000 base pairs) and relatively easy to isolate (there are perhaps 500 to 1000 mitochondria per cell, and thus, 500 to 1000 copies of a gene, compared with two copies in the cell nucleus.) Moreover, and of substantial significance, the mitochondrial genes apparently do not get rearranged over time, nor is the DNA subdivided. However, the genes undergo mutation at the random sites including some where restriction enzymes act, and herein lies the foundation of Wilson's work. When a mutation occurs at a particular restriction site, the restriction enzyme cannot function and a longer-than-normal fragment of DNA emerges. (A shorter-than-normal fragment can also occur if the mutation yields a new site where the enzyme can act.) By use of a series of different restriction enzymes, many mutated sites can be identified.

mutation

an irreversible and permanent change in the DNA of a cell.

In the early 1980s, Wilson analyzed the **rate of mutations** in mitochondrial DNA

(a)

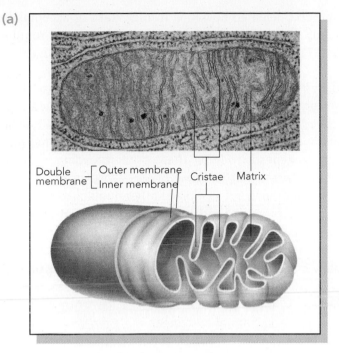

(b)

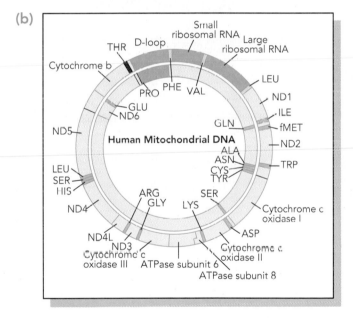

FIGURE 9.14

A mitochondrion and its DNA. (a) The mitochondrion is a cellular organelle shaped somewhat like a sausage. Internally, it contains a series of membranes called cristae and a volume of cytoplasm-like material called the matrix. Enzymes used for energy metabolism are located along the cristae and dissolved in the matrix. The DNA of the mitochondrion is found in the matrix. (b) Mitochondrial DNA occurs as a double-stranded circular unit. The inner or the outer strand may be transcribed, as noted on the diagram. The three-letter abbreviations refer to the amino acid whose tRNA molecule is encoded by that gene. Other designations refer to enzymes encoded or enzyme subunits or ribosomal fragments encoded.

and found that the DNA mutates at a rate of approximately two to four percent every million years. The next step was to analyze how the DNA differs among various populations within the species *Homo sapiens* (e.g., Asians versus Europeans versus Africans). If the mitochondrial DNA mutates at a constant rate, the amount of time needed to yield those differences from a single female ancestor could be assessed.

During the mid-1980s, Wilson's coworkers Cann and Stoneking collected data on the mitochondrial DNA from 147 women living in five geographic areas: Europe, Asia, Africa, New Guinea, and Australia. Most DNA samples were very similar, indicating that the various peoples had only recently diverged from one another. However, the DNAs from the African women were far more diverse than the DNAs from people from the other regional groups. This diversity implied that humans have lived far longer in Africa because the larger number of mutations in their DNA would take much more

New Guinea

a large island off the coast of Australia.

time to accumulate. When the group calculated how long it might take to accumulate the mutations, assuming a steady rate of change, they concluded it would require about 140,000 to 280,000 years. Thus mitochondrial "Eve" lived about 140,000 to 280,000 years ago.

The **conclusions** reached by the researchers attracted substantial attention because the implications were far reaching. For instance, the results implied that fossils of humanlike creatures dated at about a million years old are not really human ancestors as believed by many anthropologists. Also, to suggest that all modern humans possess nothing but African genes is to imply that no interbreeding occurred between species of humanlike creatures, that is, *Homo sapiens* arose in Africa as a brand-new species and spread out over the world to replace other humanlike creatures rather than genetically mixing (or interbreeding) with them. This is the **out-of-Africa theory** (presumably named for the book and movie of the 1980s.). Such an insight would contradict some biologists' view that the evolution to *Homo sapiens* occurred roughly at the same time in many places. This is the **multiregional evolution theory**, as illustrated in Figure 9.15. The latter holds that humans arose in Africa a long 2 million years ago and evolved as a single species (*Homo erectus*), with multiple populations interconnected by genetic and cultural exchanges.

In the years after 1987, critics of the out-of-Africa theory pointed out several **possible errors** in the study. They noted, for example, that restriction enzyme analysis

anthropologist

one who studies the sociology and behavior of modern and ancient humans.

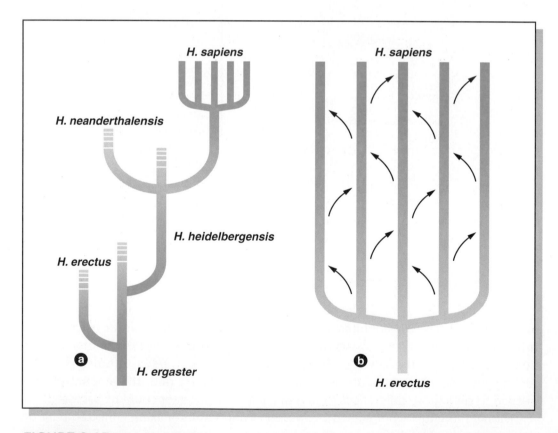

FIGURE 9.15

Two conflicting theories for the origin of humans on the basis of analysis of DNA from ancient specimens. (a) The out-of-Africa theory holds that humans arose in Africa as a new species, then spread to other parts of the world to replace other humanlike species such as *Homo neanderthalensis* (Neanderthals) and *Homo erectus*. (b) The multiregional evolution theory maintains that humans arose from *Homo erectus*, then formed many different populations that interacted by exchanging genes while remaining part of the same species.

was used to detect the mitochondrial DNA rather than the preferred method of sequencing the DNA itself; they also pointed out that African-Americans rather than native Africans were used for the study; and they questioned the method used to build the inheritance pattern. To counter the criticisms, Wilson's group published another paper in *Science* showing that they sampled a larger number of people from more diverse geographic origins (including several ethnic groups from Africa); and that they used a highly regarded computational program to deduce the evolutionary relationships. They continued to hold to their theory.

The controversy continues to this day. Some scientists question whether other evolutionary patterns of development are as viable as the Wilson proposal, and there is controversy regarding how the data and computational programs were used. Other lines of genetic and fossil evidence are offered to refute and support the theory. One study shows that the DNA segments used for DNA fingerprinting are more diverse among African peoples than any other peoples. These results imply a longer human heritage in Africa. However, human skulls located in China in 1992 have been dated to 350,000 years. This finding would indicate that humans existed outside of Africa well before Eve emerged.

And what of mitochondrial Eve's genetic consort, the **mythical "Adam?"** This man's Y chromosome (the male sex chromosome) was presumably passed to every modern male. It is the equivalent of the female's mitochondrial DNA; and it gives the male side of human history. In 1997, two groups of scientists from two American universities reported results of their concurrent studies on the Y chromosome: studies of more than 100 genetic markers from thousands of men around the world point to Africa as the birthplace of modern humans about 100,000 to 200,000 years ago. The research supports the out-of-Africa theory and indicates that the origin of genetic Adam is apparently identical to that of mitochondrial Eve.

Y chromosome

a chromosome found only in cells of the human male and containing some genes for the "maleness" trait.

There is an added twist to the scenario. In 1998, a research group led by **Michael Hammer** uncovered evidence that some of the original humans migrated out of Africa into Asia, then **returned to Africa**. Descendants of these individuals display a mutation on the Y chromosome that seemingly arose in Asia. Moreover, some African males could have interbred with Asian females, and traces of genes from these women may still exist in the nucleus of human cells. As the fascinating story of human origins continues to unfold, scientists await the next day's developments, and writers prepare to rewrite the textbooks.

■ DISPUTING THE NEANDERTHAL CONNECTION

For more than a century, many anthropologists have claimed that humans descended directly from a burly, now vanished type of human named **Neanderthal** (also called Neandertal). First discovered in 1856 by quarry workers in Germany's Neander Valley, the Neanderthal skull (and many others since found) has a protruding browridge and flat forehead; and the individual was supposedly less intelligent than modern humans. But did it lie in the direct line of human ancestry? Or was it an evolutionary dead end? In 1997, DNA researchers had an answer: a snippet of Neanderthal DNA is "very different" from the corresponding region of DNA in modern humans.

Neanderthal

Ne-an'der-thal

The research was performed by a group that included **Svante Pääbo, Mothias Krings,** and **Ralf Schmitz** pictured in Figure 9.16. The scientists were given access by the German government to the original Neanderthal skeleton, and they removed a one-eighth ounce, half moon slice of bone from the right humerus (the upper arm bone). From drilled samples of the outer bone layer, the researchers hunted for cells

FIGURE 9.16

Svante Pääbo and his research team at the University of Munich. Pääbo has pioneered the science of DNA analysis and has used it for numerous purposes in anthropology and comparative evolution, including studies on Neanderthals.

control region

a sequence of bases in the mitochondrial DNA that lacks a protein-encoding function.

and focused on a section of mitochondrial DNA known as the "**control region**." This sequence of bases does not encode any protein. It is a crucial tool for inferring evolutionary relationships because it mutates faster than most other regions of DNA and can reveal more differences between populations. Also, because it encodes nothing, it is not subject to the forces of natural selection as encoding genes are. However, like other DNAs, it begins to degrade immediately after death, as water, oxygen, and microbes attack it. Short, damaged fragments were all the scientists could hope for. And contamination from outside human and microbe sources was an ongoing problem.

Nevertheless, the research team extracted and identified a sequence of DNA containing 379 nucleotides. When compared with a corresponding 379-nucleotide sequence from a modern human, there were differences at twenty-seven positions, on average. The number of differences is highly significant because differences at only eight positions are observed on average when comparisons are made among a variety of modern humans (e.g., Europeans, Africans, Asians, native Americans, Australians, and Oceanians). Thus, the Neanderthal sequence is apparently outside the statistical range of variations in modern humans.

To ensure reliability, the research was reproduced with a fresh slice of bone tissue and at a different laboratory, that of Mark Stoneking. The identical 379-base sequence was found. In a salient summary statement, the researchers concluded that it is "highly unlikely" that Neanderthals contributed to the pool of mitochondrial DNA in modern humans.

The research supports the view that Neanderthals were a separate species of humanlike creatures that had little or no interbreeding with modern humans and eventually became extinct (possibly at the hands of *Homo sapiens*). It also adds evidence to the out-of-Africa theory because it apparently eliminates the possibility that humans were evolving in various places at the same time. (Indeed Neanderthals were once believed to be ancestors of modern Europeans, but there is no unusual similarity between Neanderthal and European DNA). However, the research leaves open the hypothesis that Neanderthals may have interbred with modern humans; or that a "population bottleneck" may have occurred to render genetic variants extinct; or that random genetic loss may have contributed to the lack of Neanderthal genes. Still, many paleontologists are now visualizing human origins as less a linear tree (with modern humans at the top) and more as a bush in which all branches but one became extinct.

Of course, the case is not closed. By 1998, for example, some researchers were pointing out that the DNA differences were not exceptional and that other primates such as chimpanzees have even greater variation of mitochondrial DNA within the species. Other investigators were awaiting results with more substantial segments of DNA than the ones assayed. A third group was preparing evidence to indicate human movements both out of and back into Africa. And consideration was being given to models of human evolution involving both expansion out of Africa and genetic interchange among populations.

THE FIRST AMERICANS

Over the years, many controversial theories have developed regarding the origin of the **first Americans**. Now the weight of DNA evidence has been added to the theory that the Americas were colonized in three waves: (1) About 30,000 years ago, a group called Amerinds came from Asia; (2) then about 20,000 years ago, a linguistic group called the **NaDene** followed the Amerinds from Asia; and (3) about 5000 to 7000 years ago, a group of Eskimos and Aleuts followed the NaDene people from Asia and settled in Canada.

The DNA research contributing to this theory was performed in part by **Antoni Torroni** and **Douglas Wallace** of Emory University. Using mitochondrial DNA from 24 tribes of peoples from Alaska to Argentina, the scientists discovered that all native Americans belong to only four variants of DNA called **haplotypes**. Because Asian peoples have numerous other haplotypes, it is apparent that migrants carried a small sample to the Americas. Furthermore, the three groups of settlers gave different lineages of DNA to their descendants: Amerinds have all four haplotypes; NaDene have the first haplotype only; and Eskimo-Aleuts have the first and fourth haplotypes. These observations suggest different migrations at different times.

But the question of the first settlers in the Americas is not completely answered. More recent evidence presented by **Andrew Merriwether** indicates that there are nine distinct genetic subtypes in addition to the four haplotypes. The subtypes show up in various peoples at random and argue against three separate migrations. In fact, contends Merriwether, the subtypes point to a single wave of migration in which the first settlers to set foot in America carried all four haplotypes. Scientists are now searching for the kin of those ancestors in northeastern Siberia and Mongolia. They are attempting to show that some of the haplotypes were lost as climate and other factors took their toll on the spreading groups of settlers. Later, they maintain, the survivors rebounded and gave rise to the NaDene and Eskimo-Aleuts peoples.

Another disquieting note is offered by geneticists who point out that studies of mitochondrial DNA reflect only the movement of women. Their movement may not reveal the migrations of the entire population, which is why studies of male Y chromosome are needed to complete the story.

OTHER STUDIES OF ANTHROPOLOGY

DNA analysis also provides clues on the lifestyles of ancient peoples. They show what people ate, what animals they domesticated, what pets they had, how they occupied their time, and how they buried their dead.

For example, evidence reported in 1997 indicates that plants ancestral to modern **wheats** were first cultivated in southeastern Turkey about 9000 years ago. Researchers were able to relate cultivated forms to those growing wild by comparing their DNA profiles. They postulated that agriculture originated in an area stretching between present-day Turkey and Iran.

NaDene

na-den′

Aleuts

inhabitants of the Aleutian islands off the west coast of Alaska.

haplotypes

subdivisions of a DNA group.

That same part of the world, according to conventional wisdom, is apparently where **cattle** were first domesticated, also about 9000 years ago. But it was not the only place. Researchers from Dublin University analyzed mutations occurring in the mitochondrial DNA of thirteen breeds of modern cattle and found that European and African cattle share the same genetic lineage, but Indian cattle are from different stock. It appears that the two lines diverged from a common ancestor about 200,000 years ago and were domesticated during two coincident periods about 9000 years ago. Anthropologists point out that the results show how people from different locales used similar ideas to achieve a settled agricultural lifestyle.

A similar analysis was performed in 1998 by a research team studying the domestication of **dogs** (Figure 9.17). Analysis of the mitochondrial DNA from 67 breeds indicates four separate lineage groups closely related to wolves but distantly related to coyotes and jackals. The research team studied mutations in the DNA as well, and they came to the astonishing conclusion that dogs were domesticated 135,000 years ago. Considering that the origins of agriculture date to about 9000 years ago, the finding would indicate that humans had pet dogs 100,000 years before they had farms. The fossil record has not yet verified the research findings.

And what was used to paint the **cave paintings**? Scientists from Texas A & M University have discovered that the pigment used in two such paintings contains DNA. The paintings, were found along the Pecos River in Texas and are between 3000 and 4000 years old. Bone marrow or blood from a bison or deer was apparently used to draw them because the DNA profiles from the paint approximate these sources. A thin mineral deposit of calcium carbonate over the paintings had offered the DNA a measure of protection against degradation. The researchers used a gene that encodes the protein histone as a marker because bacteria do not possess this gene, and they wished to eliminate the possibility of a bacterial contaminant. Comparing the "paint" gene with that of a host of animals and plants narrowed the gene to a bison or deer gene.

Even the burial patterns of ancient peoples are useful clues to how they traveled and lived (Questionline 9.3). Molecular biologists at Oxford University are studying DNA from **Anglo-Saxon burial grounds** to see whether individuals from fifteen centuries ago came to Britain from Germany (as tradition suggests) or are more closely related to the Romans who arrived before them. Other anthropologists are using DNA analyses to distinguish the remains of boys from those of girls in the burial grounds. This distinction can be made in adults by comparing the skeletons, but the skeletons are not different in children. By analyzing the DNA in the Y chromosome, technologists

histone

a protein used as a supportive framework in eukaryotic, but not prokaryotic chromosomes.

FIGURE 9.17

Variations in dogs. Research performed on the mitochondrial DNA of dogs indicates that 67 breeds of dogs now known to exist are descended from only four lineage groups. The research also points to the domestication of dogs far before scientists had estimated.

QUESTIONLINE 9.3

1. **Q.** I know that recovering fossil DNA was the basis for the movie *Jurassic Park*, but how did they do it?

    A. In *Jurassic Park*, DNA technologists recovered DNA from dinosaur blood cells preserved in insect fossilized amber. They then placed the DNA in an amphibian cell and transformed the cell to a dinosaur cell. The cell developed into a dinosaur. Could this happen in real life? Possibly. But you would need an entire set of dinosaur chromosomes, and only a minuscule fragment has been obtained until now. And that's only the first of a thousand problems that must be resolved.

2. **Q.** Has DNA ever been obtained from centuries-old human remains?

    A. Yes, it has. During the 1980s, scientists first learned how to extract DNA from human bone tissue obtained from mummies. Now they have obtained it from Neanderthal bones and other ancient specimens.

3. **Q.** What is the value in recovering DNA from fossilized specimens?

    A. DNA recoveries from fossils make good headlines, but they also have practical values. For example, DNA obtained from ancient animals can be compared with that from modern animals to determine where kinships lie. Ancient and modern humans can be compared the same way. DNA analyses can also distinguish male and female skeletons to learn about the habits of a society. And DNA recoveries can give insight into the evolution and societal patterns of our ancestors.

hope to identify males and to sudy the ancient society. For example, a graveyard may show a disproportionately high number of young male skeletons, which, together with other data, may indicate cultural selection of boys over girls.

ABE LINCOLN'S DNA

Abraham Lincoln's tall, spindly stature and mildly deformed chest (Figure 9.18) have led modern scientists to suspect that he suffered from a genetic disease called **Marfan syndrome**. Named for French pediatrician Antonin Marfan, who described it in 1896, Marfan's syndrome is a connective tissue disorder that affects more than 40,000 Americans annually. It results in abnormalities in the tissues that support the skeletal system, eyes, and cardiovascular system, and it makes the individual prone to aortic aneurysms. These dilations of the aorta can be fatal.

Marfan syndrome
a genetic disorder of the connective tissue resulting in gauntness and long limbs, among other features.

In 1990, **Darwin J. Prockop**, a molecular geneticist at Jefferson Medical Center in Philadelphia approached the National Museum of Health and Medicine for samples of Lincoln's tissue for DNA isolation and cloning. (The museum maintains two locks of Lincoln's hair, seven bone fragments from his head wound, and the blood-stained shirt cuffs from the physician who performed Lincoln's autopsy.) Prockop theorized that Lincoln's physique, especially his gauntness and long and gangling limbs, reflected Marfan's syndrome. Prockop proposed to test Lincoln's DNA for evidence of Marfan genes. The next year, an ethics panel convened by the museum gave its approval for the experiment. It cited how the investigation might help counter problems of discrimination against people with genetic diseases, while enhancing the self-esteem of individuals with a genetic disease.

Coincidentally, two months after the ethics panel rendered its approval, three teams of researchers announced that Marfan's syndrome results from damage to certain

FIGURE 9.18

Abraham Lincoln, whose DNA is to be analyzed to determine whether the gene for Marfan syndrome is present.

fibrillin

the protein believed to be lacking in individuals who have Marfan syndrome.

genes that encode the protein **fibrillin**. In an article in *Nature*, in July 1991, the investigators noted that fibrillin lends strength and stature to connective tissue and that defective genes prevent fibrillin's production. The connective tissue is now left without an essential molecular scaffold. One apparent effect is the bulging of the aorta that results in aneurysms.

Prockop's work encountered ethical and technical problems. Ethicists argued the ownership of Lincoln's DNA and questioned whether this national treasure should be depleted. They also asked whether it was proper to pry into Lincoln's life. Technical problems were encountered when numerous mutations in the Marfan gene were discovered and no one mutation could be related to all cases. (At the time, the technology for locating all mutations was not available). However, at this writing, the technology has improved considerably and the search for the Marfan gene will likely be reinstituted. It should be noted that Prockop's work has had a positive twist—the news media published enough articles on Marfan syndrome to focus attention on the disease and encourage hope for a test to detect the defective genes.

A test for Marfan syndrome would be particularly valuable for **athletes** because the rigors of sports often place persons with Marfan's syndrome at considerable risk. In 1986, for example, volleyball star Flo Hyman collapsed and died of a ruptured aorta. Unaware that she had Marfan's syndrome, she had used her unusual height (6 feet, 5 inches tall) and her unusually long arms and large hands to achieve excellence in volleyball. Had a diagnostic test been available, Hyman could have adapted her lifestyle to avoid excess rigor. Other athletes will have the opportunity to do so in decades ahead, largely because of the advances in DNA technology.

■ MISCELLANEOUS DNA SEARCHES

As noted earlier in this chapter, DNA analyses have solved numerous questions of parentage. They have also focused on one question related to the parentage of **wine**. Vintners have long wondered whether the grape of **Cabernet sauvignon**, a dry red wine was related to the grapes for Cabernet franc (a red wine) and sauvignon blanc (a white wine). The names obviously reflect a relation, but the answer to the question would be found in the genes. In 1997, researchers at the University of California at Davis provided the answer: Yes, they are related. The scientists surveyed 30 DNA segments in 51 different grapes (alicante to zinfandel) and found a definite relationship between

Cabernet

Cab'er-nay

sauvignon

sau'vign-yon

the cabernet sauvignon grape and the two suspected relatives. They guessed that a cross-pollination between the two plants yielded the new variety. Case closed.

With poaching on the rise in many states, game and fish departments have begun using DNA analyses to win convictions against illegal hunters. In 1998, for instance, Wyoming prosecutors used DNA fingerprinting to show that DNA from the carcass of an **illegally hunted antelope** matched the DNA from a mounted antelope head in the possession of a poacher. DNA tests have also been used to match entrails from a discarded carcass to the steaks in a person's house and prove cases of hunting out of season. DNA analyses can tell whether animal parts for sale are genuine or fraud. And in one case in Japan, DNA tests indicated that meat from the endangered **humpback whale** was being illegally offered for consumption.

On a more microscopic level, the hunt for DNA has strengthened the **endosymbiosis theory**. This theory holds that about 1.5 billion years ago, a primitive cell engulfed one or more bacteria, which remained in the cell and eventually evolved to the current-day mitochondria. Supporting the theory are observations that mitochondria (like bacteria) have their own DNA, divide as bacteria do, and have bacteria-like enzymes.

In 1997, more support for the theory was added by DNA technologists from Dalhousie University in Nova Scotia. A research team led by **Michael Gray** located a protozoal mitochondrion more closely related to a bacterium than any mitochondrion previously studied. The mitochondrion, found in a mud-dwelling protozoan, has sixty-two genes (versus thirteen in a human mitochondrion). Many of the genes are typical, but eighteen of the genes have never been observed in a mitochondrion, and some of the eighteen are similar to genes in a bacterium. Among the similar genes are four that encode RNA polymerase, the RNA-synthesizing enzyme. No other mitochondria have these genes, but bacteria have the genes. It is conceivable that the genes were lost when the bacteria took up residence in the cell cytoplasm and evolved to mitochondria. The new mitochondrion resembles a "missing link" between the ancestral bacterium and the modern mitochondrion.

Speaking of bacteria, it is apparent that these organisms occupy two of the **three domains** of living things. In 1997, researchers determined the DNA compositions of the entire genome of a methane-generating species of bacterium living on the ocean floor under extreme conditions. The organism, *Methanococcus jannaschii*, belongs to a group of bacteria called **archaea** (because they apparently lived during ancient times). Researchers found that only forty-four percent of the bacterium's 1738 genes resemble those from organisms outside the archaea (Chapter 12). Scientists led by **Carl R. Woese** of the University of Illinois used the data with other evidence to argue that all living things should be classified to three domains: Archaea, Bacteria, and Eucarya. The three-domain classification is gaining wide acceptance among scientists.

It is somewhat interesting that we close this chapter with the newly suggested three-domain classification of living things. DNA, a focal element of all living things, has now been used to shed light on the fundamental relationships among those living things.

endosymbiosis

en″do-sym-bi-o′sis

the theory holding that mitochondria in complex cells originated from indwelling bacteria.

Methanococcus

meth″an-o-kok′us

jannaschii

jan′ah-shi

archaea

ar-ka′ah

SUMMARY

The use of DNA probes and the development of retrieval and amplification techniques have made it possible to match DNA molecules to one another for identification purposes. In DNA fingerprinting, the presence of repeating base sequences yields DNA fragments of different lengths. Because the pattern of fragments is unique for an individual, the pattern can be used as a molecular fingerprint.

Extraction, digestion, electrophoresis, blotting, and matching with DNA probes are some of the procedures used to develop the fingerprint. Statistical evaluations of the results are important to establishing a confidence level that a suspect was at a crime scene. Paternity suits and identifications of the deceased can also be resolved using the DNA fingerprint technique. Moreover, the pathology of such infectious diseases as influenza, AIDS, and tuberculosis lends itself to DNA analysis.

The ability to retrieve DNA from ancient materials and museum specimens gives archaeologists and anthropologists a glimpse of ancient life. DNA has been cloned from extinct animals, such as the quagga and plants such as 18-million-year-old magnolia leaves. Evolutionary biologists can attempt to draw lineage patterns from the data to give a better understanding of species' relationships.

Molecular studies of mitochondrial DNA have helped solve questions concerning the origins of humans in Africa. Because an offspring's mitochondrial DNA is derived from its mother, and because this DNA represents an unbroken line of genetic information, the analysis of mutation sites in mitochondrial DNA leads one back to the mitochondrial "Eve." Such an analysis made by Wilson and his colleagues encouraged researchers to postulate the existence of a mythical woman who lived in Africa approximately 200,000 years ago. The implication was also that all human populations developed when this woman's relatives migrated from Africa, the so-called out-of-Africa hypothesis.

Anthropologists studying DNA from ancient specimens have also decided that Neanderthals were not on the direct line of succession to modern humans. In addition, they have traced the movements of human populations that first colonized the Americas, and they have studied the way of life of ancient humans, including what they ate, what pets they had, and what they used in their cave paintings. One of the more notable requests made by biochemists was for samples of Abraham Lincoln's tissue to be used for DNA analysis to test for Marfan syndrome. Other DNA studies revolve around the endosymbiosis theory and the three domains of life.

REVIEW

The major insights of this chapter have been how DNA analysis is used in medical forensics for identification purposes and how DNA can be sought and studied to answer questions in evolution, anthropology, medical history, and other subject areas. To test your understanding of these insights, answer the following review questions.

1. Explain the molecular basis of DNA fingerprinting, with reference to the variable number tandem repeats (VNTRs) and restriction fragment length polymorphisms (RFLPs).

2. Describe the major steps performed by a DNA technologist in the preparation of a DNA fingerprint. How is the statistical analysis conducted? What are some of the inadequacies and social objections that must be resolved before DNA fingerprinting is universally accepted?

3. How can DNA analyses be used to resolve paternity suits and identify some cases in which deceased individuals have been identified by DNA analysis?

4. Using specific examples, explain instances in which DNA analysis is used to study the pathological process of infectious disease.

5. Trace the search for DNA back into ancient times by listing the specimens from which DNA was retrieved and significance of each retrieval. Explain

The Agricultural Revolution  10

LOOKING AHEAD

In the decades ahead advances in DNA technology will have a major impact on agriculture. This chapter examines some places the impact will be felt. On completing the pages, you should be able to:

■ describe methods for introducing genes to plant cells, determining whether cells have incorporated the foreign genes.

■ understand how bacteria bring nitrogen into the cycle of animal and plant life and explain how DNA technology can circumvent the role bacteria play.

■ summarize how plants can be protected from freezing through the intervention of gene-altered bacteria.

■ appreciate how bacteria bring about plant resistance to pests and how DNA technology can enhance that resistance.

■ discuss the resistance that viral proteins impart in plants and indicate how it can be increased by the methods of genetic engineering.

■ describe some of the bioengineered agricultural products that will soon be appearing in the marketplace.

■ explain how plants can be gene-altered for use as carriers of vaccines.

■ denote some agricultural uses for genetically engineered chemical compounds.

⬡ INTRODUCTION

At several junctures, the technological innovations of the twentieth century have encouraged farmers to change their agricultural methods and boost their yields. In the 1920s, for example, farmers moved from horse power to mechanical power, and they measurably increased their production of agricultural goods. Then, from 1950 to 1980, another wave of growth occurred as new chemical fertilizers, new pesticides and herbicides, and new animal drugs came into widespread use. In recent years, farmers have entered a third era, the era of DNA technology. Revolutionary advances made in this period may change agriculture more profoundly than ever before.

DNA technology raises numerous possibilities heretofore unimagined. The technology provides new ways to increase a plant's resistance to insects and bacterial and fungal disease and to grow plants under conditions once believed impossible. By use of the techniques of DNA technology, productivity is improved by increasing the size of edible plant parts such as seeds, fruits, or storage roots. Furthermore, foods are made more nutritious by altering their amino acid content by means of DNA technology. Some botanists even dream of converting plants, notable for their production of carbohydrates, into producers of protein. The hunger problems of the world could conceivably be resolved overnight by such an innovation.

Plant biotechnologists have the **advantage** over their counterparts who work with animals because several important species of plants (e. g., cabbage, carrots, and potatoes) can be cultivated from a single, nonreproductive cell, as shown in Figure 10.1. A single plant cell can be modified by introducing a novel gene, then used to develop a colony of genetically altered cells. (This technique, although possible in animals, is not quite as thoroughly developed.)

To develop such a colony, the genetically altered plant cells are placed in a sterile medium of carefully balanced nutrients and hormones. Here the cells develop into a random mass called a callus. A few days or weeks later, depending on the species of plant, the forerunners of roots, stems, and leaves appear on the **callus**. Miniature plantlets can now be transferred to individual containers until they attain sufficient size and hardiness for planting outdoors. The result is a genetically altered plant, a so-called **transgenic plant** in which all the cells are derived from one cell, and all cells express the genetic alteration. By contrast, researchers working with animals usually must make their gene alterations in animal embryos. Transgenic animals are discussed in Chapter 11.

But performing experiments with plants is not quite as simple as just implied. Plants, for example, are far more complex than microorganisms and much more difficult to work with, and introducing genes into plant cells requires innovative approaches to insertion, as we will see presently. Moreover, the plant characteristics that need improving (the growth rate, for instance) are usually governed by multiple genes. Replacing a multiple set of genes can be extremely tedious, assuming, of course, that the genes have been identified and their interactions have been determined.

Of even more fundamental concern is the question: Why perform genetic engineering on plants in the first place? Opponents of DNA technology point out that new methods are unnecessary because for decades plant breeders have successfully introduced desirable genes into plants by conventional crossing techniques. DNA enthusiasts acknowledge that success, but they add that DNA technology can introduce genes from distant plant families, families not easily bred

herbicide

a chemical compound that kills certain plants such as weeds.

callus

a random mass of plant cells developing from a single cell and capable of forming a mature adult plant.

transgenic plant

a plant whose cells contain genes not usually found in the plant.

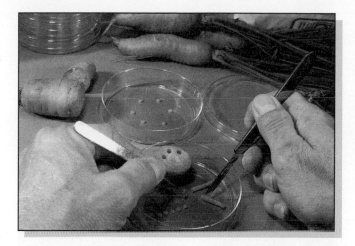

(a)

FIGURE 10.1

Cultivating an entire plant from a tissue sample. (a) A section of a carrot root is separated and placed in a plate containing an enriched medium. (b) In a few days, a cluster of cells known as a callus has formed. (c) A few days later, the callus sprouts rudimentary plant cell parts. (d) After maturity, the new plant can be moved to potting soil to continue its growth and differentiation.

(b)   (c)   (d)

to one another. Moreover, DNA technology can make substantive changes in selected cellular proteins, changes that are unlikely to occur during breeding.

In addition, DNA technology can eliminate many problems by introducing desirable genes directly into a plant. For example, DNA techniques bypass the uncontrolled and random mixing of genes that occurs during a crossing experiment. They also preclude the repeated crosses that must be performed to obtain the desired gene in the offspring, and they sidestep the incompatibility between plants leading to pollen's failure to fertilize the ovule. Taking these advantages into account, plant breeders have generally been quick to see the potential in DNA technologies.

To date, there have been some impressive examples of applied DNA technology in plants, as we will discuss in the pages ahead. The products of DNA technology in plants have fostered an extremely positive outlook. Together with startling advances made in animal technologies (Chapter 11), they have ushered in what has been called the new agricultural revolution.

pollen

the counterpart of the sperm cell in the plant.

DNA INSERTION METHODS

Inserting fragments of DNA into plant cells requires the proper instrumentation, a high degree of dexterity, and an element of luck. The method selected must have a reasonable expectation of success and must inflict minimal damage on the cell (Questionline 10.1).

One widely used procedure for DNA insertion is **microinjection**. In this procedure, a microscopic syringe penetrates the cell wall and plasma membrane and propels DNA

microinjection

a process for introducing DNA to cell nuclei using microscopic syringes.

QUESTIONLINE 10.1

1. **Q.** **What is the advantage of using DNA technology in agriculture?**

   A. Plants provide the basic stock of food for all complex creatures on Earth, and DNA technology holds promise for increasing agricultural yields or enhancing plant resistance to disease. DNA technology can also be used to dramatically increase the protein content of plants.

2. **Q.** **Are plants easier to work with than animals?**

   A. Yes. Modifying an animal biochemically generally requires thousands of cells, but performing a genetic manipulation on a plant involves a DNA insertion into only a single plant cell. The plant cell can then be cultivated to produce an entire plant.

3. **Q.** **How does working with plants compare with working with bacteria?**

   A. Working with plants is much more difficult because plants have extensive mechanisms of gene control not found in bacteria. Also, the internal environment of a plant cell is much more complex than that of a bacterial cell, and the increased complexity reflects a more complex gene pattern. Finally, plants operate on a tissue and organ level, whereas bacteria are self-contained in a single cell. Plants thus present a greater challenge to the DNA technologist.

---

fragments into the nucleus. Despite its straightforward approach, microinjection is a laborious procedure with a high failure rate. Moreover, only one cell at a time can be injected, and 10,000 or more injections may be required before there is a successful DNA incorporation to the cell's nucleus.

An alternative method for DNA insertion was developed in the mid-1980s by scientists at Cornell University. The method uses a type of genetic shotgun named the **biolistic** (for biological ballistic). The biolistic consists of a cylinder with a cartridge containing a nylon projectile. The projectile carries millions of microscopic tungsten spheres, each coated with DNA. When the cartridge is fired at a colony of cells, as shown in Figure 10.2, the projectile shoots down the cylinder, and the gene-carrying spheres enter the cytoplasm of the cells. Unfortunately, the cellular damage can be great, and the method is quite costly. Moreover, there is no way of predicting how many cells will incorporate the DNA into their chromosomes.

A modification of the biolistic is the gold-particle **electric-discharge gun**. In this apparatus, DNA molecules are attached to beads of gold, which are then suspended in water droplets together with a colony of plant cells. An electrical discharge is activated and finely tuned to determine the correct conditions for cell penetration. The electrical discharge propels the gold beads into the cellular cytoplasm, where the DNA is released. The method is called **electroporation** because electricity is used to drive the beads through cellular pores. In 1989, researchers used electroporation to insert marker genes into plant seeds and into the skin cells of frogs. In both cases, the target cells remained alive after the treatment.

Still another DNA insertion method uses needlelike fibers, or "**whiskers**" of silicon carbide. When combined with cells and fragments of DNA in a laboratory mixer, the fibers punch innumerable holes in the target cells. These holes permit passage of DNA fragments into the cells. In 1990, whiskers were successfully used to insert DNA fragments into tangerine-like tissues grown in laboratory culture.

One of the most widely used mechanisms for DNA insertion is the plasmid from

biolistic

a cylinder containing a cartridge with a nylon projectile carrying spheres that ferry genes into plant cells.

electroporation

a method of gene insertion to plant cells whereby electric currents force gold beads carrying DNA into the cells.

FIGURE 10.2

A DNA insertion method. A DNA vector carrying desired genes is coated onto microscopic tungsten spheres. The spheres are loaded into a plastic case (foreground) and positioned in the gene gun (at rear). When detonated, the spheres are propelled down the gun barrel and into the target.

Agrobacterium tumefaciens. This bacterium dwells in the soil and is a parasite of plants. When it passes through the plant surface through a scratch or cut, *A. tumefaciens* infects the tissue and induces formation of a plant tumor called **crown gall**, as illustrated in Figure 10.3. Crown gall develops when the bacterium releases its **Ti plasmid** (Ti for "tumor-inducing") into the plant cell cytoplasm. A segment of the Ti plasmid called the T-DNA leaves the Ti plasmid and inserts into a plant cell chromosome. Here it encodes the production of proteins and supplementary food materials needed by other *A. tumefaciens* cells. As the bacteria multiply and continue the infection, a visible mass of bacteria and plant material accumulates and the crown gall develops.

For experiments in DNA technology, Ti plasmids are modified to carry desirable genes into the plant cells. First the genes for tumor induction are experimentally removed from the Ti plasmids. However, the "insertion genes" are left intact. Then foreign genes are spliced onto the plasmids at the promoter site used by the bacterial

Agrobacterium tumefaciens

ag″ro bac-ter′e-um toom-e-fac′e-enz

a rod-shaped bacterium that causes crown gall by inserting its DNA into plant cells and transforming the cells to form tumors.

FIGURE 10.3

A crown gall on a tobacco plant. The crown gall was generated by cutting off the top of the plant and inoculating with *Agrobacterium tumefaciens*, which produces tumor tissue in the crown gall. Note the somewhat normal plant tissues arising from the tumor.

protein genes. The recombined plasmids are now returned to fresh *A. tumefaciens* cells, and the bacteria are encouraged to infect healthy plants.

When using Ti plasmids, it is difficult to know which plant cells have been infected. To solve this problem, a **marker gene** is combined with the foreign genes. If evidence of the marker gene's activity is displayed by the plant cells, it may be assumed that the foreign genes have also been inserted.

The use of marker genes and the mechanism of gene transfer are worthy of further elaboration because they demonstrate some of the technology for a successful gene insertion. For gene transfer, scientists do not attempt to infect the entire plant with *A. tumefaciens*. Instead, they incubate individual cells from a plant leaf together with *A. tumefaciens* in a nutritious culture medium as illustrated in Figure 10.4. The medium also contains an **antibiotic** deadly to plant cells. For marking purposes, an antibiotic-resistant gene is chemically bound to the foreign genes. Cells that acquire the antibiotic-resistant gene will survive the antibiotic's effect and grow to form a callus. Because the marker gene is bound to the foreign genes, it may be assumed that the callus cells have taken on the foreign genes and the resistance gene.

The value of the Ti plasmid system for DNA insertion is underscored by the observation that foreign genes show up in all plant cells derived from the callus. But, suppose the marker genes must be removed from the plant cells because they pose a hazard to the plant and spread to unrelated plants nearby? To perform a deletion, DNA technologists insert a special segment of DNA called a **recombinase target** on either side of the marker gene. Further away, they insert another gene that encodes a deletion enzyme, together with a regulatory sequence. Now the entire combination is inserted into a plant cell by means of a Ti plasmid system. Later, when scientists wish to excise the marker, they activate the regulatory gene, and the latter stimulates the enzyme gene to produce its deletion enzyme. The enzyme attaches to the two recombinase targets and excises the gene between them. The system works so well that by 1997, researchers were removing genes from isolated cells and mature plants. The technique can also be used to remove the foreign genes to prevent their spread to other plants.

Over the years, botanists have learned that plants treated with Ti plasmids form normal seeds, and they have developed methods for spotting the foreign genes in the

Ti plasmid

the plasmid of *Agrobacterium* that induces tumor formation in plant cells and is used as a vector in DNA technology.

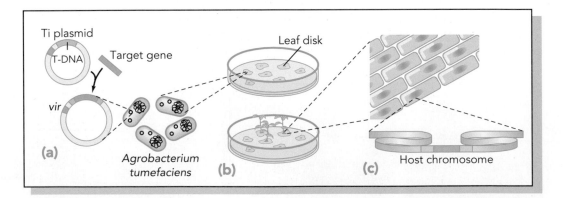

FIGURE 10.4

Use of *Agrobacterium tumefaciens* to ferry genes into a plant cell. (a) The target gene is inserted to the transforming region (T-DNA) of the Ti plasmid. For the insertion, the *vir* genes on the Ti plasmid encode essential enzymes for the excision of the T-DNA and the integration of the target gene. (b) In a plate of nutrient medium, *Agrobacterium* cells containing the recombined Ti plasmid are combined with leaf discs (each disc called a callus), and the Ti plasmids insert to the plant genomes. (c) Examination of cells from the growing plant reveals that the target gene has integrated into the host cell chromosome.

seeds. Unfortunately, *Agrobacterium* species do not usually infect such important crop plants as corn, rice, and wheat; but biotechnologists have approached this dilemma with imaginative solutions as we will see in the pages ahead.

THE NITROGEN CONNECTION

Nitrogen is an essential element of nucleic acids, amino acids, and numerous other chemical compounds with biological significance. Even though **nitrogen** is the most plentiful gas in the atmosphere (about eighty percent of the air), neither animals nor plants can use gaseous nitrogen to synthesize their biological compounds. Instead, they must depend on the microorganisms to bring nitrogen into their biochemical systems.

We can appreciate the roles played by microorganisms by examining a typical **nitrogen cycle** taking place in the soil. In such a cycle, nitrogen enters the soil with the deposits of dead animals and plants in the soil. Nitrogen also reaches the soil from urea in animal urine. Certain microorganisms, through a complex biochemical process, then release much of the nitrogen in simple chemical groups—plants recycle this nitrogen back into living systems. However, microorganisms release most nitrogen into the atmosphere as an unbound gas, where it is useless to plants and animals. A reverse trip back to living things is absolute necessary if life as we know it is to continue. (Figure 10.5 depicts the cycle.)

urea

a waste product of protein metabolism in the liver that is excreted in the urine and accounts for the return of much nitrogen to the soil.

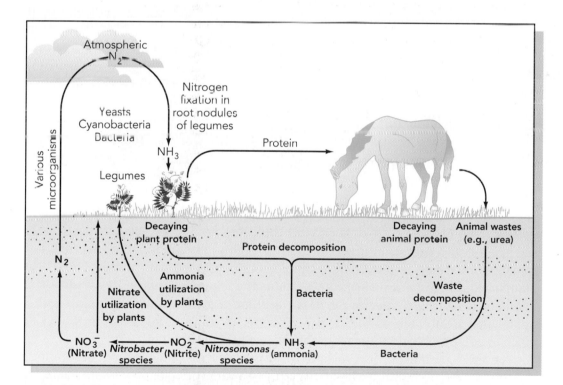

FIGURE 10.5

A simplified nitrogen cycle. Plant and animal metabolic wastes are decomposed by bacteria into ammonia and other products. The ammonia may be used by plants, but it may also be converted to nitrate, which can be used by plants. Bacteria of the genera *Nitrosomonas* and *Nitrobacter* can make this conversion. In many cases, the nitrate is broken down to yield atmospheric nitrogen. The nitrogen is then returned to the cycle of life by nitrogen-fixing bacteria on the roots of leguminous plants. Animals consume the leguminous plants as food, and the cycle is complete.

nitrogen fixation
the biochemical process in which certain bacterial species remove nitrogen from the atmosphere and incorporates it to amino acid molecules.

Rhizobium
ri-zo′be-um
a rod-shaped bacterium that lives symbiotically with leguminous plants and fixes nitrogen into organic compounds useful to the plant.

legume
a plant that bears its seeds in pods; examples are beans, peas, clover, and alfalfa.

symbiotic
living together in harmony.

amino acids
nitrogen-containing molecules used to synthesis proteins.

Enter the microorganisms. Species of bacteria and other microorganisms living in water and soil have the enzyme systems that trap atmospheric nitrogen and use it to synthesize various compounds. The process by which nitrogen returns to the cycle of living things is called **nitrogen fixation**.

Two types of microorganisms have the capacity for nitrogen fixation: one type lives freely in soil and water and includes several species of bacteria and fungi. The organisms extract nitrogen from the air and produce nitrogen-rich compounds that enrich the soil when the microorganisms die. The second type of microorganism lives together with plants in a mutually beneficial relationship called **symbiosis**. The major nitrogen-fixers of this type are species of bacterial rods in the genus *Rhizobium*. Species of *Rhizobium* infect pod-bearing plants such as peas, beans, soybeans, alfalfa, peanuts, and clover. Plants such as these are known as **legumes**.

The relationship between *Rhizobium* species and legumes is distinctive. On their roots, legumes have a number of swellings (nodules) in which *Rhizobium* species thrive. Although complex factors are involved, the central trade-off of the symbiotic relationship is that rhizobia trap atmospheric nitrogen and synthesize nitrogen-rich compounds used by the legume, while the legume contributes important nitrogen compounds to the metabolism of the rhizobia. Legumes use the nitrogen compounds to construct amino acids and proteins. These nutrients are used by the plant for its own metabolism and are passed on to animals when they eat the plant.

Humans have long recognized that **soil fertility** can be enhanced when legumes are cultivated in the field. Indeed, so much nitrogen is captured through symbiosis that the net amount of nitrogen compounds in the soil (and consequently, the worth of the soil) increases significantly after a crop of legumes has been harvested. Moreover, no nitrogen fertilizer needs be added when legumes are cultivated in the field or when crops such as alfalfa or clover are plowed under. Humans are indebted to *Rhizobium* species for the protein in such edible plants as peas and beans, as well as for many indirect protein products of nitrogen fixation such as milk, steaks, and hamburgers (Figure 10.6).

The importance of nitrogen-fixing bacteria to the **world's food supply** has not escaped the interest of DNA technologists. They foresee the day when gene alterations can improve the nitrogen-fixing chemistry in rhizobia and increase the efficiency of the bacteria-plant interaction. Other agricultural scientists are seeking ways to transfer the genes for nitrogen-fixation from *Rhizobium* species to another bacterium such as *Agrobacterium tumefaciens*. If *A. tumefaciens* could be modified to be a nitrogen-fixer, scientists could extend nitrogen fixation to such plants as tobacco, petunia, and tomatoes (because *A. tumefaciens* technology is developed with these plants). The plants would then become as protein-rich as the legumes. Another possibility is to engineer a *Rhizobium* species so it can assume a symbiotic relationship with a nonlegume plant such as wheat, rice, or corn. An advance such as this would do much to solve the world's hunger problems (Questionline 10.2).

The most optimistic DNA technologists predict that genes for nitrogen fixation can be removed from rhizobia and transferred to a variety of plants or even to animals, including humans. Then the bacterial link in the nitrogen connection could be severed, and plants, animals, and humans could extract their nitrogen directly from the atmosphere. It tugs the imagination to envision animal cells that can use nitrogen to synthesize amino acids they cannot now synthesize. What, for instance, would life be like if we did not need meat, fish, cheese, or other protein-rich food on the daily menu to supply our amino acids?

FIGURE 10.6

Alfalfa and clover are leguminous plants growing in the field and used as food by cattle. The nitrogen-rich compounds in these plants can be used to synthesize amino acids and proteins. Milk and meat products contain this protein, and the nitrogen is passed on to humans in this form.

ENHANCING RESISTANCE IN PLANTS

Decades ago chemical pesticides were the cheap and effective way of waging war on the insects that plague plants. Times have changed, however, and the cost of chemicals has multiplied (in some cases as much as 2000 times). Moreover, insects have developed resistance to the chemicals, and an educated public has become very concerned about chemical damage to the ecosystem.

DNA technology holds great promise for resolving some of the concerns about chemical pesticides as researchers seek to enhance the natural biological controls that exist in the environment. In the newly revitalized science of biocontrol, scientists attempt to marshal the capabilities of beneficial insects and soil microorganisms as a practical alternative to chemicals. In addition, they use genetic engineering techniques to help a plant resist cold temperatures, as we will see in the paragraphs ahead.

THE ICE-MINUS EXPERIMENTS

Ice forms in nature under a complex set of chemical and biochemical conditions. In many instances, a number of **ice-nucleating bacteria** are involved. In these bacteria, crystals of ice form when molecules of protein coalesce with water on the outer membrane of the cell. This process takes place at a temperature of about 32°F.

Ice-nucleating bacteria are commonly found on diverse plants, including cereals, beans, fruit and nut trees, and a wide variety of vegetable crops. The density of bacteria along the plant surface ranges from less than 100 bacteria per gram of fresh weight of plant to more than one billion bacteria per gram.

ice-nucleating bacteria

bacteria that synthesize proteins, which coalesce with water molecules to form ice crystals at 32°F.

QUESTIONLINE 10.2

1. **Q.** Occasionally I will read about a Ti plasmid in DNA technology. Exactly what is a Ti plasmid?

    A. A Ti plasmid is a loop of DNA (a plasmid) derived from a bacterium called *Agrobacterium*. When the plasmid is released inside a plant cell, it attaches to the plant chromosomes and becomes a tumor-inducing (hence "Ti") factor. For DNA technology, the tumor-inducing ability is removed, and the plasmid is used to ferry new genes into the plant chromosome.

2. **Q.** How can scientists tell whether the Ti plasmid has successfully inserted itself and its genes into the plant chromosome?

    A. To determine successful insertion, scientists attach an antibiotic-resistant gene to the Ti plasmid. If the Ti plasmid inserts, the resistance gene will also be inserted. When the plant cells are placed into an antibiotic solution, they will resist the effects of the antibiotic. If insertion was unsuccessful, the plant cells will die.

3. **Q.** Why is it so important to insert genes for nitrogen fixation into plants?

    A. Plants are well-known carbohydrate producers, but very few plants are protein producers. (Animals must be eaten to obtain protein for structural materials and cellular enzymes.) The ability to produce protein is intimately linked with the ability to extract (or "fix") nitrogen from the atmosphere. This ability occurs in certain bacteria, but not in plants. If the nitrogen-fixing genes could be obtained from the bacteria and inserted into plants, the plants could become nitrogen fixers and protein producers. The world's starvation problems could conceivably be resolved this way because people could obtain their protein from readily available plants.

---

ice-minus bacteria

gene-altered bacteria unable to produce the proteins that participate in ice crystal formation.

Pseudomonas

soo-do-mo'nas

syringae

si-rin'ja

Biochemists have tested several approaches to protecting plants by controlling ice crystal formation stimulated by bacteria. One approach involves treating the plants with copper-containing compounds to kill the bacteria; another uses urea solutions or copper salts to interfere with ice formation. Perhaps the most controversial (and most publicized) approach is to replace ordinary ice-forming bacteria with mutants of bacteria forming ice at a much lower temperature. Such mutants are referred to as **"ice-minus" bacteria**.

In the 1970s, **Steven Lindow** and his colleagues at the University of California discovered that the bacterium ***Pseudomonas syringae*** is one of the more prevalent ice-forming organisms in nature. (Lindow is pictured in Figure 10.7.) With DNA technology, Lindow and his group successfully removed the gene that directs synthesis of ice-related bacterial proteins from *P. syringae* cells. They thus produced the first ice-minus bacteria, that is, bacteria lacking the ice gene. The researchers proposed to spray their gene-altered (transgenic) strain of bacteria onto young plants, hoping that ice-minus *P. syringae* would crowd out the normal ice-nucleating strains of *P. syringae*. The replacement with ice-minus bacteria would presumably give the host plants a few extra degrees of frost tolerance. Indeed, with the ice-minus bacteria in place, the plants might be able to withstand temperatures as low as 23°F because dew can cool to that point before it changes to frost. With frost protection in place, the growing season could be extended considerably and crop yields raised proportionally.

By 1982, Lindow's group was ready to spray its ice-minus strain of *P. syringae* onto plants growing in an experimental field. This was to be the first-ever **deliberate release experiment** involving genetically engineered microorganisms. However, the experiment

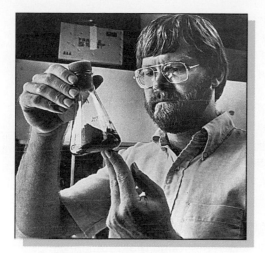

FIGURE 10.7

Steven Lindow, the plant pathologist who performed the first deliberate release experiments, in which gene-altered bacteria were sprayed on plants outside the laboratory environment.

was repeatedly delayed by court challenges brought by opponents of DNA technology who argued that safety issues had not been considered. A number of questions arose about federal regulation of deliberate release experiments, and people who lived near the test site voiced their fears about spreading "bacterial mutants."

The **rebuttals** from DNA technologists were long and loud. Researchers argued that no new genetic information had been placed into *P. syringae* (rather, genetic information had been removed). They showed that the organism was simply being used to replace a closely related organism already in the environment, and they highlighted the valuable contributions previously made to medicine and science by experiments in DNA technology. They spoke of their record of social responsibility and pointed out that DNA technology was merely a tool, no different than fire or electricity. Five years passed while the courts considered the merits of deliberate release experiments.

Finally, in 1987, the deliberate release experiments took place. Lindow and his colleagues sprayed their ice minus bacteria onto a field of potato plants. Concurrently, a research group led by **Julie Lindemann** sprayed a strawberry field with a second strain of bacteria also genetically altered to retard frost damage. The second strain was produced by a California biotechnology company and was commercially known as Frostban. With these two experiments, a scientific logjam was broken; they were the first to involve tests with transgenic bacteria outside the laboratory. The experiments yielded encouraging results but nothing worthy of screaming headlines. The bacteria reduced the frost damage on the treated plants and remained within the confines of the test plot. Experiments continue at this writing.

Frost damage can be retarded by another method reported in 1998. Scientists led by Michigan State's **Michael Thomashow** reported the development of a cold-hardy strain of **Arabidopsis**, a member of the mustard family and the key test plant of botanical DNA technology. The researchers added to the plant a repertoire of **cold-tolerant genes** that come on gradually when the plant is exposed to slowly declining temperatures. *Arabidopsis* plants, as well as wheat plants, have more than twenty-five such genes. The genes normally activate in sequence, but the researchers found a coordinating gene that turns on all the genes simultaneously when the temperature drops quickly. This **coordinating gene** encodes a protein (a transcription factor) that binds to regulatory sequences at the cold-tolerant genes and activates them.

The researchers surmised that if they could turn on the coordinating gene continuously and get it to overproduce the transcription factor, the cold-tolerant genes

deliberate release experiments

experiments in which gene-altered bacteria are released to the environment to assess their effects.

Arabidopsis

a′′rab-i-dop′sis

might come on all at once, regardless of whether the temperature dropped slowly or rapidly. They attached the coordinating gene to a regulatory gene to ensure continual activation and inserted the gene combination to *Arabidopsis*. To their delight, the plant's cold-tolerant genes were activated under normal conditions, and they protected the plants when the leaves were frozen.

How the cold-tolerant genes protect the plant remains a point of debate, although most evidence points to activity at the cell membrane. Scientists believe that cold-protecting proteins guard the cell against disruption resulting from freezing or other dehydrating conditions where water molecules are lost from the area. The implications of the research are noteworthy because late frosts and cold snaps cause substantial crop damage annually.

■ RESISTANCE TO BIOLOGICAL AGENTS

In the early 1900s, the German scientist G. S. Berliner found that moth larvae could be killed by a chemical product of a certain bacterium. Berliner named the bacterium **Bacillus thuringiensis** after the European province Thuringia where he lived. The organism remained in relative obscurity until the present generation when the need for novel insecticides prompted renewed interest in its activity.

Biochemists now know that *B. thuringiensis* produces **toxic crystals** as a normal part of its metabolism. Figure 10.8 shows how these crystals appear under the electron microscope. Crystal formation is rapid while bacteria are producing their highly resistant spores, although the relationship between crystal formation and spore production is unclear. The toxin is an alkaline protein deposited on plant leaves. It is ingested by caterpillars such as corn borers, cotton bollworms, tobacco hornworms, and gypsy moth larvae. In the caterpillar's intestine, the toxin dissolves substances that bind together the cells of the gut wall. As the cementing substances liquefy, bacteria penetrate between the cells and spread infection. The caterpillar soon dies.

B. thuringiensis has been widely used as an **insecticide** because it is relatively specific for an insect pest or group of insects. In addition, it acts rapidly, is stable in the environment, and is easily dispensed. Sprayed onto foliage, the bacteria are so effective against caterpillars that use of *B. thuringiensis* has rapidly expanded in recent years, especially among those who wish to avoid chemical sprays.

But bacteria sprayed onto plants can soon wash off, so the protective effect can be limited. DNA technologists have realized that long-term protection can be provided by inserting the genes for toxin production directly into plants. To date, they have identified and isolated the gene for toxin production and have spliced it into plant genes using the Ti plasmid of *A. tumefaciens* as the carrier (or vector).

---

thuringiensis

thur-in-jen´sis

toxin

a poisonous substance usually associated with disease.

B. thuringiensis

a species of bacteria that systhesizes toxic proteins, which affect various insect caterpillars.

FIGURE 10.8

An electron micrograph of *Bacillus thuringiensis* (x 60,300), the bacterium used as an insecticide. The organism's endospore (ES) and toxic crystal (TC) can be seen in this view. Also note the multiple coverings of the spore and the cell membrane (CM) and cell wall (CW) of the organism.

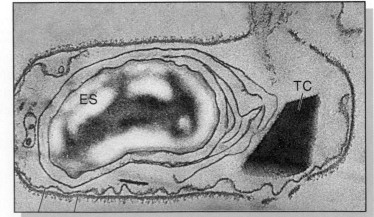

In 1991, scientists at Monsanto Chemical Company reported successful gene insertions to cotton plants. Plants so-treated suffered only minor damage from cotton bollworms compared with 70 to 100 percent damage in nontreated control plants. The tests were conducted at six locations in five states. Although both natural and artificial insect infestations were encouraged, the transgenic plants consistently resisted disease as if sprayed with bacterial insecticide.

The first crops with inserted *B. thuringiensis* (Bt) genes were planted in 1996. Corn, cotton, seed potatoes, and other crops were planted on more than three million acres in the United States. Although the crops displayed increased resistance to pests, scientists noted an emerging resistance to Bt toxin in caterpillars. In caterpillars of cabbage plants, for example, the resistance was related to a recessive gene passed along to successive generations by typical Mendelian patterns. Nevertheless, the plantings continued, and by 2000, over half the soybean crop in the U.S. was planted with Bt-engineered plants.

Another approach to gene-related plant resistance has been taken to protect corn plants against **corn borers**. Figure 10.9 shows some of the experimental results. Researchers at Monsanto isolated the gene for toxin production from *B. thuringiensis* and spliced it into a bacterium that lives harmlessly with corn plants. The researchers then placed a colony of gene-altered bacteria together with corn seeds in a pressurized container. Here the bacteria penetrated the seeds through naturally occurring cracks. When the seeds were sown, the toxin-producing bacteria flourished along with the plant, and when an insect ate the plant, it consumed the toxin. This approach is advantageous because only the insect attacking the plant is subjected to the toxin; other insects in the vicinity are spared. Field tests of the transgenic bacteria have been taking place since 1988 with the hopes of limiting the estimated $400 million of corn borer damage that occurs in the United States annually.

toxin
a poisonous substance.

FIGURE 10.9

DNA-engineered resistance in plants. The upper corn plant has been transformed with the genes of *Bacillus thuringiensis* and is producing toxins to resist the European corn borer. The lower plant is a normal plant, has no such genes, and is susceptible to the borer's effects.

baculovirus

bac″u-lo-vi′rus

lepidopteran

belonging to the
Lepidoptera order of
insects and including
scaled-winged insects such
as butterflies and moths.

**tomato mosaic
disease**

a viral disease of tomatoes
in which the leaves
become shriveled and
assume a mosaic
appearance as a result of
the viral infection.

GFLV

the grape fan leaf virus
that infects leaves of
grape plants and causes
then to shred and "fan."

rice stripe virus

a virus that attacks rice
plants and causes
stripelike discolorations in
the plants.

To be sure, the Bt toxin is not the only natural insecticide available. The American Cyanamid Company, for example, has experimented with a **scorpion toxin**. The toxin, normally found in the scorpion's venom, paralyzes the larvae of moths and other insects. The toxin gene is attached to a **baculovirus**, a virus with a high affinity for lepidopteran tissues. Then the virus is sprayed on lettuce and cotton plants infested with moth larvae. At the conclusion of the field trial, the plot is sprayed with one percent bleach to destroy any remaining viruses. Experiments have been conducted since 1994, and development continues.

Resistance of a different sort has been demonstrated in DNA-altered tomato and tobacco plants. These plants are infected by viruses. The viruses mottle the leaves of the plants, giving them a mosaic appearance; hence, the so-called **tomato mosaic disease** and **tobacco mosaic disease**, respectively. The responsible virus for both diseases consists of a helical strand of RNA surrounded by a protein coat.

In 1983, **Roger Beachy** and his colleagues at Washington University transformed plants genetically so they would resist viral disease. They isolated the gene that encodes the protein capsid of the virus, and they attached the gene to the Ti plasmid of *A. tumefaciens* and inserted the plasmid into tomato and tobacco cells. Soon the plant cells were producing viral capsids. The transgenic cells were coaxed to grow and multiply in laboratory dishes before cultivation in soil.

By 1987 Beachy and his colleagues were ready for field trials. They set out transgenic tomato plants and normal plants and inoculated both with tomato mosaic viruses. At the conclusion of the growing season, the crop yields from the transgenic plants were twenty-five to thirty-five percent higher than from control plants. Moreover, in experiments in which no viral inoculations were made, the crop yields were substantially identical. The gene alteration had increased the crop yield when tomato mosaic virus was present, but in the absence of virus, the crop yields from transgenic and normal plants were virtually identical. Beachy's group hypothesizes that the presence of viral protein in the plant cells blocks viral replication, possibly viral uncoating. It is also possible that plant-viral proteins bind to cellular receptors and prevent pathogenic viruses from binding there.

Viral genes have also been used to protect **grape vines**. In 1993, French biotechnologists announced the successful incorporation of genes from the **grape fan-leaf virus (GFLV)** to champagne grape vines. This virus is transmitted by a nematode and is endemic in the soils of many French regions. It causes malformation of the plant's leaf ("fan-leaf") and induces the plant to lose chlorophyll and become yellow. To protect the vines, researchers inserted the genes for viral capsids into *A. tumefaciens* and infected the plants with this bacterium. Soon the cells were producing viral capsid proteins, and they became resistant to the virus. Wine from the new transgenic plants is expected to reach market by the year 2000.

Success with capsid proteins as protective devices has spurred other teams to adapt the research to other plants. As of 1993, researchers were conducting field tests to determine whether capsid proteins could protect potatoes against **potato leaf-roll virus** and cantaloupes against **cantaloupe mosaic viruses**. In 1993 a research team from Japan announced incorporation of genes for the capsid of **rice stripe virus** into rice plants with successful plant protection. A disquieting note was sounded in 1994, however, when researchers from Michigan State University presented evidence that viral genes inserted into plants could recombine with genes from infecting viruses in the environment and produce new gene combinations, a potential source of new viruses.

In some cases, a disease-resistance gene can be found in one crop strain and used to engender resistance in another strain. Rice, for example, suffers infection by the bacterium *Xanthomonas oryzae* and plants wilt in a few days. This bacterial blight can devastate half a farmer's field. But certain wild strains of rice possess genes that confer natural resistance to the bacterium. Researchers at Cornell University have located the resistance genes, cloned them, and inserted them into rice cells. In laboratory experiments, plants derived from these cells resisted infection by twenty-nine of thirty-one strains of *X. oryzae* obtained from eight different countries. As of 1998, the researchers were hoping that their rice strain would be the first transgenic disease-resistant crop available to farmers for cultivation.

Research is also continuing with a naturally occurring gene that will protect a plant against **nematode (roundworm) infestation**. Microscopic nematodes are estimated to destroy $100 billion of the world's annual crop yield. European investigators announced in 1997 that they cloned a nematode-resistance gene from wild beet plants. This, they claimed, is the first gene ever located for resistance to an animal pest. The discoverers postulated that the gene works by encoding a protein to detect a pest's chemical signals and trigger a defensive reaction in the plant. The reaction may consist of a chemical compound that degenerates the feeding structures of the nematode.

The European researchers hope to use the resistance gene to protect the commercially important **sugar beet**. However, they point out that sugar beets are notoriously difficult to cultivate from gene-altered cells. Also, the various strains of nematodes may resist the protective device developed against one strain. And, without a good understanding of the resistance mechanism, it may be difficult to understand the activity of the gene product.

It might be a stretch to consider **tree farming** a form of agriculture, but DNA technologists are less concerned about semantics than about their work. And when their work results in a tree's increased resistance to defoliating insects, they are proud indeed (Figure 10.10)

Traditionally, trees have been very difficult to work with because of their large size, slow growth patterns, and immense chromosome numbers. But in the late 1980s, scientists managed to introduce *B. thuringiensis* genes into poplar and white spruce trees, and painstakingly they developed mature trees that killed any caterpillars feeding on their leaves. Other researchers genetically engineered poplars to express a protein that inhibits an enzyme in the digestive tract of a beetle, thereby causing the beetle to starve to death. And in 1995, scientists announced the successful incorporation of a flower-development gene called *LEAFY* (or *LFY*) into a **European aspen tree**. Normally, aspens take more than a decade to produce their first flower, but in the genetically altered trees, flowers appeared during the first year of growth. Getting a tree to flower early accelerates breeding experiments by a large factor.

It should be noted that many years will pass before DNA technologists develop trees that grow faster or display improved resistance to pests. One problem is the difficult identification of genes that encode desired traits. A solution may lie in the development of identifying markers that help pinpoint the gene's presence without waiting for the gene to express itself in the mature tree. Arborists can then propagate the embryo or seedling, thereby speeding up the breeding process. Eventually, researchers would like to clone the desired genes and insert them into trees lacking them, a sort of gene therapy for plants.

Biological controls such as those discussed are dramatically different than chemical controls, which have been predominant in agriculture since World War II. Biological research has languished until recent decades. One reason is the limitations inherent in biological controls. For example, a microorganism such as *B. thuringiensis* thrives

Xanthomonas
zan"tho-mon'as

oryzae
o-ri"za

nematode
a relatively simple worm having a complete digestive tract and commonly found in the soil.

(a)   (b)

FIGURE 10.10

DNA technology in trees. (a) Aspen trees normally grow very slowly and take about ten years to produce their first flowers. The aspen shoots on the left have been altered with the *LEAFY* gene from *Arabidopsis* to encourage the seedling to produce flowers during the first six months of growth. Note the solitary flowers in the axils of leaves as well as the terminal cluster of flowers. The shoot on the right is unaltered and displays no flowers after an equal amount of time. (b) A field of young loblolly pine trees, the most widely planted commercial tree in the United States. This tree is a prime target of DNA technologists who seek to improve its resistance to disease.

outdoors in one part of the country but not another; moreover, a biological agent does not bring about a dramatic kill, and a biological agent is generally useful against only one pest at a time and often on only one crop. A chemical agent, by contrast, works in all weather, kills dramatically, and eliminates a broad spectrum of pests.

Despite the inadequacies of biological controls in agriculture, interest has resurged for a number of reasons: Many scientists and environmentalists believe that chemical pesticides are irreversibly contaminating the environment. In addition, they point out that chemical production depends too heavily on energy from limited fossil fuels. Also, some chemicals are losing their effectiveness as pests emerge with resistance to the chemicals. And an aroused public, fearing dangerous chemical residues in foods, has begun to voice its opposition to chemical pesticides (Questionline 10.3).

DNA technology can do much to enhance the development of biological controls, but it is also worth noting that the proper use of biological controls is inseparable from a broader strategy of manipulating the microenvironment of the soil and field. Therefore, many scientists are focusing their research on understanding the communities of microorganisms that inhabit the soil and whose interactions can determine the health of crops. In this regard, the practical uses of DNA technology depend on the more fundamental principles of pure research.

herbicide

a chemical compound used to kill weeds.

■ RESISTANCE TO HERBICIDES

On another front, scientists are working to insert into plant cells a set of genes that will protect plants against **weed-killing herbicides**. In modern agriculture it is common

QUESTIONLINE 10.3

1. **Q.** Is it safe to release gene-altered bacteria into the agricultural environment?

   A. For many years there has been an ongoing concern about releasing gene-altered bacteria into the environment for the purpose of improving agricultural yields. Finally, in 1987, the first deliberate release experiments took place. No damaging effects or health-related problems have been noted in any such experiment since that date.

2. **Q.** How can the methods of DNA technology improve plant resistance to disease?

   A. Certain bacteria have the ability to produce toxins that kill insect larvae (caterpillars) without having any effect on the plant. The bacterial genes for toxin production have been identified and cloned, and experiments are continuing to see whether the genes can be inserted into plant cells so that they can produce the toxin. So transformed, the plants would survive insect damage, and crop yields would increase.

3. **Q.** When will gene-altered foods be ready for market?

   A. The so-called gene altered foods are the products of genetically engineered plants. One such food is a tomato that contains a gene to delay rotting; another is a potato that resists bruising; still others are animals that provide more milk, better quality meat, and improved wool. Many such products are already available to consumers.

---

practice to clear the land before sowing crops. There are always unwanted seeds landing on the ground, however, and the weeds that spring up not only compete with crop plants for nutrients but also impede their growth by taking up necessary space. Applying an herbicide kills the weeds, but the herbicide also reduces growth of crop plants because it cannot distinguish weed from crop plant. If plants could be protected against the herbicide, use of the herbicide could be expanded.

One commonly used herbicide is **glyphosphate** (commercially available as Roundup and Tumbleweed). This herbicide inhibits the activity of the plant enzyme used to synthesize certain amino acids found in the chloroplast and essential to plant growth (the enzyme's name is 5-enolpyruvylshikimate-3-phosphate synthetase, or EPSPS). It happens that the common intestinal bacterium *Escherichia coli* contains genes that confer resistance against glyphosphate activity by making the organism much less sensitive to glyphosphate activity. The *E. coli* has a gene that encodes a mutant EPSPS enzyme. The enzyme is less sensitive to glyphosphate than the normal EPSPS enzyme, and a plant with the mutant enzyme would be less sensitive than the one with the normal enzyme.

To engineer **glyphosphate resistance** into plants (Figure 10.11), DNA technologists begin with *E. coli* cells and isolate the mutant *EPSPS* genes. Then they attach the genes to the Ti plasmid of *Agrobacterium tumefaciens* and transfer the plasmids to the cells of plants such as tobacco and petunia. The plant cells now produce the mutant enzyme and develop resistance to glyphosphate. When glyphosphate is sprayed on the tobacco or petunia field during the growing season, it effectively eliminates the weeds but has a minimal effect on the plants because they can resist the herbicide's activity. The crop yield is thus increased.

A similar experiment has been performed with cotton plants and the herbicide **bromoxynil**. In 1994, the U. S. Department of Agriculture deregulated the transgenic cotton plant to permit field testing of the plants resistant to the herbicide. No further

glyphosphate
gli-fos′phate

bromoxynil
bro-mox′i-nil

a bromine-based herbicide that attacks weeds in an agricultural setting.

FIGURE 10.11

Glyphosphate resistance in soybeans. All the soybean plants shown were sprayed with the herbicide glyphosphate. The plants in the foreground fared poorly, and many died. However, the plants in the background were genetically engineered to encode an enzyme that destroys the herbicide. The plants overcame the effects of the herbicide and thrived.

USDA permits would be required. This step signaled a measure of approval of the bromoxynil-tolerant cotton.

As with other forms of genetic engineering, some controversy has arisen in the scientific community about the possible consequences of introducing gene-related herbicide resistance. Critics of the process suggest that herbicide-resistance genes might move naturally from crops to weeds, thus rendering the herbicide ineffective. A 1997 report seemed to confirm these concerns. A consultative group representing a consortium of sixteen research centers acknowledged that such gene transfers could occur and urged those preparing the bioengineered crop to assess how long the gene will serve its intended role.

BIOENGINEERED FOODS

Food and Drug Administration

a federal agency that oversees the safety of foods and drugs in humans.

On May 26, 1992, the U. S. Food and Drug Administration (FDA) announced its landmark decision that food and food products altered by genetic engineering will be regulated no differently than foods created by conventional means. So ended a decade of waiting by food manufacturers to see whether they would have to pass unusually stringent guidelines to market their DNA-altered products. The announcement also heralded a potentially vast change in the food supply because it implied that bioengineered foods may be placed on supermarket shelves without FDA-mandated testing or special labeling.

THE FLAVR SAVR TOMATO

One of the first examples of a genetically altered food to reach market was the **Flavr Savr tomato**. Developed by the California firm Calgene, Inc., the tomato contained a gene that delays rot. The value of such a tomato was that it did not have to be picked hard and green for shipping because it decays slowly. The tomato could be left on the vine to ripen for several days longer and shipped without refrigeration (shipping without refrigeration also helps retain flavor). The tomato resisted rot for more than three weeks, or about twice as long as the conventional tomato.

polygalacturonase

pol″e-ga-lac″to-ur′on-ase

The Flavr Savr tomato worked this way: tomato plants produce the enzyme **polygalacturonase** (or "PG"), a chemical substance that digests **pectin** in the cell walls of tomato plants, as Figure 10.12 displays. This digestion induces the normal decay ("rotting") that leads to plant propagation in the wild. Researchers at Calgene

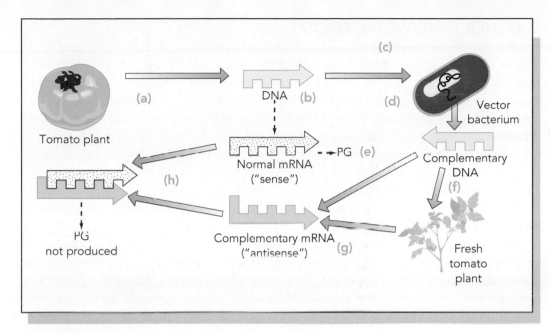

FIGURE 10.12

Producing a rot-resistant tomato. (a) DNA technologists begin by isolating from the tomato plant the gene that encodes polygalacturonase (PG), the "rotting enzyme." (b) In normal tomato plants, this gene encodes normal mRNA (the "sense" molecule), which is translated to PG. (c) DNA technologists transfer the PG gene to a vector bacterium and (d) induce it to produce a complementary DNA molecule. It is a complementary copy of the PG gene. (e) The complementary gene encodes an mRNA molecule that complements the normal mRNA molecule. The complementary mRNA molecule has "antisense." (f) Now the complementary DNA is inserted to a fresh tomato plant. (g) Here the DNA encodes the antisense mRNA molecule. (h)The antisense mRNA unites with the sense mRNA molecule, and thereby neutralizes it. The message in the normal mRNA molecule is not translated, and PG is not produced. Without PG, rotting does not occur in the tomato.

identified the gene that encodes PG, removed the gene from plant cells, and inserted it into a vector organism. Then they induced the vector to produce a complementary copy of the gene.

Continuing the process, researchers isolated the gene copy and inserted it into fresh tomato plant cells. Here it encoded an mRNA molecule (an antisense molecule) that unites with and inactivates the normal mRNA molecule (the sense molecule) for PG production. With the normal mRNA inactivated, the plant could not produce enzyme and, therefore, pectin digestion did not occur. And without pectin digestion, rotting slowed considerably. This use of an antisense molecule is similar to that used in drug research (Chapter 6).

The Flavr Savr tomato debuted in 1995 to great fanfare. Unfortunately, it did not fulfill its promise. Apparently, the company had not predicted some of the **problems** it would encounter: The company had failed to develop tomato varieties that would grow in adequate quantities in different regions of the country; it erred in assuming that existing equipment geared for picking hard, green tomatoes would work for its ripe, softer product; it had exhausted company resources gaining approval from the FDA; and it had failed to account for skeptical people reluctant to pay premium prices for an unknown quantity. In attempting to become the first company to market a biotech product, the company had overstepped its bounds. By 1996, the Flavr Savr tomato was forgotten, and the company was owned by industry giant Monsanto Inc.

antisense molecule

an RNA molecule that reacts with and neutralizes the mRNA molecule used for synthesis of a particular protein, effectively halting the production of that protein.

■ OTHER INNOVATIVE FOODS

But the technology developed by Calgene and the expertise of its scientists survived, and soon, a new product, **canola oil**, was ready for market. Canola oil is a vegetable oil extracted from canola seeds and used for such diverse purposes as detergents, face creams, and ice cream. Scientists have transformed the canola plant by introducing a gene from the California bay laurel tree that encodes the enzyme **thioesterase**. This enzyme synthesizes **lauric acid**, a fatty acid not normally found in the canola oil. The transgenic canola plant now produces a specialty oil containing up to forty percent lauric acid (laurical). The oil has the low saturated fat content of olive oil and is attractive to health enthusiasts. Furthermore, it does not break down under heat and is favored by chefs, as well as homemakers. The first commercial acreage of laurate canola was planted in 1995 under contract to Calgene.

Also anticipated is a **biotech potato** that resists color changes when peeled. In this vegetable, DNA technologists have eliminated the enzyme that promotes brown color changes by removing the responsible gene. It will be joined by a variety of **yellow squash** called Freedom 2. The vegetable looks like a normal squash, but it has been genetically engineered to resist two common plant viruses (Figure 10.13). Already on the market is a **mini-pepper** about the size and shape of a jalapeno pepper but with a mild, sweet flavor. (Strictly speaking, this pepper is not a product of DNA technology—it was produced by cell cloning from a mutant specimen.)

Although not strictly a form of agriculture, transgenic plants are being used to crank out increased amounts of vitamins, other organic substances, and minerals. One product of a transgenic plant is **vitamin E**. Scientists at the University of Nevada are coaxing *Arabidopsis thaliana* plants to increase their production of **alpha-tocopherol**, a useful form of the vitamin. The plant normally produces a compound close to alpha-tocopherol, but it cannot produce the enzyme that makes the final conversion. DNA technologists have located the gene that encodes the enzyme in a bacterium, then found its equivalent lying unexpressed in *Arabidopsis*, shown in Figure 10.14. They found a regulatory sequence that would activate the dormant gene and inserted it into the plant cells. The plant now produced a tenfold amount of vitamin E. With the theoretical basis for producing the vitamin in place, scientists could envision soy

thioesterase
thi″o-es′-ter-ase

jalapeno
ha-la-pen′yo

Arabidopsis
a″rab-i-dop′sis

thaliana
thal-i-an′ah

alpha-tocopherol
al′fa-to″co-fer′ol

FIGURE 10.13
Transgenic squash altered with genes to resist the deleterious effects of two viruses. The squash is appropriately called Freedom 2.

FIGURE 10.14

Arabidopsis thaliana, a member of the mustard family of plants and a commonly used specimen in plant-related DNA technology. *Arabidopsis* is sometimes called the "mouse of the botanical world."

plants, already known for their production of vitamin E, as major suppliers for health-conscious individuals.

Scientists have dubbed their emerging discipline **nutritional genomics**. Research studies are already underway to insert bacterial genes that encode carotenoid into rice plants and induce the latter to synthesize vitamin A. Other researchers are attempting to enhance the iron content of edible plants by enhancing their ability to absorb iron from the soil.

And consumers can expect to see other bioengineered foods in the near future: Researchers from Auburn University have spliced trout genes into **catfish** so the latter will reach maturity months earlier; genes from a chicken are being spliced to **potatoes** to make the potato more protein rich; and "ice-minus" genes are being incorporated to **ice cream** and cake products to prevent ice crystal formation during freezing. Figure 10.15 depicts an artist's whimsical approach to biotech agricultural products.

To help allay the fears of consumers, the FDA has not given carte blanche to bioengineered foods. Rather, its new policy allows new foods to be treated as conventional foods as long as they meet three conditions: their nutritional value has not been lowered; they incorporate new substances (e. g., proteins or carbohydrates) that are already a part of the human diet; and they contain no new allergenic substances. So constructed, the new foods are considered fit for human consumption.

■ VEGETABLE VACCINES

Before leaving the topic of bioengineered foods, we will consider transgenic plants as carriers of **vaccines**. Plants are desirable carriers because they are inexpensive to store and readily accepted in developing countries, where vaccines are needed most.

Researchers at New York's Boyce-Thompson Institute were among the first to splice microbial genes into plant cells. In 1997, a research group reported successful incorporation of genes from a strain of *Escherichia coli* into a **potato** plant. When chunks of the "veggie vaccine" were eaten by volunteers during trials at the University of Maryland, the gene-encoded protein provoked an antibody response against the protein. Unfortunately, the potatoes were uncooked, and they turned up the volunteers' noses as well.

To resolve this dilemma, the researchers tried cooked potatoes, but they observed a significant reduction in the amount of bacterial protein. Colleagues at Loma Linda University attempted to amplify the amount of bacterial protein produced by the

nutritional genomics

the study of genes with the purpose of using the information to enhance the study of nutrition.

vaccine

a substance that will stimulate an immune response when introduced to the body.

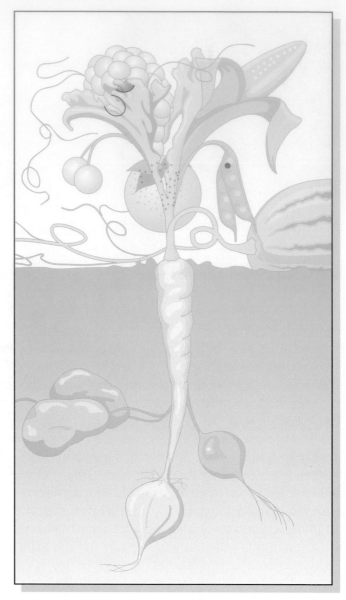

FIGURE 10.15

An artist's fanciful conception of a superplant of the future produced through the processes of DNA technology.

potato and administer "booster" feedings, both with limited success.

Future research efforts will use more palatable plants such as tomatoes and bananas. **Tomatoes** are appealing vaccine carriers because of their short growing season and the wealth of experience DNA technologists have with transgenic tomatoes. **Bananas** have attracted interest because children like them (children are the main recipients of vaccines); the fruit grows in many tropical countries (where vaccines are essential to public health systems); and bananas do not have to be cooked before consumption. Indeed, a research group has recently reported successful incorporation of the *E. coli* genes to bananas (Figure 10.16).

Using transgenic vegetables has its unique **problems**: Investigators must address the problem of how to get the antibody-stimulating antigens through the acid and enzymes of the stomach and intestine, respectively. Because these antigens are proteins, they may combine with food proteins to produce allergic substances. And there is the problem of getting the antigens from the gastrointestinal tract to the blood where the immune reaction occurs. Despite these obstacles, DNA scientists see great hope for the new wave of transgenic vaccines. Their hope is accompanied by many years

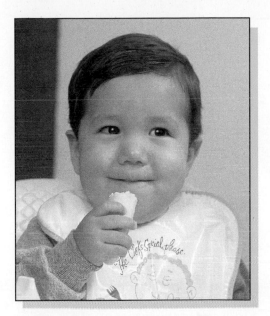

FIGURE 10.16

Researchers have genetically altered a banana plant so its fruit produces the antigens associated with the bacterium *E. coli*. In the human body, the antigens stimulate an immune response, and the banana can be used as a vaccine carrier.

of laboratory tinkering and many clinical trials before the banana replaces the needle as a vehicle for vaccine delivery.

The final achievement we will mention occurred in 1998. Researchers from Agracetus Inc. proudly announced that a field in Wisconsin was home to a corn crop where **human antibodies** were being produced. The scientists had altered corn seeds with genes for anti-cancer antibodies, and the corn plants were now synthesizing the so called **plantibodies**. Company scientists hoped that radioactive killer substances could be tagged to the antibodies for delivery to cancer cells. They emphasized the inexpensive and allergy-free characteristics of their astounding achievement. But, they cautioned, clinical trials have yet to begin.

plantibodies

plants engineered with genes to encode human antibodies

"PHARM" ANIMALS

Although most of the excitement in agriculture has been generated by work with plants, much innovative experimentation has been performed with animals. Since the early 1980s, animals have been treated with a variety of genetically engineered pharmaceuticals, such as hormones to increase their efficiency as producers of valuable food and other products. Over the years such animals have acquired the catchy name of **"pharm animals."**

The dairy industry may be among the first to feel the benefits of DNA technology. As early as 1983, biochemists identified the gene for **bovine growth hormone (BGH)**, cloned it, inserted it into bacterial cells, and modified the cells to produce the hormone as Figure 10.17 indicates. In the cow, BGH promotes general growth of the bone and muscle by stimulating protein synthesis and regulating other growth-related activities. When genetically produced BGH was injected into dairy cows, researchers reported a ten to twenty-five percent increase in milk production compared with control cows.

Another step forward occurred in 1990. That year, the FDA declared that milk produced by cows injected with BGH is as safe to drink as any other milk. In effect, the FDA concluded that BGH is biologically inactive in humans and poses no threat. Two years later, however, a government report urged caution: Experimental cows are more prone to infectious disease. To treat and prevent disease, the experimental cows must be treated with large doses of antibiotics, a practice that leads to antibiotic

"pharm" animal

an animal treated with genetically engineered drugs to produce agriculturally desirable products.

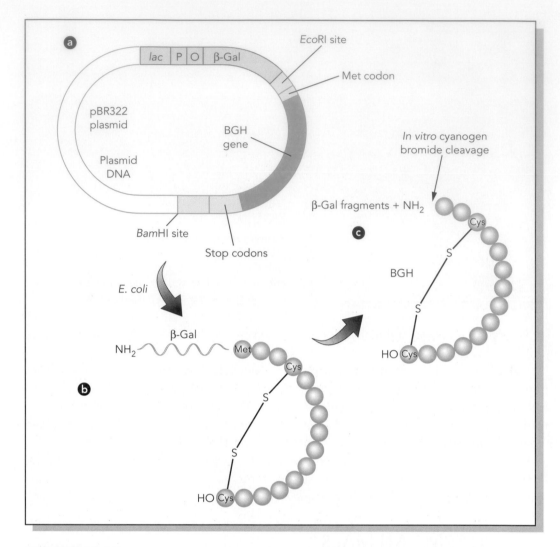

FIGURE 10.17

Some of the underlying biochemistry involved in bovine growth hormone (BGH) production.
(a) The BGH gene is attached to a plasmid containing genes of the *lac* operon. The plasmid is named pBR322. It contains promoter (P) and operator (O) genes for ß-galactosidase production. Stop codons and a special methionine codon (Met) are included, and attachment at two sites is accomplished by restriction enzymes *Eco*RI and *Bam*HI, respectively. **(b)** Incubated in *E. coli* cells, the plasmid encodes BGH linked to the enzyme ß-galactosidase. This molecule is cleaved by cyanogen bromide at the methionine molecule after test tube isolation of the gene product. **(c)** The cleavage results in active BGH, a small protein composed of fourteen amino acids.

appearance in the milk. Nevertheless, in 1994, the FDA approved the use of genetically engineered BGH in cows.

Also in the future is a pig with more meat and less fat. Experiments with genetically engineered **porcine growth hormone (PGH)** have resulted in a notable reduction in untrimable fat within the pig's muscle. In other experiments, egg cells from female pigs are being injected with genes that encode PGH. The egg cells are then united with sperm cells in vitro and implanted in female pigs until birth. Pigs, genetically altered in this way, have been shown to produce a fifth to a tenth of the fat normally found in a pig. Unfortunately, the pigs are weak, lethargic, and plagued by arthritis and gastric ulcers. And they are sterile.

Intriguing results have been attained by researchers seeking to improve the ease of **wool collection** from sheep. DNA technologists have isolated the gene for

porcine growth hormone

a hormone that encourages muscle growth in pigs.

sheep epidermal growth factor (EGF) and have inserted it into bacteria to produce large quantities of the synthetic hormone. Injected into a sheep when the fleece is ready to harvest, the hormone weakens the hair follicles, causing them to loosen their grip on the wool. About a week after the injection, the fleece comes off the sheep in a single, unmutilated sheet and gathers in a net fitted about the animal. To this writing, there have been no reports of deleterious effects of the hormone, and in 1998, an Australian company was ready to launch the hormone commercially as Bioclip.

Researchers conducting experiments with the new agriculture have alternately felt emotions running from euphoria to despondency and back to euphoria as failures follow successes follow failures, and so forth. They have learned that nothing comes easy in science. But it is important to continue to dream—to dream of radishes as big as potatoes, of skim milk directly from cows, of carrots that taste like apples, of plants that mimic beefsteaks, and of countless other possibilities. Many years ago Luther Burbank reformed agriculture with his innovative experiments in plant genetics; DNA technologists do the same today by manipulating what Burbank could not manipulate: the genes that make plants work.

SUMMARY

Some of the more innovative applications of DNA technology have come in plant biotechnology and have forecast a new agricultural revolution. Although plants are more difficult to work with than bacteria, gene insertions can be made to single plant cells, and the cells can be cultivated to a mature plant. Microscopic syringes, genetic projectiles, and silicon carbide fibers represent the physical methods for gene insertions, while the gene-altered bacterium *A. tumefaciens* remains the major biochemical insertion method.

DNA technologists have sought to increase the efficiency of plant growth by increasing the plant's capacity for nitrogen fixation. Although nitrogen fixation is normally a bacterial function in the nitrogen cycle, the process can conceivably be transferred to plants (and even animals) by transferring the genes from the nitrogen-fixing bacterium *Rhizobium*. Other possibilities are to encourage *Rhizobium* to assume symbiotic relationships with a larger variety of plants and to transfer the nitrogen-fixing genes to bacteria that infect plants. The implications of these experiments are far-reaching.

The fruits of DNA technology research have been seen in experiments involving plant resistance. For example, the so-called ice-nucleating bacteria have been replaced in plants with ice-minus bacteria, a strain of bacteria that permits plants to better withstand cold temperatures. DNA technologists have also learned to control cold-tolerant genes in wheat plants.

To resist biological pests, genes from the bacterial insecticide *B. thuringiensis* have been isolated and inserted into plants and into bacteria that live with plants. The toxin is effective against caterpillars of insect pests because it dissolves the cementing substances in the gut. And to protect against viral disease, plants have been modified to produce proteins for viral capsids. Nematode resistance genes have been found as well, and work has been done to protect trees against defoliating insects.

Genetically altered foods typify the practical uses of DNA technology. One of the first foods to be marketed was to be a rot-resistant tomato. Altered by the addition of a gene to produce an antisense mRNA molecule, the tomato does not produce the enzyme that encourages rotting. It was the forerunner of numerous agricultural products of DNA technology such as canola oil rich in lauric acid, bruise-resistant potatoes, and vaccine-containing vegetables.

sheep epidermal growth factor

a protein that loosens the fleece of a sheep and hastens the harvesting of wool.

REVIEW

These pages have explored some of the imaginative uses of DNA technology in agriculture. To test your understanding of the chapter contents, answer the following review questions.

1. What are some advantages of DNA technology with plants compared with that with animals? What are some disadvantages compared with animals?

2. List three "nonbacterial" methods for inserting genes to plant cells and explain how each works.

3. Describe the value of *Agrobacterium tumefaciens* in plant DNA technology and show how marker genes are used to confirm gene insertion.

4. Discuss the role of bacteria in a typical nitrogen cycle occurring in the soil.

5. Indicate some methods by which DNA technology can enhance the incorporation of nitrogen into the life cycles of living things.

6. What are "ice-minus" bacteria and how can they be used to help plants withstand the damaging effects of cold temperatures?

7. How does *Bacillus thuringiensis* function as an insecticide, and how can its genes be used to enhance plant resistance?

8. Describe the imaginative genetic methods that have been used to protect plants against the damages caused by viral disease.

9. Why is it important for a plant to be resistant to an herbicide, and how can DNA technology increase that resistance?

10. Discuss how rotting occurs in tomato plants, and show how rotting can be prevented in marketable products. Explain how vegetables can be used as carriers for vaccines.

FOR ADDITIONAL READING

Adler, T. "Mothra Meets His Match." *Science News* 146 (1994): 154–157.

Greenberg, J. "The Great Gene Escape." *Discover* (May, 1998).

Karcher, S. J. "Getting DNA Into a Cell: A Survey of Transformation Methods." *American Biology Teacher* 56 (1994): 14–20.

Kling. J. "Could Transgenic Supercrops One Day Breed Super Weeds?" *Science* 274 (1996): 180–182.

Mestel, R. "Altered Vegetable States." *Discover* (January, 1993).

Milot, C. "Plant Biology in the Genome Era." *Science* 281 (1998): 331.

Moffat, A. S. "Moving Forest Trees Into the Modern Genetics Era." *Science* 271 (1996): 760–763.

Raloff, J. "Taters for Tots Provide an Edible Vaccine." *Science News* 153 (1998): 149.

Ronald, P. "Making Rice Disease-Resistant." *Scientific American* (November, 1997).

Transgenic Animals

 CHAPTER OUTLINE

 LOOKING AHEAD

Transgenic animals are another startling application of DNA technology, as the pages ahead will illustrate. On completing your study of this chapter you should be able to:

- understand the nature of a transgenic animal and appreciate some of the technology in its development.

- summarize the various transgenic animals that have been bred to date and indicate how each is genetically different from its unaltered counterpart.

- explain some of the problems in adapting the methods used for laboratory animals to research in larger, barnyard animals.

- describe a number of pharmaceutical products derived from transgenic animals and show how each is useful in medicine.

- conceptualize how an animal can be made more resistant to infectious disease through transgenic DNA technology.

- recognize methods by which transgenic animals can help interrupt the spread of disease in nature.

- identify some of the major researchers who spurred the development of transgenic animals.

INTRODUCTION

The well-worn cliche says: "Build a better mousetrap and the world will beat a path to your door." Now, in the age of DNA technology, the newer saying is "Build a better mouse!" And that is precisely what scientists have done: a mouse that develops pathological conditions not normal to its species, a mouse with a human immune system, and a mouse whose organs can be "knocked out" on command.

In the barnyard outside the laboratory, there is an equally stunning group of large animals once existing only in the imaginations of DNA technologists: a pig that produces human hemoglobin; a cow that synthesizes human iron-binding proteins; goats that manufacture medications to relieve cystic fibrosis; and animals that resist diseases once thought untreatable.

These are the **transgenic animals**, animals in which one or more extra genes have been introduced to the cells. Although technically complex, the scheme for producing a transgenic animal is relatively straightforward: The foreign genes are injected into the nuclei of fertilized eggs, and the latter are cultivated and screened for a successful implantation of DNA. (Amazingly, the chromosomes take up the genes as if they were broken bits of DNA that need to be repaired.) Then the embryos are transferred to surrogate mothers for development and birth. In those rare instances when all the pieces fit into the jigsaw puzzle, a transgenic animal results (Questionline 11.1).

To be sure, Old MacDonald never envisioned moos, bahs, and oinks from the animals now inhabiting the farms of biotechnology companies. He could not foresee the exhilarating promises of transgenic animals that make human proteins. The health and well-being of the animals have not been disturbed, and the extra genes behave like all other genes present in the animal's genome, including passage to the animal's offspring. The eventual hope is that animal-derived proteins can be produced at levels presently unattainable and at reasonable prices.

cystic fibrosis

an inherited disease in which sticky mucus accumulates in the airways and impedes the flow of air.

transgenic animal

animals whose somatic cells have one or more foreign genes.

surrogate

alternative.

QUESTIONLINE 11.1

1.   **Q.** Exactly what is a transgenic animal?

A. A transgenic animal is an animal that has been genetically engineered to contain one or more extra genes in its cells. It now has a new, inheritable property of some type.

2.   **Q.** How are these genes placed into the cells of the transgenic animals?

A. A straightforward process called microinjection is used to install new genes into an animal's cells. DNA fragments such as plasmids are loaded into an ultrathin syringe then injected into the nucleus of a sperm or egg cell or into a fertilized egg cell nucleus. The cell is then implanted in the reproductive system of an animal and allowed to develop to a full organism.

3.   **Q.** How can technologists be sure that the desired genes were installed into the cells?

A. Some sort of marker must be used to ensure that the genes were installed properly. One marker is the enzyme thymidine kinase. Recombined cells are made to depend on this enzyme and grow in certain specified medium when they have incorporated the genes for the enzyme plus the desired genes. Other cells, the ones that did not take up the desired genes, die.

Since the 1970s, DNA technologists have been working their way through the kingdoms of living things, beginning with the bacteria (Chapter 6), continuing through the fungi and protozoa, and on to the plants (Chapter 10). They have breached the barriers of humans with gene therapy (Chapter 8) and are well on their way with transgenic animals. They have already built a better mouse, and the world is beating the proverbial path to their door.

PLACING DNA IN MAMMALIAN CELLS

To produce a transgenic animal, novel methods must be developed to introduce DNA to mammalian cells (compared with those for introducing genes to bacterial cells). Until the mid-1970s, the only way to make this introduction was by the technique of **somatic cell hybridization**. In this technique, two unrelated cells (one containing experimental DNA) fuse together, and the hybrid cell is encouraged to grow and multiply. Unfortunately, when the nuclei of the cells fuse, the cell often eliminates some of the chromosomal material and should this material contain the experimental DNA, the incorporation process is short circuited.

During the mid-1970s, an alternate method using calcium phosphate was developed by researchers from University of Leiden in The Netherlands. The researchers mixed foreign DNA fragments with calcium phosphate and formed **DNA-calcium precipitate**. When mixed with cultivated mammalian cells, the latter took up the precipitate efficiently, possibly because of endocytosis performed by the cell. Endocytosis is the process in which the cell forms a membrane around food particles in its environment and pinches off the new vesicle into its cytoplasm. Unfortunately there is no way of predicting which cells take up the DNA-calcium particles, and any transformed mammalian cells are usually lost within the morass of nontransformed cells.

The **microinjection** process now widely used is much more simple and direct, as Figure 11.1 displays. A thin glass pipette is placed in a flame and its tip becomes liquefied. Quickly, the technologist draws out the tip to form an ultrathin tip about one-half to a few micrometers in diameter (roughly a hundredth the diameter of a human hair). The micropipette that results is a syringe with an ultrathin tip. It is

endocytosis

the biochemical process in which a cell surrounds a particle or volume of fluid and takes it into its cytoplasm.

microinjection

injections made into microscopic objects such as cells.

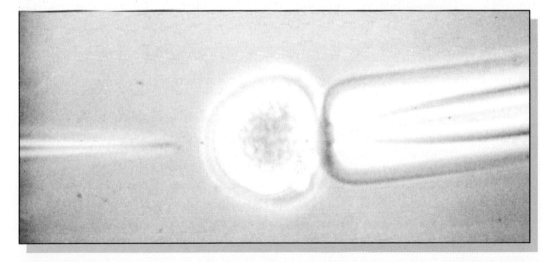

FIGURE 11.1

The microinjection process. The cell in the center of the photograph is held in place by the object to the right. A microsyringe containing the DNA for insertion is about to penetrate the cell from the left. Integration of DNA to the genome of the mammalian cell can occur at a high or low rate, depending on the conditions used.

placed in a microinjection apparatus and filled with DNA fragments or with vectors containing genes. The experimenter controls the micropipette while watching through the oculars of a microscope and restraining the target cell with a second micropipette. (A suction device in the second micropipette keeps the cell in place while the injection proceeds.)

Microinjection procedures are time-consuming, costly, and labor intensive. However, they are **advantageous** because less DNA is lost compared with calcium-based uptake methods and the method does not require marker genes to show which cells have taken up the DNA. Moreover, the rate of DNA integration into the mammalian cell nucleus can be relatively high, depending on the conditions used. As we will see in the pages ahead, this method of DNA integration has become a favored tool for developing transgenic animals.

For methods that require a marker, a popular one in mammalian cell experiments is the gene that encodes the enzyme **thymidine kinase (TK)**. The TK enzyme functions in mammalian cells by extracting thymidine molecules from previously used DNA and reincorporating them to new DNA molecules. Normally, mammalian cells do not need TK enzyme because they have an alternate method for accomplishing the same goal. By use of a biochemical inhibitor, DNA technologists can kill the cells that use the TK enzyme and focus the thymidine metabolism on the "alternative method."

The cells that survive the selection process (the "alternative" cells) obtain their thymidine molecules by synthesizing them, not by salvaging them. To synthesize thymidine, the cells must obtain raw products from the local environment. However, if hypoxanthine and aminopterin are present in the environment, the two chemicals will inhibit thymidine synthesis, and the cells will not grow. Moreover, if thymidine is present in the environment, one would expect the cells to take it up and use it. This does not happen, however, because the cells lack the TK enzyme necessary to salvage thymidine and incorporate it. Thus, the cells do not grow in a laboratory medium containing hypoxanthine (H), aminopterin (A), and thymidine (T), a so-called **HAT medium**. Figure 11.2 illustrates this biochemistry.

Now the DNA technologist inserts a gene for TK enzyme into the DNA fragment for insertion into the mammalian cell. If both DNA fragment and TK gene are successfully inserted into the cell, the TK gene encodes TK. The cell can use this enzyme in HAT medium to obtain thymidine from the medium and incorporate it into new DNA for its growth. Thus, the growing cells are the ones with the TK genes and, by inference, the other genes of the DNA fragment. By contrast, the dead cells could not take up the TK gene, nor could they produce thymidine kinase. They could not survive in the HAT medium.

In 1977, **Michael Wigler** and his colleagues at Columbia University successfully used the TK marker to demonstrate the incorporation of viral genes into mammalian cells. With calcium phosphate as a carrier, they took the genes from herpes simplex viruses, bound them to TK genes, and exposed the DNA-calcium precipitate to cells derived from a mouse. The mouse cells were transformed. The experiment was among the first to demonstrate the efficacy of the **TK marker gene** in mammalian cells.

Other gene markers were soon developed, among them the ***Alu* marker**. This is a repetitive sequence of base pairs, so named because the restriction enzyme *Alu* 1 operates at this marker. About 300,000 copies of the *Alu* sequence exist in the human genome; the number of copies in the cells of other mammals varies. By using radioactive gene probes to pinpoint the number of repetitive sequences present, DNA technologists can determine which type of DNA is present. For example, the presence of human DNA in a mammalian cell genome can be determined by locating the human

hypoxanthine

hi-po-xan'then

aminopterin

am"in-op'ter-in

thymidine kinase

an enzyme that cells use to obtain thymidine for use in DNA synthesis.

hypoxanthine

hi-po-xan'then

aminopterin

am"in-op'ter-in

TK gene

a marker gene inserted to DNA for incorporation to a mammalian cell; if the DNA is inserted, the gene will encode thymidine kinase.

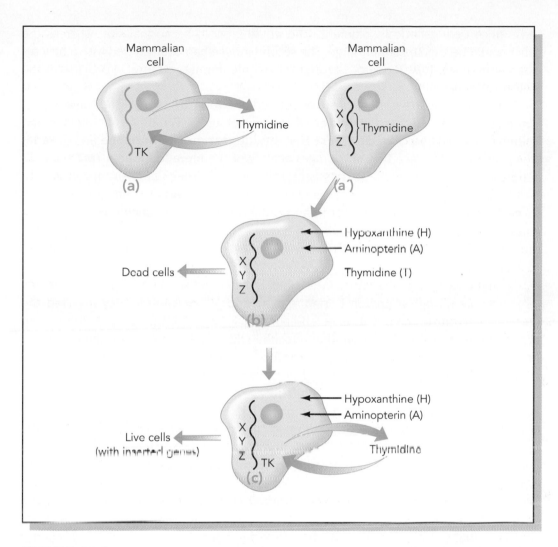

FIGURE 11.2

Use of the thymidine kinase (TK) marker to determine whether gene insertion has taken place. (a) Certain mammalian cells use the enzyme TK to obtain thymidine from the growth environment and incorporate it to DNA for new cells. These cells are killed by inhibiting the activity of TK. (a´) The surviving mammalian cells obtain their thymidine by using certain genes (X, Y, and Z) to encode thymidine in the cytoplasm. These cells are used for gene insertion. (b) The cells are placed in a medium containing hypoxanthine (H) and aminopterin (A), which inhibit thymidine synthesis by interfering with the activity of the X, Y, and Z genes. Without thymidine, the cells die even though thymidine is contained in the growth environment. (c) The gene for TK is bound to the DNA fragment to be inserted to the cell. If the insertion is successful, the TK gene will function and TK will be produced. Despite the presence of H and A in the growth environment, thymidine will be taken up and incorporated into DNA for new cell production. The cells will thrive, and the DNA technologist can assume that the desired genes have been inserted.

Alu marker and determining how many copies exist in the experimental cells.

The DNA incorporation methods and the use of selective markers were essential to the DNA technology experiments that developed transgenic animals. The remaining pages of this chapter will focus on the present and future of this innovative technology.

▨ CUSTOM-MADE ANIMALS

In 1983, two seemingly normal littermates were born to a female mouse. Both mice were gray and had black eyes, and to all appearances they were interchangeable. But these mice were genetically different. One would grow to be normal size, whereas the other would become extremely large. The second mouse had been altered with a new set of growth genes. It was one of the first transgenic animals. Figure 11.3 displays the historic littermates.

The pioneering experiments were performed by **Ralph Brinster** of the University of Pennsylvania and **Richard Palmiter** of the University of Washington. Brinster and Palmiter isolated and cloned the gene that encodes **rat growth hormone** and linked it to a promoter sequence that encourages activity of the **metallothionein (MT)** gene (a gene that encodes a metal-binding protein). This promoter switches on the growth hormone gene when the MT promoter is activated by a metal in the environment such as **cadmium**. Thus, to trigger hormone production within a specific tissue, all the researcher need do is add cadmium. An inducer (the metal) is used to stimulate the promoter (the MT promoter), which encourages the foreign gene (the hormone gene) to express itself.

With the techniques available at that time, Brinster and Palmiter prepared a chimeric plasmid containing rat growth hormone gene and MT promoter. They inserted the plasmids into one of the pronuclei of fertilized mouse eggs. **Pronuclei** are the sperm and egg nuclei that fuse with one another after the sperm cell and egg cell unite. (Twelve hours after mating, fertilized egg cells can be collected from the oviduct of a female mouse before the cells have undergone division and before the nuclei have fused.)

After insertion of plasmids, the fertilized eggs were surgically implanted into the Fallopian tubes of a surrogate female mouse. To simulate a "natural" pregnancy, the female was mated with a vasectomized male mouse. After birth, scientists studied the littermates for the presence of rat growth hormone gene by testing various tissues for gene expression. Where genes were incorporated, the addition of cadmium to the environment induced the MT promoter and growth hormone was produced. Indeed, certain littermates grew to twice their normal size. (They attained the size of the normally larger rat.)

The initial success, however, was tempered when it became clear that bigger is not necessarily better. The transgenic mice encountered a number of physiological problems such as reproductive failure and organ diseases. Also, the mice did not

metallothionein

met″tal-othi′o-nin

chimeric plasmid

a plasmid that has been biochemically modified to contain one or more foreign genes.

oviduct

the tube leading from the ovary to the uterus.

vasectomized

having had the vas deferens cut so that sperm cannot be delivered during mating.

FIGURE 11.3

The first transgenic animal. The two mice in the photograph are littermates. The mouse on the right is normal, whereas the mouse on the left has the gene for rat growth hormone in its cells. The size difference is apparent.

convert food to muscle any more efficiently than normal mice, even though they were of larger size. Thus, the initial expectation of leaner cattle and meatier pigs still remained elusive.

■ THE HUMAN MOUSE

The experiments of Brinster and Palmiter encouraged others to pursue their experiments with transgenic animals. One result was the so-called human mouse, a mouse with a **human immune system**. This animal was not microinjected with genes, so strictly speaking, it is not a product of DNA technology. Instead it was exposed as an embryo to human cells that grew together with the mouse tissues and became part of the mouse's system.

The "human mouse" experiment was performed in 1988 by researchers at Stanford University. Led by **Joseph M. McCune**, the group selected mice having severe combined immunodeficiency (SCID), a condition in which no immune system cells develop in the animal. (The condition is somewhat analogous to the situation arising from ADA deficiency discussed in Chapter 8). Researchers took thymus, lymph node, and liver tissues from an aborted human fetus and placed them under the capsulelike membrane surrounding the mouse's kidney. A week later they transplanted immature immune system cells from a human fetus into the mice hoping that the cells would mature in the mouse to typical cells of the immune system. Two weeks after the transplant, the mice displayed an **immune response** characteristic of both T-lymphocytes and B-lymphocytes. The mice had acquired an immune system, a human immune system. (Figure 11.4 summarizes how the experiment was performed.)

Having a mouse with a human immune system is a valuable asset to many forms of research, particularly to AIDS research. AIDS researchers are seriously hampered by the lack of good animal models for laboratory tests. For example, no animal except the chimpanzee contracts AIDS. The use of an animal with human cells will permit researchers to determine whether an experimental vaccine is eliciting a response. Moreover, pharmaceutical companies will be able to observe the effects of new drugs on the ability of human immunodeficiency virus (HIV) to infect immune system cells.

■ THE ONCOMOUSE

Another animal to make news in 1988 was the so-called **oncomouse**. This transgenic mouse was the creation of Harvard University's **Philip Leder**. Leder produced a mouse that is highly prone to breast cancer. Such a mouse permits researchers to test carcinogens and cancer therapies more accurately and efficiently than with normal mice. In addition, the oncomouse will increase understanding of how oncogenes contribute to cancer.

Leder's group worked with an **oncogene** called c-myc. Early in their studies the researchers noted that the gene is involved in certain types of lymphomas occurring in children. However, in the presence of the **mouse mammary tumor (MMT) virus**, the c-myc gene induces breast cancer in animals. With this knowledge, the investigators synthesized chimeric plasmids consisting of the c-myc gene and sections of MMT virus. They injected the chimeras to the pronuclei of fertilized mouse egg cells and transferred the cells to surrogate females. The mice that developed did not display breast cancer as adolescents, but at maturity, the MMT genes were activated, thereby disrupting the normal activity of c-myc genes. Breast cancer followed spontaneously. And the trait was passed to the offspring.

On April 12, 1988, history was made when the U. S. Patent Office granted to

SCID

severe combined immunodeficiency; a condition in which immune system cells fail to develop in an animal or human.

thymus

a organism in the human neck area; site of immune cell maturation.

oncogenes

genes associated with tumors and cancers.

MMT virus

a virus that appears to cause tumors in the mammary glands of mice.

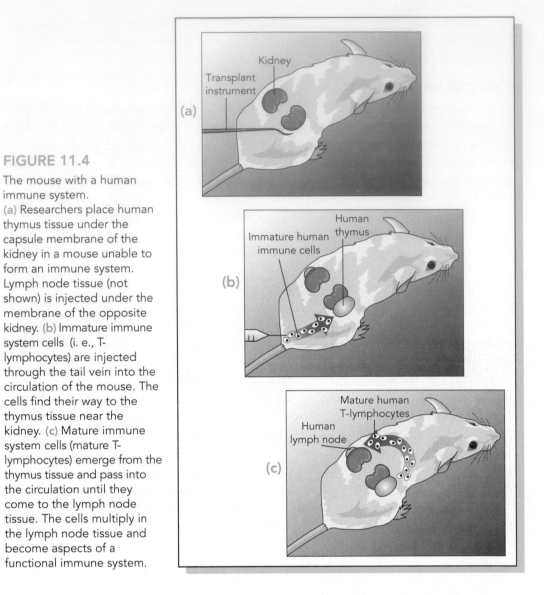

FIGURE 11.4

The mouse with a human immune system.
(a) Researchers place human thymus tissue under the capsule membrane of the kidney in a mouse unable to form an immune system. Lymph node tissue (not shown) is injected under the membrane of the opposite kidney. (b) Immature immune system cells (i. e., T-lymphocytes) are injected through the tail vein into the circulation of the mouse. The cells find their way to the thymus tissue near the kidney. (c) Mature immune system cells (mature T-lymphocytes) emerge from the thymus tissue and pass into the circulation until they come to the lymph node tissue. The cells multiply in the lymph node tissue and become aspects of a functional immune system.

Harvard a **patent** for the cancer-prone transgenic mice shown in Figure 11.5. The mice and their offspring officially became "Number 4,736,866." As such, they were the first complex animals patented. (The first ever patent for a transgenic organism was granted for a bacterium in 1980). The patent application claimed rights to any gene-altered nonhuman mammal, "preferably a rodent such as a mouse," whose cells have been engineered to contain an activated oncogene sequence.

The second patent for a complex animal was also awarded to Leder and his Harvard group. In 1991, the researchers developed a mouse that develops cancer of the **prostate gland**. The prostate gland is a donut-shaped gland that surrounds the urethra of males and supplies sperm-activating fluids to the semen. In mature men, the gland often becomes cancerous, and surgery is generally required to remove the gland. (In mice, the condition does not normally occur unless induced by an environmental agent.)

Leder's research team developed their newly patented mouse in much the same way as the oncomouse. They spliced the oncogene *int*-2 to a viral promoter and inserted the chimeras to fertilized eggs. The mice that resulted, the so-called **prostate mice**, displayed enlarged prostate glands with predictable accuracy.

prostate gland

a donut-shaped gland that surrounds the urethra in males and adds secretions to the semen.

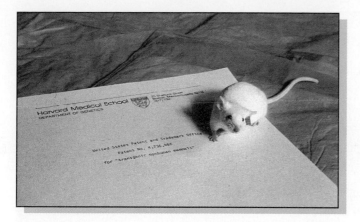

FIGURE 11.5

The first animal ever patented. This mouse, the so-called oncomouse, has been genetically modified so that its cells are prone to cancer. Harvard University received a patent for the transgenic animal in 1988.

■ THE ALZHEIMER'S MOUSE

Still another custom-made animal is the mouse with **Alzheimer's disease**. Alzheimer's disease is a brain disease marked by severe memory loss, decline in judgment and thinking, and mood swings (Chapter 7). Postmortem analyses of brains of Alzheimer's patients consistently reveal "plaques" of dead nerve cells enmeshed in fibers of a protein called **amyloid**. Several years ago, researchers purified amyloid and determined its amino acid sequence. Soon thereafter, a precursor protein was identified, and its gene sequence was subsequently determined.

Through all these experiments, the question of amyloid's involvement in Alzheimer's disease was left unresolved. Was it a cause of the disease or was it debris left by the disease? Experiments reported in 1991 shed light on the question. That year DNA technologists led by **Jon Gordon** of New York's Mount Sinai Hospital engineered a transgenic mouse having the signs and symptoms of Alzheimer's disease. Gordon's team introduced the amyloid precursor gene to the fertilized egg cells of mice and observed that adult mice were synthesizing human amyloid protein. As amyloid concentrations built up in the animal brain, there appeared the typical brain configurations of **amyloid plaques** seen in Alzheimer's patients. The model replicated the disease process accurately and offered a practical understanding of the disease.

■ THE KNOCKOUT MOUSE

One of the hottest commodities in contemporary molecular genetics is the so-called **knockout mouse**, a mouse in which the genes for a single organ or organ system have been "knocked out." The key to knockout mice traces to a discovery made in 1985 at the Fox Chase Cancer Center in Philadelphia (Questionline 11.2). That year, researchers discovered the **SCID mouse**, later used by McCune to produce the mouse with the human immune system. The SCID mouse has no immune system because a defect in a single gene eliminates its ability to produce the B-lymphocytes and T-lymphocytes of its immune system. In later years, **Frederick Alt** of Children's Hospital in Boston and **Susumu Tonegawa** of MIT pinpointed the gene defect and developed methods to convert the normal gene to the defective gene. This discovery permitted them to destroy on command the mouse's ability to produce an immune system on command. Moreover, the gene defect could be passed on to the next generation.

Knockout mice have now been developed so that a **single gene** can be eliminated. To develop a knockout mouse missing one gene, scientists begin with very early embryonic cells separated from the embryo. They remove the cellular DNA and then excise a functional gene using DNA-cutting restriction enzymes. Now they recombine the two outer bits of DNA using ligases and drive the DNA back into fresh cells using

Alzheimer

alz-hi′mer

amyloid

am′i-loid

a protein associated with fibrils that form in the brain tissue of patients with Alzheimer's disease.

knockout mouse

a mouse in which the genes for an entire organ or organ system have been eliminated.

embryo

the developmental stage in which the organs are not yet developed; precedes the fetal stage.

QUESTIONLINE 11.2

1. **Q.** In the last few years, I've heard references to strange and unusual kinds of mice. One is the "supermouse." What is that?

   A. The supermouse is a very large mouse whose cells have been altered to contain rat genes that encode growth hormones to spur overall growth. So changed, the mouse, ordinarily a small animal, grows to the larger size of the rat.

2. **Q.** How about the "human mouse?"

   A. The human mouse is yet another product of modern science. This is a mouse that contains human immune system cells. Strictly speaking, it is not a product of DNA technology, however. The mouse originally had no immune system cells of its own, and its body was tricked into treating human immune system cells as mouse cells. The human cells survived and grew into a human immune system, harbored within the mouse's body.

3. **Q.** Still another strange mouse that the papers have talked about is the "oncomouse." What's that?

   A. The oncomouse is a mouse that is highly prone to having cancer develop ("onco" refers to cancer). In this mouse, the cells have been transformed to contain a gene that induces cancer when a second gene is also present. Ordinarily, the development of cancer cannot be predicted easily, but with both genes in the cells, the mouse will develop a cancerous condition at a specified time in its life. Researchers can then conduct experiments under controlled conditions.

electric currents. At the nucleus of some cells, the altered DNA replaces ("**knocks out**") its analogous segment of DNA, and the altered DNA is passed on to daughter cells during cell division. (Markers such as the TK marker point up the genetically altered cells.) Next, the cells are added back to a developing mouse embryo, where they act as normal embryo cells, growing and dividing to become a mouse. Finally, by selectively breeding the partially altered mice, scientists can develop a line of mice in which the normal gene is almost completely eliminated.

In 1994, scientists at the University of North Carolina used a knockout mouse missing one gene to verify a basic tenet of science. Allergy researchers have long believed that certain body cells have on their surface a molecule used as a docking site (a receptor) for the IgE antibodies that trigger most **allergic reactions**. The DNA technologists developed a mouse whose gene for encoding the receptor was "knocked out." Unable to synthesize the receptor, the cells failed to bind the allergy antibodies, and the mouse did not suffer allergic reactions when challenged. (Ordinary mice with the receptor displayed the typical allergic reaction on challenge.) With the importance of the receptor verified, researchers are continuing work on a therapy to block the receptor and provide relief to the 40 to 50 million Americans who have allergies.

The knockout mouse has made possible a new type of **transplantation experiment** in which rejection is not a problem. Indeed, because the experimental animal has no immune system, it cannot reject any transplanted tissue. In addition, there seems to be no recovery of immune system activity. (This is important because in the original SCID mouse, some immune system activity appears to return on occasion, thus jeopardizing the ongoing experiment.)

Ralph Brinster and Richard Palmiter and their colleagues were back in the news in

allergic reaction

an immune system process in which antigens unite with antibodies on the surfaces of basophils and mast cells and release mediators that encourage smooth muscle contraction.

1992, this time using knockout mice without an immune system and with a gene to destroy the animal's liver. The researchers developed a **suicide gene**, one that would selectively kill a mouse's liver cells from within the cells themselves. After the elimination of the liver cells, a sample of human liver cells could be transplanted to replace the mouse's liver. Normally, the immune system would reject these cells, but the knockout mouse has no immune system. In effect an entire organ could be replaced in this brave new world of DNA technology.

An even more controversial idea is to raise farm animals for their **transplantable organs**. Surgeons have long dreamed of relieving the perpetual shortage of human donors by harvesting hearts, kidneys, livers, pancreases, and lungs from other animals. And the pig appears to be the favorite subject for such **"organ farms."** Their organs approximate human organs in size; their physiology is similar to human physiology; they carry few infectious diseases transmissible to humans; they have a politically correct image for slaughter; and their genetics have been studied in depth, as evidenced by the work with hemoglobin to be discussed presently.

Nor does it end there. By the end of the 1990s, knockout mice were almost as common in DNA technology as the test tube. Tonegawa's group, for example, had developed a technique for knocking out a gene in one group of cells in the brain (rather than the entire brain). They eliminated the ability of cells of the **hippocampus** to produce receptors for the neurotransmitter glutamic acid. In doing so, they eliminated the mouse's ability to remember.

And at Lawrence Berkeley National Laboratory, scientists developed sickle cell mice: knockout mice having mutated genes for encoding mutated human hemoglobin. These are the genes that encode the deformed hemoglobin in patients with **sickle cell anemia** (Chapter 3). Animal models could now be used for testing new drug and gene-therapy strategies compared with the test tube blood samples formerly used. It is safe to assume that knockout animals and the genetic alchemy of DNA technologists hold great promise for radical improvements in science.

ANIMAL BIOREACTORS

Although transgenic laboratory animals are experimentally interesting and have some medical benefits, the techniques of DNA biochemistry must be applied to farm animals if any large-scale commercial benefit is to be derived. Working with large domestic animals is more difficult, however. Large animals do not produce as many offspring as smaller laboratory animals, and implanting manipulated embryos to surrogate females is more difficult because of the animal's larger size. Also, the eggs of many domestic animals have opaque cytoplasm, and it is impossible to see the nuclei without resorting to special techniques. For example, centrifugation may be required to separate cytoplasmic contents from the nucleus.

Despite these problems, almost any protein the human body produces can also be made in other animals, as long as the animal's genes are programmed correctly. This potential gives the animal the ability to become the **living equivalent** of a laboratory **bioreactor**. In essence, the animal becomes a drug factory or, as whimsically stated in the press, a **"pharm"** animal (Figure 11.6) .

A generation ago, the possibility of obtaining unlimited amounts of scarce human proteins at low cost was little more than a dream. Researchers used building-sized stainless steel vats containing genetically engineered cells in their attempts to produce the proteins. The proteins were extremely expensive and subject to the vagaries of culture sensitivities. Transgenic livestock now provide low-cost alternatives as

suicide gene

a gene that kills cells by encoding protein toxins to poison the cell.

hippocampus

an area at the midsection of the brain that is believed to function in memory processes.

sickle cell anemia

an inherited disease in which red blood cells assume a distorted appearance because of the collapse of the hemoglobin within them.

bioreactor

a living animal whose cells are engineered to produce desired proteins.

bioreactors whose numbers can be increased by breeding; that require only routine attention to ensure healthful living conditions; and that produce many more times the desired protein than the metal vat. Scientists envision tidy stalls occupied by healthy livestock modified to carry several human genes.

The process of **creating a transgenic animal** begins by splicing together two genes, each cloned separately. One gene encodes the desired protein, whereas the other comes from the organ or gland where the protein will be produced. For example, if the protein is to be isolated from the milk, the mammary gland is the production site, and genes specific for this organ are included. In addition, the second gene controls other biochemical "bookkeeping" for the protein production.

hybrid DNA

a DNA molecule containing foreign DNA as well as its own.

Next comes **microinjection** of the hybrid DNA into fertilized egg cells or embryos. In typically five to ten percent of cases, the DNA inserts randomly into an embryo's genome. All the eggs are then transferred to females. Once the young animals are born, they must be tested for presence of the hybrid gene. In the case of milk-borne proteins, the successful incorporation of genes takes longer to test because females must be bred before they begin producing milk. And if the offspring animal happens to be a male, then the process takes even longer. However, once a high expressing **"founder animal"** is identified, it can be used to build a herd without any further microinjections. Its offspring will serve as bioreactors

■ HUMAN HEMOGLOBIN FROM PIGS

hemoglobin

the red-pigmented protein of red blood cells that transports oxygen from the lungs to the tissues.

One example of a living bioreactor is the transgenic pig that produces **human hemoglobin**, the essential oxygen-carrying component of red blood cells. (Figure 11.6 pictures this animal.) The animal is a milestone in the effort to find a blood substitute useful for all types of transfusions and capable of storage for months (compared with weeks for donated blood). Because of its relatively short generation time (twelve months) and large litters (ten to twelve piglets), the pig has become an **"agricultural mouse."**

In 1991, researchers at a New Jersey biotechnology company announced the development of three **transgenic pigs**. Day-old embryos received injections of two genes for hemoglobin production, and the embryos were then implanted to surrogate mothers. Only about five in a thousand injections succeeded, and only three pigs survived to birth. Researchers found that about fifteen percent of the pigs' red blood cells contained human hemoglobin (the other eighty-five percent had normal porcine hemoglobin.)

porcine

relating to the pig.

To isolate pure human hemoglobin, researchers draw blood from the pigs and put it through a purification system that distinguishes human and porcine hemoglobins by

FIGURE 11.6

The pig in this photograph is a transgenic animal that is able to produce human hemoglobin. Researchers from a New Jersey biotechnology company called DNX Inc. produced a litter of the pigs in 1991.

their different molecular charges. The human hemoglobin is then extracted by adsorbing it to a special compound material and separating the hemoglobin-compound combination from the mixture. The hemoglobin is released from the compound and purified. The genetically engineered hemoglobin binds to and transports oxygen molecules similar to authentic human hemoglobin.

When free in the bloodstream, porcine-derived hemoglobin breaks down in a period of several hours to a few days, depending on how it is kept. However, the hemoglobin can be stabilized by cross-linking its proteins. Naked hemoglobin could probably not be used when there is massive blood loss, but when five to six pints of blood are needed (as during many surgical procedures), the naked hemoglobin could be used to help patients through the crisis period.

A key **advantage** of animal-derived hemoglobin is its freedom from human pathogens (Questionline 11.3). The problem of blood contamination was pointed up dramatically in the 1980s by the presence of HIV in blood. Before the advent of HIV testing in 1985, many transfusion recipients contracted the virus. And as late as the early 1990s, it was possible to contract various types of hepatitis from transfused blood. Moreover, using naked hemoglobin would preclude the need for determining blood types of donor and recipient because the surface antigens of erythrocytes are not there. A potential **disadvantage** is the contamination by animal viruses or cellular debris that could induce allergic reactions.

hepatitis

a viral disease that affects the liver and may be transmitted by infected blood.

■ OTHER TRANSGENIC ANIMAL PRODUCTS

The successful development of the transgenic pig has raised the expectations for what gene insertions into animals can lead to. Another animal being studied is a cow that produces **human lactoferrin** in its milk. Lactoferrin is a major protein that functions in iron uptake in nursing infants. The protein also has strong antibacterial properties because it removes iron from the environment that bacteria need for growth. Because lactoferrin increases iron binding in the body, it could also be used to relieve iron-deficiency anemia.

lactoferrin

lac-to-fer´in

a protein found in animal milk that has the ability to bind iron molecules.

QUESTIONLINE 11.3

1. **Q.** Do transgenic animals have any direct benefit to humans?

   A. Yes, they do. For example, DNA technologists have been able to produce a transgenic pig that produces human hemoglobin. Purified hemoglobin from this animal can one day be used in place of human blood for transfusions during many surgical procedures.

2. **Q.** Are there any other products possible from transgenic animals?

   A. Human hemoglobin is just one of many possible products from transgenic animals. Research work is currently progressing in methods to obtain compounds that may be used to treat emphysema, lessen the symptoms of cystic fibrosis, and dissolve blood clots.

3. **Q.** How about improving the quality of the animal itself? Can DNA technology be used for that purpose?

   A. Again the answer is yes. Transgenic cows, for instance, could one day produce a higher quality milk, one that is more similar to human milk. And the forecast is for chickens and cattle that synthesize antiviral proteins to encourage disease resistance in their bodies. Such animals, would require less antibiotic therapy, and there would be less possibility of antibiotic carryover to humans.

As of 1993, a Netherlands biotechnology company called GenPharm International had produced a number of **transgenic dairy calves** having the lactoferrin gene, and researchers were steadily improving the technology for improving extraction and purification. Figure 11.7 displays the notable animals. Because cows can produce thousands of liters of milk per year (far more than goats or sheep), the possibility of a lactoferrin-producing cow is very appealing.

Elsewhere in the barnyard is the product of a Scotland-based firm, a **transgenic sheep** that manufactures the protein **alpha-1-antitrypsin**. This protein helps cell membranes retain their elasticity and thereby encourages membrane passages of gases, nutrients, and waste products. Currently extracted from human blood serum, the protein helps emphysema patients breathe more easily. Scottish researchers reported that their transgenic sheep could secrete thirty-five grams of alpha-1-antitrypsin per liter of milk, more than a fifth of the dose to treat a patient for a year.

In an attempt to treat cystic fibrosis, a Massachusetts biotechnology company has used **transgenic goats** to produce **cystic transmembrane conductor regulator (CTCR)**. The CTCR protein is an essential component of cell membranes that transport ions, as Chapter 8 describes. Cystic fibrosis patients lack the protein, so therapeutic doses of the protein could conceivably make up the deficiency and help resolve the symptoms of cystic fibrosis. A transgenic goat (shown in Figure 11.8) has also been developed with the ability to synthesize **tissue plasminogen activator (TPA)**, an important clot-dissolving enzyme used to treat myocardial infarction. The same company has provided a transgenic goat that manufactures **human antithrombin III** to regulate bood-clotting.

But preventing disease with vaccines is even more basic than treating it, and transgenic technology is being developed as a **vaccine-producing device**. One of the first such achievements was reported in 1998, when scientists produced a transgenic goat whose milk contains proteins normally found on the surface of the parasite that causes **malaria**. The proteins elicit an antibody response in test animals and could conceivably be used to immunize against malaria. It stirs the imagination to think that 500 million people could one day be protected against the devastating effects of malaria simply by drinking a glass of goat's milk.

◼ OTHER BIOREACTORS

Pigs, cows, and goats dominated the transgenic headlines in the 1990s, but other animals have also found value in the DNA research laboratory. Among the four-footed pharmaceutical factories now drawing attention is the **transgenic mouse**.

---

emphysema

a lung disorder in which the lung tissues lose their elasticity and become inefficient at obtaining oxygen.

cystic fibrosis

a genetic disease in which mucus buildup in the bronchioles makes breathing difficult.

myocardial infarction

blood clots occurring in the muscle of the heart.

malaria

a protozoal disease transmitted by mosquitoes and accompanied by destruction of the body's red blood cells.

uroplakins

u-ro-pla´kins

FIGURE 11.7

Herman, a young transgenic bull whose cells contain the genes for human lactoferrin. The bull was bred in the early 1990s by researchers at Pharming Group N.V. Since then, it has sired numerous transgenic dairy calves all of which carry the gene for human lactoferrin.

FIGURE 11.8

A transgenic goat. In its milk, this goat produces a protein called human tissue plasminogen activator (TPA). The protein is an important clot-dissolving enzyme.

Biotechnologists are not so much interested in its milk, but its **urine**. Despite its unappealing image, urine offers at least three advantages over milk: both males and females produce it; production starts soon after birth; and, because there is less protein in urine than milk, processing costs are reduced.

Producing therapeutically valuable urine became possible in 1995 when **Tung-Tien Sun** of New York University demonstrated that certain genes are active only in the bladder. The genes encode proteins called **uroplakins**, which are found exclusively in the bladder lining. Soon thereafter researchers attached to the uroplakin gene the gene for human growth hormone (HGH) and included a promoter sequence to activate the gene. Mice containing the gene now produced several hundred nanograms of HGH per milliliter of urine. Although this yield is relatively low compared with that from milk, the process may be worth developing for products that damage mammary tissues.

To treat human disease, DNA technologists have produced transgenic mice that synthesize **human antibodies**. Human antibodies can be produced in cell cultures but with great difficulty and at great expense. Under normal circumstances, animal antibodies can be used to treat disease in humans, but there is danger of potentially fatal anaphylactic (allergy) reactions. Antibodies from transgenic mice, by comparison, can be synthesized in unlimited amounts and do not stimulate allergic responses.

Antibodies are exquisitely specific in their ability to locate one, and only one, kind of molecular target. Produced by lymphocytes in the immune system, antibodies are protein molecules that help neutralize infectious disease agents, cancer cells, and other foreign substances in the tissues. A single antibody molecule contains two "light" chains of amino acids (about 200 amino acids per chain) and two "heavy" chains (about 400 amino acids per chain). To produce human antibodies, DNA technologists modified one strain of mouse with the human genes for the light chain and a second mouse strain with human genes for the heavy chain. Next they bred the mice to one another and produced a third strain, producing both light and heavy chains. The mouse's genetic machinery then united the chains to produce the complete antibody. The achievement signaled the beginning of a new field of therapeutic medicine with far-flung implications in public health.

Transgenic rabbits are being investigated for their ability to produce the enzyme **alpha-glucosidase**. This enzyme helps relieve the symptoms of **Pompe's disease**, a genetic disorder in which the liver cells fail to break down glycogen molecules and release their glucose molecules for the body's energy needs. The glucose deficiency leads to a broad spectrum of diabetes-like symptoms related to energy deficiency.

uroplakins
u-ro-pla´kins

nanograms
billionths of a gram.

lymphocytes
A type of single-nucleated, nongranular white blood cell involved in the immune reaction.

alpha-glucosidase
al˝pha-glu-cos´i-daz

A Dutch biotechnology firm has announced the identification, cloning, and insertion of the alpha-glucosidase gene into rabbits. The transgenic females produce about a gram of the enzyme per liter of milk. According to the company, 200 rabbits should meet the annual worldwide demand. In the body the transgenic alpha-glucosidase replaces the defective enzyme encoded by the mutated gene. Rabbits are useful animals because they mature quickly and produce milk in quantities sufficient to extract enzyme.

In addition to urine and milk, DNA technologists are focusing on another animal product: the egg. They see the oviduct of the **chicken** as a bioreactor that could incorporate high amounts of human protein (e. g., antibodies, blood proteins) into egg protein. Moreover, they envision "**designer eggs**" that contain proteins to enhance human nutrition, especially in developing countries.

One method for delivering genes to a chicken is the **avian leukosis virus** (to be discussed presently). This virus was developed in France to transport genes into embryonic cells. Once transformed, the embryonic cells are added to an egg through a microscopic hole in the shell. Scientists have achieved a thirty to forty percent hatching rate with the modified injection process, but a transgenic chicken remains a difficult objective to reach.

avian leukosis

a disease of birds in which viruses replicate and induce uncontrolled multiplication of the white blood cells of the circulation.

■ "DOLLY" AND THE TRANSGENIC CLONES

In 1997, the scientific community and general public were electrified by the announcement that a lamb had been cloned from a single adult sheep cell. The lamb's name was **Dolly**. She was the first mammal ever cloned; she had a mother, but no father.

The stunning achievement was the work of **Ian Wilmut** and **Keith H. S. Campbell** of the Roslin Institute, near Edinburgh, Scotland. Dolly was born in July 1996, but the researchers decided to wait until she was more mature and all the scientific papers had been carefully prepared. Thus, Dolly was introduced in March 1997 with the inevitable whimsical headlines; "The Lamb that Roared," "Ewe Again?" "Spring Cloning," "Send in the Clones," and "Mary Had a Little...Clone."

cloning

a process in which a population of genetically identical cells such as in an organism is produced from a single adult cell of the parent organism.

Broadly defined, **cloning** is a type of asexual reproduction resulting in an offspring genetically identical to the parent organism. (In laymen's terms, cloning is manipulating a cell from an animal so it grows into an exact duplicate of that animal.) Cloning has been the subject of much speculation in recent decades, and it was the basis for the 1978 movie *The Boys from Brazil*, where several boys develop from cells cloned from Adolph Hitler. In the more recent 1996 movie *Multiplicity*, Michael Keaton is cloned with comic results. Indeed, cloning has been going on for many years as long as the procedure of **embryo splitting** is included in the definition. Here, researchers take a very young embryo and divide it in halves, thirds, or quarters, then encourage each portion to develop into a normal animal. Twins, triplets, or quadruplets result.

embryo splitting

the laboratory procedure in which the cells of an embryo are separated then reimplanted in surrogate mothers.

But Wilmut and Campbell performed a cloning procedure more in keeping with the traditional view of **adult cloning**. They obtained a mammary gland cell from a pregnant, six-year-old **Fin Dorset ewe**; then they fused it to an egg cell, minus the nucleus, from a **Poll Dorset ewe**. After several divisions in a Petri dish, the embryo was implanted in a surrogate mother of yet a third breed, a **Scottish Blackface ewe**. Five months later, Dolly the lamb was born (Figure 11.9). It was apparent that she was strikingly similar to the Finn Dorset and strikingly unlike the Poll Dorset or Scottish Blackface. Wilmut and Campbell had used three different breeds so it would be clear that Dolly had received her genes from the adult six-year-old Finn Dorset.

To many observers, the work of Wilmut and Campbell was the end result of an orderly progression of events starting with the 1952 nuclear transfer experiments in

(a)

(b)

FIGURE 11.9

The scientist and his work. (a) Ian Wilmut, the Roslin Institute researcher who led the study that resulted in Dolly, the first cloned mammal. (b) Dolly, the cloned sheep who had a mother, but no father.

frogs. It soon became clear that, although embryonic nuclei could be successfully transplanted, adult nuclei could not. The results seemed to confirm that by the adult stage, the genes have lost their **totipotency**. In biological terms, totipotency is the ability of a cell's genes to encode liver cells, brain cells, epidermal cells, or any other cells of an adult. As the embryo develops, cells lose totipotency and specialize, much as a CD album plays only one track over and over again. Although every cell of the adult contains the full complement of 100,000 genes, only the genetic "melody" for liver is actually played in a liver cell. The genes have lost their full potential; they are no longer totipotent.

Restoring totipotency was the major breakthrough achieved by the Roslin investigators. Experiments with mice proved fruitless because the nuclei refused to be reprogrammed. (Scientists speculate that methyl groups added to the DNA adult cells deactivate most genes.) But sheep cells did not appear to have the same limitation, and Campbell had a possible solution to the problem: deprive the mammary gland cells of almost all nutrients for five days. This procedure seemed to induce the cells to abandon their normal cycle of growth and enter a dormant stage of metabolic torpor, where very few genes are active (akin to the quiescence of an unfertilized egg). The **starvation** was the genetic equivalent of a Fountain of Youth, and the genes regained their totipotency, as Figure 11.10 illustrates.

Now the procedure continued. Wilmut and Campbell removed the nucleus from a Poll Dorset egg cell and fused it to the starving mammary gland cell, using a pulse of electricity. The electricity broke down the cell's outer membrane and allowed the egg cell to envelope the nucleus. Soon the proteins, mRNA molecules, and other factors of the egg cell cytoplasm were activating the dormant genes and restoring the vitality of the nucleus. As a result of still unidentified signals, the genes expressed their totipotency and began constructing a new lamb. In separate experiments, Wilmut and Campbell fused 277 adult mammary gland cells with 277 egg cells, but they achieved

totipotency

to´te-po-ten´´cy

the quality of a cell by which its genes have the ability to give rise to any possible chemical substance, cell, tissue, or organ produced by an organism.

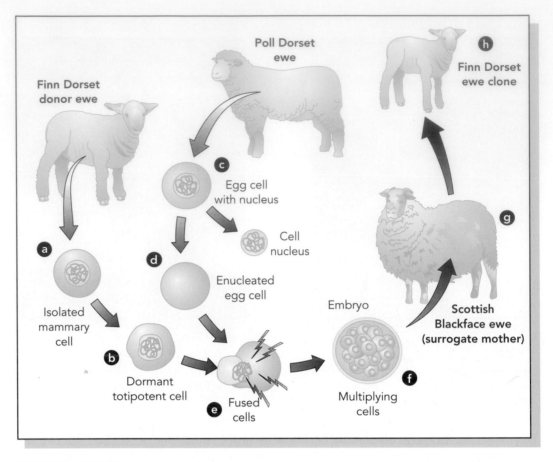

FIGURE 11.10

The cloning process that led to Dolly, the first cloned mammal. (a) A Finn Dorset donor ewe contributes a mammary gland cell. (b) Researchers deprive the cell of nutrients for five days to encourage it to enter a dormant stage and to confer totipotency on the genes. (c) A Poll Dorset ewe contributes an egg cell, and (d) the nucleus of the egg cell is removed. (e) The dormant cell and the enucleated egg cell are fused by a pulse of electricity. (Now the proteins, mRNA molecules and other factors of the egg cell activate the genes of the dormant cell.) (f) The cells are placed in a Petri dish to undergo several divisions, (g) then the fused cell is implanted in the uterus of a surrogate mother, a Scottish Blackface ewe. (h) Five months later, a lamb is born. It is a clone of the donor ewe that contributed the original mammary gland cell. A single cell has been cloned to an adult mammal.

only thirteen pregnancies, and only one pregnancy resulted in a live birth. But that one offspring, Dolly, was the blockbuster of science in 1997.

Even as photographers snapped their pictures, the **questions** were emerging: in view of her six-year-old nucleus, what was the biochemical age of this newborn? Did something in the egg cell reverse the damage associated with aging, or would Dolly age prematurely? Considering the egg cell origin of her mitochondria, were Dolly and her parent identical twins? Of course, there were the human questions and the specter of cookie-cutter clones grown for spare parts and other potential uses currently in the realm of science fiction.

Researchers saw the successful cloning as a major step in developing transgenic animals to synthesize valuable products. As discussed in previous sections, animals can be valuable bioreactors in which hard-to-obtain drugs are synthesized. The more traditional method is to randomly inject gene copies and hope for uptake by the cells. Scientists speculated that in the cloning method, the genes could be added before the nuclei are transplanted into egg cells, thereby increasing the odds of a commercially

mitochondria

submicroscopic organelles of the cytoplasm of a cell where energy metabolism takes place.

useful animal. Cloning could then be used to create a herd of transgenic animals.

Indeed, transgenic animals were in the near future. Four months after the presentation of Dolly, Roslin Institute researchers introduced five other lambs developed by the slightly different process of **fetal cell cloning**. Instead of adult cells, the lambs originated from fibroblasts, a type of totipotent muscle and tendon cell arising early in the fetal stage. The cells were altered with genes for human blood clotting proteins and biochemical markers, then they were subjected to the same cloning strategies as for Dolly. Fetal cells were used because the success rate with these cells is higher than for adult cells. One lamb, Polly, was a **transgenic sheep**, having a human gene in all her cells.

Soon, other cloned animals were appearing. For example, a Wisconsin biotechnology company unveiled a **Holstein bull calf** (appropriately named Gene) that was developed by **stem cell cloning**. To create the calf, DNA technologists used fetal stem cells, a series of totipotent cells appearing early in life. They took **stem cells** from a month-old fetus, stimulated their growth, and fused one cell with an egg cell minus its nucleus. After several divisions, a cell was removed from the mass and fused with another enucleated ("no nucleus") egg cell. After seven days in a nutrient-rich environment, the embryo was placed in a surrogate mother to gestate. Some weeks later she gave birth to Gene, the cow whose originating cell came from the month-old fetus. Gene was a transgenic calf because it carried foreign genes. Before the fusion process, researchers had inserted genetic markers into the stem cells, and now those markers could be located in cells of the calf. The calf demonstrated that stem cells could be manipulated by inserting new genes.

The transgenic livestock exemplified by Polly and Gene (and their successors) galvanized efforts to create herds of animals acting as living factories. Scientists soon returned to and further developed the earlier techniques for **nuclear transfer technology**. Here they could alter the DNA of fibroblasts or stem cells, then remove the nucleus and implant it to a totipotent cell for development to an embryo and transfer to a surrogate mother. This technique is illustrated in Figure 11.11 as it is used to clone transgenic calves. Originally used with fully formed nuclei, the method was improved in 1998 by researchers who inserted hepatitis antigen genes to unfertilized bovine egg cells **arrested in metaphase** (where the chromosomes are without their nuclear membrane). This approach makes insertion of foreign genes more efficient and increases the chances that they appear in all cells of the animal resulting from the egg cell's fertilization, including the reproductive cells.

Nuclear transfer technology is attractive to DNA technologists because it permits them to select as donors only the cells that express a transplanted gene. By contrast, injections to embryonic cells must await the birth of an offspring to see whether transplanted genes have been incorporated. Moreover, nuclear transfer technology and cloning provide as many animals as the researcher needs in the first generation, whereas cell injection technology requires further breeding.

On the **negative side**, nuclear transfer technology results in high mortalities and low birth rates. Moreover, the offspring are often plagued by abnormally large size, undeveloped lungs, and other conditions contributing to ill health. Implantation problems and instances of poor placental development are common. And there is the difficulty of controlling the final destination of the transplanted genes and the number of gene copies taken up.

Ethical problems associated with cloning abound as well. Shortly after Dolly's birth, the emotional impact of copying a mature animal persuaded President Bill Clinton to announce a ban on the use of federal funds for human cloning research.

fetal cell cloning

development of a clone of cells from a totipotent fetal cell such as a fibroblast.

stem cell cloning

development of a clone of cells from stem cells that develop very early in life.

metaphase

the stage of mitosis in which the chromatids align at the equatorial plate of the multiplying cell.

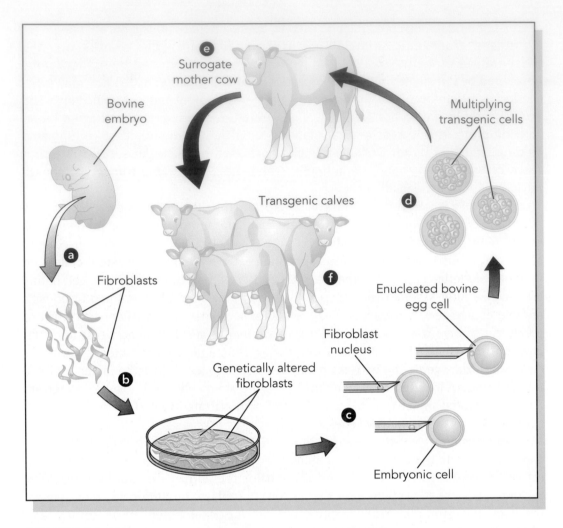

FIGURE 11.11

Transgenic cattle produced by cloning with fetal cells. (a) Fibroblasts are obtained from a fifty-five-day-old bovine fetus. The fibroblasts are totipotent muscle and tendon cells arising early in the fetal stage. (b) The fibroblasts are cultivated in nutritious medium Petri dishes and modified with foreign genes. (c) Then the nucleus, with its genetically altered DNA, is removed from the cell, and the nucleus is implanted into an egg cell lacking a nucleus. (d) The egg cell with its new nucleus is encouraged to multiply and form an embryo. (e) Embryos are implanted to surrogate mothers, and (f) some months later, transgenic calves are born. They are clones because they have originated from single cells, and they are transgenic because all their cells bear foreign genes.

Still the questions persisted: Would it be acceptable to copy a child to obtain a source of tissue to save its life? Or to copy an embryo to avoid a genetic disease related to damaged mitochondria? Or for a couple to replace a dying child? To be sure, cloning could bring enormous changes in animal husbandry and biomedicine, but cloning also raises legal questions regarding procreation and philosophical questions of how we view children and how we think of families and relationships; in short, what it means to be human.

BUILDING A BETTER ANIMAL

To many DNA technologists, the important objective is not what drugs an animal can produce, but rather the animal itself. These technologists seek to improve the quality of the animal by enhancing its ability to produce valuable foods and by increasing its resistance to disease.

Scientists have been attracted to a gene that encodes **whey acidic protein (WAP)**, a group of acidic proteins found in the fluid portion (the whey) of the milk. The gene has attracted attention because it is active only in the mammary gland and its alteration would not affect other body organs. Normally found only in rodents, the gene for WAP was introduced into the fertilized egg cells of pigs in 1991. The transgenic pigs produced the fluid protein in their milk, and, as anticipated, the gene was inactive in the remainder of the pig's body.

The ability to transfer protein-encoding genes to a large animal's mammary gland raised the possibility of enhancing the nutritional value of milk by increasing its content of **casein**. Casein is the protein component from which cheese is made. Boosting casein content by twenty percent might yield an estimated $200 million profit increase to the cheese industry annually. Another possibility is to reduce the lactose content of milk for those who have lactose intolerance. And, as one observer notes, it may even be possible to genetically modify a cow to produce skim milk

A better animal may also be in the future for research laboratories as DNA technologists begin using animals that light up from the inside. In 1997, for example, **John Morrey** of Utah State University developed a transgenic mouse with a **luciferase-encoding gene** in its cells. Luciferase is an enzyme that breaks down the firefly compound **luciferin**, with the release of visible light, a process called bioluminescence. When the gene is activated and luciferin is present, the cells and local tissues give off an eerie green glow. The system could be used to determine when a gene is turned on or off during animal development by simply noting whether the tissue is glowing or not. Also, gene therapies could be tracked in the living animal by watching for the light.

Another light-emitting system attracting interest centers about a **green fluorescent protein** from jellyfish. Japanese investigators have successfully incorporated the gene into transgenic mice, thereby developing animals that are completely green when exposed to blue light (Figure 11.12). The system is presumably better than the luciferase system because no outside substance (i. e., no luciferin) is required to activate the gene; the gene is perpetually active, and the mice are perpetually green. Scientists hope to track cancer cells in the mice, while waiting for their minds to conjure other possible uses for these novel laboratory tools.

whey

the fluid portion of the milk that remains after the protein has been removed.

casein

the major protein found in milk.

luciferase

lu-cif'er-az

bioluminescence

a phenomenon in which visible light is released as a result of an enzyme-catalyzed reaction.

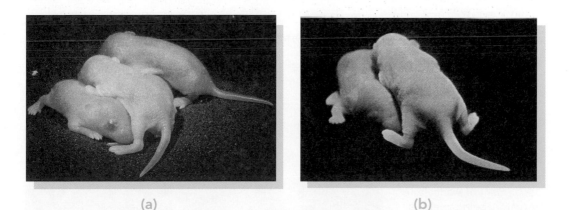

(a)   (b)

FIGURE 11.12

Transgenic mice that glow. (a) The animals in the first photograph are normal mice. (b) The mice in the second photograph were genetically altered with a jellyfish gene that encodes a green fluorescent protein. The mice glow with a green luminescence when they are illuminated by blue light.

■ INCREASING ANIMAL RESISTANCE

Among the many dreams of DNA technologists is the intriguing possibility of farm animals that are genetically altered to resist infectious disease. This concept has been investigated in plants (Chapter 10), and it may be possible to develop a similar animal.

To pursue disease resistance in animals, DNA technologists have spliced viral genes into animal cells. Chickens, for example, are susceptible to the **avian leukosis virus (ALV)**, a retrovirus containing RNA. In the early 1990s, DNA technologists identified and cloned the *env* gene that encodes the ALV envelope. Then they injected the gene into the nucleus of fertilized chicken eggs, whereupon the RNA encoded a double-stranded DNA provirus. The provirus inserted into the nuclear DNA, and soon the egg cells were producing glycoproteins normally associated with the ALV envelope. Chickens developed from these fertilized eggs displayed resistance to ALV when injected with active viruses (Figure 11.13). By comparison, fertilized eggs injected with the genes of complete viruses expressed complete viruses, and chickens developed from these eggs experienced the effects of ALV infection.

As in plants, the **mechanism of resistance** induced by viral proteins is not well understood. One possible theory is that the viral proteins react with receptor sites present on the host's cells (where viruses would normally dock). Because the sites are neutralized by the proteins, they are not available to pathogenic viruses, and infection does not occur. Such a mechanism has not escaped the interest of researchers who are working on human vaccines.

Similar work is ongoing with cows. At the University of Colorado, researchers have introduced genes from the **bovine leukemia virus (BLV)** to the nuclei of fertilized eggs of cows. The cells are then incubated overnight to permit cell division to begin. Next, they are transferred to the oviduct (Fallopian tube) of a female rabbit, where they reach the 100-cell stage (after about six days). Finally they are flushed out, recovered, and a few cells are tested for successful BLV gene insertion. Gene-altered embryos are reimplanted, but this time to the uterus of a surrogate female cow.

Another approach to augmenting disease resistance in animals is to enhance the activity of inborn genes that **encode antimicrobial substances**. For instance, DNA technologists are investigating the possibility of boosting the activity of genes that encode **interferon**, the well-known antiviral protein (Chapter 5). Transgenic cows have now been produced with higher-than-normal interferon levels in their tissues. Tests

leukosis

uncontrolled replication of white blood cells.

provirus

the DNA form of an RNA virus; integrated to nuclear material of a host cell.

bovine leukemia virus

a virus that induces uncontrolled replication of white blood cells in cows.

interferon

in-ter-fer′on

an antiviral protein produced by animal cells on stimulation by replication viruses.

FIGURE 11.13

Chickens at the U. S. Department of Agriculture laboratory being injected with avian leukosis virus to test whether transgenic chickens could produce viral glycoproteins and acquire protection to the viruses.

are ongoing to determine whether they can resist viral pathogens normally dangerous to cows. Other researchers have worked on gene-based methods to improve the efficiency of antibody production.

ENVIRONMENTAL REPLACEMENTS

Another dream of DNA technologists is using replacement organisms to interrupt disease cycles in nature. For example, a **transgenic snail** could conceivably interrupt the life cycle of the parasite that causes **schistosomiasis**. The parasite, a flatworm belonging to the genus *Schistosoma*, breeds in the tissues of the snail before infecting the blood of humans who stand or swim in contaminated water. French investigators are now attempting to develop a transgenic snail that resists invasion by *Schistosoma* species. Released into the environment, this snail could crowd out natural snails and thus break the life cycle of the parasitic worm. Each year, approximately 100 million people worldwide have the fever, chills, intestinal ulcerations, and diarrhea of schistosomiasis develop.

A slightly different strategy is being used by DNA technologists at the University of California, where insects are being endowed with "suicide genes." The researchers are working with cotton bollworms, the caterpillars that live in cotton bolls. They are attempting to produce **transgenic bollworms** by inserting into laboratory-reared bollworms a gene that activates a **suicide gene** in the insect's offspring. The plan is to raise millions of worms to adult moths and release them to the environment, where they will compete with the normal moths and mate with wild bollworm moths. The mating will produce caterpillars that die from the effect of the gene and save the cotton crop from caterpillar-related damage.

Then there is the **transgenic mosquito**. Of particular interest are members of the genus *Anopheles*, which transmit the protozoal parasites that cause **malaria**. Researchers have already identified certain critical genes that give anopheline mosquitoes the ability to harbor and transmit malarial parasites. Presumably, those genes can be altered, and mutated insects can be produced. Released in large numbers, the mosquitoes could dilute or overwhelm native mosquito populations and break the chain of disease transmission.

At the center of interest regarding insect biotechnology is understanding the foundations for **vector incompetence**, that is, the reasons why a vector is able to sustain a parasite and how to change that ability to make the vector inhospitable. It is possible, for example, that an organism is unable to establish itself in insect tissues because of some immunological mechanism operating in the insect's gut. Another possibility is the parasite's failure to develop properly because of a nutritionally un-sound environment. Still another possibility is the failure of the parasite to migrate to the insect's salivary gland for discharge. Indeed, researchers at Notre Dame University have found that mosquitoes with vector incompetence have smaller pores in their salivary glands, effectively keeping out the virus so it is not injected with saliva into a human host. A single gene or two may account for the critical difference, thereby providing a mechanism for producing transgenic-resistant mosquitoes, such as shown in Figure 11.14a.

But suppose transforming the insect itself was unnecessary? Suppose it was possible to transform a bacterial population living within the insect and thereby transform the insect's intestinal environment? That has been the object of University of Bristol scientists as they attempt to create a **pseudotransgenic tsetse fly**. The researchers are working with the insect that transmits **African sleeping sickness**, a protozoal disease of the blood and nervous system. They have discovered that the

schistosomiasis

a blood disease caused by the flatworm *Schistosoma*, which develops in snails and penetrates the skin tissue when contacted in water.

Schistosoma
shis-to-so'mah

Anopheles
an-of'e-les

vector incompetence

the characteristics of a vector whereby it becomes incapable of sustaining the life of a parasite normally found in its tissues.

African sleeping sickness

a protozoal disease transmitted by the tsetse fly and often accompanied by coma.

disease-causing protozoa are killed by a protein secreted by the insect's gut wall. Their experiments are directed toward identifying the protein-encoding gene and inserting it to the bacteria within the gut. Then the bacteria would produce the antiprotozoal protein and the tsetse flies would no longer harbor or transmit the disease agents.

Researchers at Yale University have adopted a variation on the same theme. They are hoping to coax the bacteria to express a gene from a mammalian immune system. In theory, the gene will encode an antibody molecule that attacks a key protein of the protozoan and kills it. The Yale researchers, however, face the same problem as the British group: how to spread the genetically engineered bacteria among the wild population of tsetse flies.

The hope of genetically disarming an important agricultural pest received a boost in 1995 when Greek scientists produced a **transgenic medfly** (Figure 11.14b). Medflies (i.e., *Medi*terranean fruit *flies*) are notorious destroyers of fruit and coffee crops throughout the world. Researchers used techniques learned from long years of experience with laboratory fruit flies (e.g., *Drosophila* species), and they found a DNA sequence that transports a foreign gene into an arthropod's sperm and egg cells. They attached a marker gene for eye color and found the marker expressed in the offspring. The next step is to insert genes useful for controlling the pest.

The transgenic snail and insects represent unique approaches to disease intervention. Traditional modes of dealing with disease have included cures for sick individuals, vaccines for disease prevention, insecticides for the vectors of disease organisms, and environmental changes that encourage disease transmission. As we have seen, the genetic approach to dealing with disease is embodied in the word "replacement." Rarely in medical history has this approach been attempted. But never before has the world been home to an oncomouse, a sheep without a father, or a pig that produces human hemoglobin.

medfly
the Mediterranean fruit fly that parasitizes fruit and coffee crops.

(a)

(b)

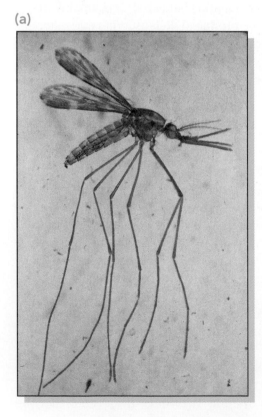

FIGURE 11.14

Transgenic arthropods. (a) *Anopheles*, the genus of mosquito vector that transmits *Plasmodium* species, the malarial parasite. Transgenic mosquitoes with vector incompetence are being used to replace naturally occurring mosquitoes in an attempt to interrupt the life cycle of the malarial parasite. (b) The Mediterranean fruit fly (medfly), a notorious destroyer of fruit and coffee crops and the object of several environmental replacement studies.

SUMMARY

A wide variety of experiments are currently in progress to assess the various uses of transgenic animals. Transgenic animals are animals in which one or more genes have been introduced to the somatic cells. Among the first transgenic animals were the so-called custom-made animals. As early as 1983, Brinster and Palmiter inserted into mice the gene for growth hormone and achieved an animal of larger-than-normal size. In later years, a mouse with human immune system cells was developed (although not by transgenic methods), and transgenic mice with certain forms of human cancers were developed and patented. Transgenic mice also were used to show the importance of amyloid protein in Alzheimer's disease. Among the major developments is a knockout mouse in which one or more genes has been eliminated so that a system such as the immune system fails to develop. Rejection mechanisms to transplants do not develop in these animals.

As living bioreactors, transgenic animals can be genetically altered to produce valuable products. The methods involve microinjection of DNA into fertilized egg cells of embryos and transferal to surrogate females, where the animal develops. In this way a transgenic pig has been produced that can synthesize human hemoglobin for use as a blood substitute. A transgenic cow has also been bred with the ability to produce lactoferrin, an iron-building milk protein and a potential antibacterial agent. A transgenic sheep is now available that can synthesize a protein to keep cell membranes more elastic and help treat emphysema patients. And finally, a transgenic goat has been developed with the capability of synthesizing the protein lacking in cystic fibrosis patients.

Among the notable development of the 1990s was Dolly, the cloned sheep. This animal was produced by returning a mammary gland cell to totipotency and combining it with an enucleated egg cell. The fused cell was implanted into a surrogate ewe for development. Researchers used the technology to develop transgenic animals and adapted it for fetal cell cloning and stem cell cloning.

Transgenic methods can also be used to develop better animals. The nutritional value of milk can be enhanced, for example, by increasing the ability of a cow's mammary gland cell to secrete casein. Another possibility is to develop transgenic animals that can resist viral disease by modifying their genes to produce viral glycoproteins. Animals can also be bred to increase their production of the antiviral substance interferon. In nature, it may be possible to introduce transgenic snails and mosquitoes to crowd out the natural vectors that transmit disease during epidemics. It is clear that the applications of DNA technology in animal development are innovative and imaginative.

REVIEW

This chapter has focused on the development and uses of transgenic animals as an imaginative application of DNA technology. To test your knowledge of the chapter contents answer the following review questions.

1. Explain the methods used by Brinster and Palmiter to develop the first transgenic animal, one that expresses a foreign growth hormone.

2. What is a "human mouse" and how is it beneficial to AIDS research?

3. Can transgenic animals be patented? Give two examples of where this has been accomplished.

4. Explain how a transgenic animal can be used to show the relationship between amyloid protein and Alzheimer's disease.

5. Explain the meaning of a knockout mouse, describe how it is produced, and indicate some uses for this animal in DNA technology.

6. Describe the transgenic pig that produces human hemoglobin and explain the value of having this hemoglobin.

7. Name four products of transgenic animals used as bioreactors and show how each product is medically beneficial. Name animals other than barnyard animals that have been used to produce transgenic animals, and indicate how they have been used.

8. Describe the process used to develop Dolly, the first cloned mammal. Explain some applications of this process in the development of transgenic cloned animals.

9. How can genes from avian leukosis virus be used to enhance resistance to the virus in transgenic chickens? Explore the benefits of increasing interferon production in transgenic cows.

10. Show how transgenic strains of snails, mosquitoes, and medflies can be used to break the cycles of disease in nature.

FOR ADDITIONAL READING

Aldous, P. "Malaria: Focus on Mosquito Genes." *Science* 261 (1993): 546–549.

Beckage, N. "The Parasitic Wasp's Secret Weapon." *Scientific American* (November, 1997).

Begley, J. "Little Lamb, Who Made Thee?" *Newsweek* (March 10, 1997).

Mestel, R. "The Mice Without Qualities." *Discover* (March, 1993).

Multiple authors "The Perils and Promises of Cloning." *The Sciences* (September/October, 1997).O'Brochta and Atkinson, P. "Building a Better Bug." *Scientific American* (December, 1998.)

Pennisi, E. "Will Dolly Send in the Clones?" *Science* 275 (1997): 1415–1417.

Richards, F. "An Approach to Reducing Arthropod Vector Competence." *ASM News* 59(10) (1993): 509–514.

Roush, W. "Knockout Mice Point to Molecular Basis of Memory." *Science* 275 (1997): 32–35.

Travis, J. "A Fantastical Experiment: The Science Behind the Cloning of Dolly." *Science News* 151 (1997): 214–215.

Velander, W. H., et al. "Transgenic Livestock as Drug Factories." *Scientific American* (January, 1997).

Wilmut, I. "Cloning for Medicine." *Scientific American* (December, 1998.)

The Human Genome Project

12

LOOKING AHEAD

The concluding chapter of this book deals with the monumental effort to pinpoint the locations of all the genes in the human genome and to learn their base sequences. On completing the pages that follow, you should be able to:

- understand the enormity and complexity of the Human Genome Project.

- describe how knowing the nature of the human genome applies to various branches of biology and specify which other genomes are to be deciphered.

- appreciate the administrative and financial underpinnings of the Human Genome Project.

- describe how DNA technologists go about developing gene linkage maps and physical maps of the human genome.

- summarize the processes by which the sequence of bases is determined in a particular gene.

- identify the organisms whose genomes have been sequenced and denote the importance of each successful sequencing.

- understand many of the issues attending the Human Genome Project and how those issues may be resolved.

INTRODUCTION

On a January afternoon in 1989, a group of biologists, ethicists, industry scientists, engineers, and computer experts gathered in a conference room at the National Institutes of Health in Bethesda, Maryland. They sat around a long oval table in a conference room and listened as molecular biologist Norton Zinder of New York's Rockefeller University rapped his gavel to bring the meeting to order. "Today we begin," declared Zinder. "Today we are initiating an unending study of human biology. Whatever else [happens],…it will be an adventure, a priceless endeavor."

With these words, Zinder launched a monumental effort that has rivaled in scope both the Manhattan project, which resulted in the atomic bomb, and the Apollo program, which landed an American on the moon. Indeed, the effort may exceed both of them in importance because the goal is to map the human genome. The intention is to spell out for humanity the entire genetic message and to identify the sequence of every single base in the DNA of a human cell.

human genome

the repertoire of the approximately 100,000 genes of a human cell.

The **human genome** is the repertoire of the approximately 100,000 genes contained in each cell nucleus (red blood cells excepting). It is the human book of life, encoding in its pages all the biochemical information to build and maintain our bodies. Passed on to each new generation during reproductive processes, the human genome encompasses both our past and our future.

To map the human genome is a project of epic proportions. To get a feel for how epic this is, recall that about three billion nucleotide pairs encode all human traits. By comparison, there are only about 4.7 million nucleotide pairs in the *Escherichia coli* chromosome. If its base sequence were written out in capital letters (e. g., CTTAGGATACCAAT…), the *E. coli* chromosome would occupy 300 pages of a thousand-page telephone book. By comparison, the base sequence of a fruit fly's genome (eight chromosomes) would take up ten telephone books, each of a thousand pages; and the base sequence of the human genome (forty-six chromosomes) would use up 200 telephone books (200,000 pages) or twenty times the amount of the fruit fly genome (Figure 12.1). For those who remain unimpressed, it should be noted that through the first eight years of the project (through 1998), only three percent of the human genome had been sequenced.

Geneticist **Eric Lander** (Figure 12.2) has invoked the following analogy to describe the Herculean endeavor of deciphering the complete human genome: take six complete sets of the *Encyclopedia Britannica* and shred them onto the floor. Now try to reconstruct the books.

human genome project

the effort to learn the sequence of bases in the DNA of the entire human genome.

The objective of the **Human Genome Project** is so colossal that the discovery of the cystic fibrosis gene (Chapter 7) seems minuscule by comparison. Over the next several years the project will be uppermost in the minds of thousands of DNA technologists worldwide, while demanding billions of research dollars. A total expenditure of $3 billion is anticipated over the ten to fifteen years of the project. (To appreciate the magnitude of $3 billion, consider that if you were to begin counting to three billion today at a nonstop rate of a dollar per second, you would not be finished until about 2090, ninety years from now.)

DECIPHERING THE HUMAN GENOME

The Human Genome Project is one of the most ambitious projects ever undertaken by scientists. Once the genome has been learned, it is expected to be the ultimate source book for understanding human biology and inherited diseases. The atlas of the

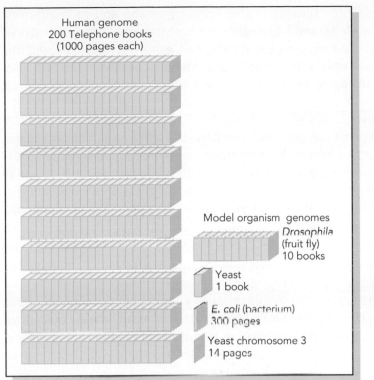

Human genome
200 Telephone books
(1000 pages each)

Model organism genomes
Drosophila
(fruit fly)
10 books

Yeast
1 book

E. coli (bacterium)
300 pages

Yeast chromosome 3
14 pages

FIGURE 12.1

The genomes of several organisms compared with a telephone book or series of telephone books. The magnitude of deciphering the human genome as is clear from these comparisons.

FIGURE 12.2

Molecular geneticist Eric Lander and his associates and students at the Whitehead Institute in Cambridge, Massachusetts. Lander has invoked various analogies to describe the Herculean endeavor of deciphering the human genome.

human genome will revolutionize biological research into the twenty-first century, much as maps of the world revolutionized travel in centuries past.

In **medicine**, an understanding of the molecular basis of inherited diseases will lead to improved diagnoses and treatments. Already, researchers have identified genes associated with Duchenne's muscular dystrophy, retinoblastoma, and cystic fibrosis (Chapter 7). As research progresses, investigators will add to this list, while uncovering the mechanisms of disease caused by genes interacting with their environment. For instance, genetic predispositions have been implicated in many major disabling diseases such as heart disease, diabetes, and cancer. More effective preventive measures and therapies will result from a discovery of the proteins involved in these diseases.

Determining the underlying biology of the human genome and its regulatory controls will also help **developmental biologists** understand how humans progress from single fertilized eggs to adults. Biologists will gain better insight into why the process sometimes is misdirected and how fetal abnormalities develop. They will also gain a better insight into the aging process by studying how genes express themselves during development.

Evolutionary biologists stand to gain from the genome project because DNA regions mutate at a constant rate, and DNA comparisons using different organisms provide a historical record of evolutionary changes over time (Questionline 12.1). Comparisons such as these have already been used to search for the mitochondrial "mitochondrial Eve" the mythical first woman or population of the first women of the species *Homo sapiens* (Chapter 9). In the future, the deluge of data will help evolutionary biologists better understand the trees of life because they will have whole families of genes to compare, rather than the single genes now available. Among the questions to be settled are how complex genes arose from simple ones, an issue at the foundation of the origin of life.

Comparative biologists will reap the benefits of genome research by studying proteins encoded by the genes. To be sure, all humans are basically alike. Differences arise in those few proteins that distinguish one individual from another, and studying those proteins will give insight into those differences. Moreover, studying the genomes of various living things will suggest new laboratory test organisms. It has been discovered, for example, that tomato cells and mouse cells have many genes in common, and each has the same number of chromosomes and the same number of arms in the chromosomes. One day, the tomato may replace the mouse as a laboratory specimen of choice.

Moreover, there is the benefit of understanding an organism like never before. Knowing an organism's genome tells us the basic structure of every one of its proteins;

QUESTIONLINE 12.1

1. **Q. What is meant by the human genome?**

    A. The nucleus of a human cell is composed of forty-six chromosomes functioning as about 100,000 units known as genes. Those 100,000 genes compose the human genome.

2. **Q. Why is knowing the nature of the human genome important?**

    A. Knowing the base sequences of a human gene will help scientists understand the nature of the protein encoded by that gene. Working with the protein will be invaluable to the diagnosis, treatment, and cure of genetic diseases such as cystic fibrosis, Alzheimer's disease, Lou Gehrig's disease, sickle cell anemia, and others. Fields such as developmental biology, evolution research, and comparative biology also stand to benefit.

3. **Q. How long is the genome project expected to last?**

    A. The Human Genome Project began in late 1989. It is expected to span a period of fifteen years and cost a total of about $3 billion. A federal organization called the National Human Genome Research Institute is directing the project in the United States, and much of the supervision is being supplied by the U. S. National Institutes of Health and the Department of Energy.

every enzyme in its metabolism; every signal-bearing hormone coursing through its body; and every receptor poised to receive those signals on the surface of its cells. We would, in effect, have the owner's manual of that organism.

In addition, the Human Genome Project will spur the development of **biotechnology**, particularly in information storage and DNA analysis. New technologies are a practical necessity because the current technology is time-consuming and intellectually mundane, as well as repetitive and unusually boring. DNA sequencing tends to be "factory work," having limited appeal to creative DNA technologists. (With tongue in cheek, one biochemist has recommended substituting laboratory penal colonies for the current prison system.)

Adding to the negative view of DNA sequencing is the fact that only about five percent of the human genome consists of exons, the genes used to encode proteins. The remaining ninety-five percent consists of introns, the "genetic gibberish" that does not appear to function in protein synthesis. Convincing an eager DNA technologist to learn the base sequence of the "genetic junk" is difficult. It is yet another reason for encouraging and supporting technological advances in DNA work.

■ ORIGIN OF THE PROJECT

No one is certain who was the first to suggest sequencing the human genome, but the first serious proposal was made in 1985 by **Robert Sinsheimer**. Sinsheimer, a molecular biologist by training, was Chancellor of the University of California, and he brought together a group of scientists to discuss the possibility of the project. **Renato Dulbecco**, a Nobel laureate in biochemistry, was one of those in attendance. After returning to New York's Cold Spring Harbor Laboratory, he spoke glowingly about how knowledge of the human genome would aid cancer research (a high priority of Cold Spring Harbor Laboratory). James D. Watson (pictured in Figure 12.3) was director of the laboratory, and as he listened to Dulbecco, his mind stirred.

Meanwhile, scientists at the federal **Department of Energy (DOE)** were made aware of Sinsheimer's proposal, and **Charles DeLisi**, head of DOE's Division of Health and Environmental Research, became an avid supporter of the project. In 1986, DeLisi convened a meeting of about eighty scientists with expertise in DNA research and offered the U. S. laboratories at Livermore and Los Alamos as places to carry out the project. Rumors of the DOE-led project spread through the scientific community and molecular biologists, including Walter Gilbert and Paul Berg (Chapter 4), became advocates. However, they argued that DOE was not the organization to lead the project because its experience in genetics research was minimal and because DOE was more

metabolism
the sum total of all the chemical reactions occurring in a living thing.

exons
sequences of chromosomal DNA that encode proteins.

FIGURE 12.3

James D. Watson, one of the early supporters of the human genome initiative and the first director of the Center of Human Genome Research.

concerned with physical sciences than with biological sciences. The National Institutes of Health (NIH) appeared to be the logical leader.

The **National Academy of Sciences** was next to intervene. At the urging of its members, the academy appointed a committee to decide whether the federal government should pursue the project and, if so, what course should be taken. The committee issued a report entitled, "Mapping and Sequencing the Human Genome," with a unanimous recommendation to proceed. The United States would work with any other nations willing to take part in the project and would seek to sequence not only the human genome but also a variety of other genomes as well. The committee urged technological improvements to support the effort and set a goal of $3 billion in expenditures. A strong genome advisory board led by a distinguished scientist was deemed indispensable; on the matter of supervision, the committee suggested that both DOE and NIH should be involved, as long as there was a single advisory board.

In 1987, Congress was ready to appropriate funds for the new Human Genome Project. Under the leadership of director **James Wyngaarden**, NIH secured $17.4 million for the project and convened a temporary organizing committee. The latter suggested the formation of a new **Office of Human Genome Research** to be headed by a director who was an active scientist so the public, the Congress, and the scientific community would be reassured that science, not politics, was uppermost in the researchers' minds. The first director of the office was **James D. Watson**, the scientist who with F. H. C. Crick first proposed DNA's double helix structure in 1953. The one who proposed the existence of the steps in the helix was now put in charge of deciphering the nature of the steps.

One of the first tasks of the Office of Human Genome Research was appointing a **Program Advisory Committee on the Human Genome**. Twelve members were chosen, including three members from industry and a number of university scientists. **Norton Zinder**, shown in Figure 12.4, was appointed Chairman. On a January afternoon in 1989 he gaveled the first meeting of the advisory committee to order and set history into motion.

As a footnote to history, the Office of Human Genome Research became a "Center" in 1990 (**The National Center for Human Genome Research [NCHGR]**). Although the change appears cosmetic, the functions of a center are different because a center has the authority to make grants, whereas an office does not. In 1998, the NCHGR became the **National Human Genome Research Institute**, a branch of the National Institutes of Health. Its current director, **Francis Collins** (of cystic fibrosis research), oversees a budget

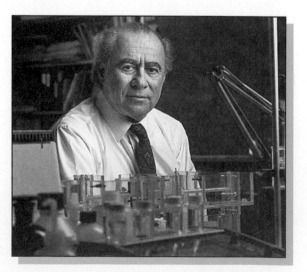

FIGURE 12.4

Norton Zinder, the Rockefeller University researcher, who chaired the first Program Advisory Committee on the Human Genome and remains active on the project today.

of hundreds of millions of dollars and supervises the project described as the "Holy Grail of Biology."

■ WHOSE GENOME?

One of the intriguing questions of the Human Genome Project is whose genome will be deciphered? Will it be the genome of the President of the United States (William J. Clinton, at this writing)? Or the director of the project (formerly James Watson; currently Francis Collins)? Or some unknown individual off the street?

The answer is that many genomes from many organisms have been decoded or are being worked on simultaneously in different laboratories throughout the world (Table 12.1). The organisms cover the interests of all biologists. For the microbiologist, DNA technologists have worked out the genome of *Escherichia coli*, the common intestinal bacterium; the genome of the fermentation yeast *Saccharomyces cerevisiae* has been pursued by other technologists partly because it is such an important research tool; for developmental biologists, the genome of the nematode worm *Caenorhabditis elegans* is the object of study because its 959 cells have been traced from fertilization to the adult worm and all the circuits of the 302 cells of its nervous system are known; the fruit fly *Drosophila melanogaster* is being analyzed for its importance in genetics research; and botany is represented by studies on two plants: the member of the mustard family *Arabidopsis thaliana* and the corn plant *Zea mays*.

Another organism whose genome is being studied is the common laboratory mouse *Mus musculus*, depicted in Figure 12.5. Decoding the genome of the mouse will provide important comparative information because humans and mice are biologically and genetically similar. For instance, analyses of the respective genomes will point up regions that have been conserved during evolution and are therefore important to evolutionists. Sections of mouse DNA similar to sections present in human DNA can be studied in the laboratory for their physiological importance, and the results can be applied to humans. Mouse models could be developed to study human diseases such as diabetes and human cancers, much as the "oncomouse" is now a valuable laboratory tool (Chapter 11).

For the human genome, the nuclear DNA can be gleaned from virtually anyone. Indeed, two people of the same sex differ in their DNA by only one tenth of one percent. Thus, the deciphered genome will apply to 99.9 percent of every human on earth. (It will also be ninety-nine percent similar to a chimpanzee's genome and eighty percent similar to a mouse's.) However, the 0.1 percent difference can add up to three million base pairs, enough to account for the individual and racial variations in the human species. Thus, it will be important to pursue the genomic sequences of various

Caenorhabditis
se″nor-hab′di-tis

Drosophila
dro-sof′il-ah

melanogaster
mel″a-no-gas′ter

Arabidopsis
a-rab″i-dop′sis

thaliana
thal-i-an′a

Mus musculus
the species of common laboratory mouse.

TABLE 12.1
A Comparison of the Genomes of Several Organisms.

ORGANISM	GENOME (ESTIMATED NUMBER OF BASES)
Human	3,500,000,000
Mouse	300,000,000
Drosophila (fruit fly)	170,000,000
Caenorhabditis elegans (nematode)	110,000,000
Arabidopsis thaliana (plant)	100,000,000
Aspergillus nidulans (fungus)	25,000,000
Saccharomyces cerevisiae (yeast)	15,000,000
Escherichia coli (bacterium)	4,720,000
Bacillus subtilis (bacterium)	4,000,000

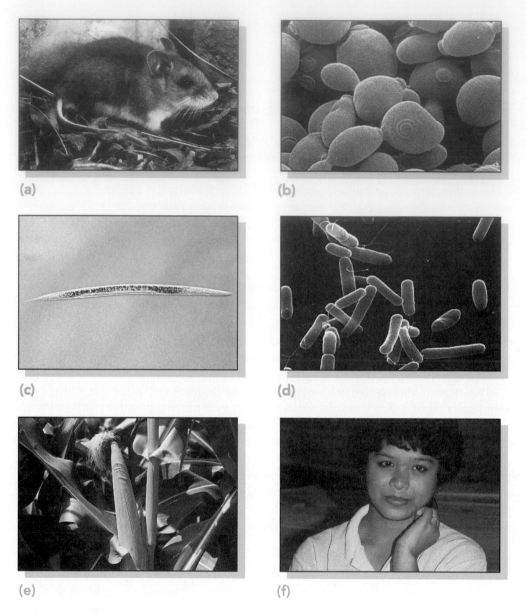

(a)   (b)   (c)   (d)   (e)   (f)

FIGURE 12.5

Several organisms whose genomes are being sequenced or already have been sequenced as part of the human genome project. (a) the mouse; (b) yeast cells; (c) a nematode worm; (d) the bacterium *E. coli*; (e) the corn plant. (f) a human being.

humans to understand the behavioral, cultural, and other differences that make each of us unique.

For the present, much of human genome work is being performed on chromosomal material supplied by the **Center for Human Polymorphism** in Paris, France. Led by Nobel laureate **Jean Dausset**, the institute has collected cell lines from more than sixty different French families each spanning three generations (including all four grandparents, two parents, and at least eight grandchildren). Thus, the institute can provide high-quality DNA of known lineage to investigators throughout the world for use in genome research. Moreover, the genome most heavily studied has been male. This is necessary because only males have the complete set of human chromosomes, that is, the forty-four autosomes and one X and one Y chromosome. Females, by contrast have two X chromosomes and no Y chromosome. In summary, it is likely that the first genome to be completely decoded will be from a French male.

autosomes

the forty-four chromosomes of the human genome that are not involved in sex determination.

■ WHOSE MONEY?

There are approximately three billion base pairs in the forty-six chromosomes of the human genome, and about $3 billion will be spent finding out what those base pairs are. A major share of the investment for genome research is being borne by the United States government through its **National Institute of Health (NIH)** and **Department of Energy (EOE)**. The NIH is involved because genome research has implications related to health, and the DOE has an interest in the project because of the energy-related technologies and equipment for gene sequencing.

Monies distributed by NIH are overseen by its National Human Genome Research Institute headed by **Francis Collins**, pictured in Figure 12.6. The center funds grants and contracts in genome technology and acts as the contact point with other agencies such as the U. S. Department of Agriculture and the U. S. Department of Health and Human Services. The first congressional appropriation for the center was for fiscal year 1988 in the amount of $6.3 million dollars. For 1994 the appropriation was $170 million. Estimates are that the appropriations will average $200 million for the fifteen-year period from 1990 to 2005. An interesting component of genome funding is the monies set aside for **ethics research**. As first director of the center, Watson saw to it that three percent of monies were earmarked to fund grants to explore the project's ethical implications: For example, should an insurance company know that you carry the gene for an incurable disease? And would you want to know that working in a chemical plant could predispose you to a genetic disease? The research into the human genome will make available much detailed and sensitive information that may affect personal and private lives. The funds for ethical research, recently increased to five percent, will help anticipate problems arising from genome knowledge.

In addition to NIH and DOE, a number of other federal institutions are also overseeing genome monies. The U. S. Department of Agriculture and National Science Foundations are examples. In addition, the privately funded Howard Hughes Medical Institute has contributed considerable sums to the research. And since 1993, many private corporations have joined the project. For a fee, some took over the work of sequencing; others assumed an interest in developing therapeutic agents from genome research; and still others were developing and merchandising DNA sequencing technology. Among the prominent companies are The Institute for Genomic Research headed by J. Craig Venter (to be discussed shortly) and Human Genome Sciences under the tutelage of William Haseltine.

The movement into the private sector and away from universities, research

FIGURE 12.6

Francis Collins, the current director of the Center for Human Genome Research.

institutes, and government laboratories was both criticized and welcomed in the scientific community. Critics questioned the conflict of interest that might confront researchers wishing to work in both university and private sectors; on the other side, scientists welcomed the new infusion of monies to spur the genome project to completion. There is also the issue of patent rights, as we shall explore next.

■ GENE PATENTS

complementary DNA

a segment of DNA whose nitrogenous bases are complementary to those of the parent DNA.

In 1991, a furor erupted over the right to patent genes when the NIH advised one of its scientists to apply for a patent. The scientist, **J. Craig Venter** of the National Institute for Neurological Disorders and Stroke, was determining the sequences of hundreds of complementary DNA (cDNA) gene fragments whose activities were then unknown. Because the fragments could be useful as probes, the NIH Office of Technology Transfer advised Venter to apply for patents on 347 of the genes detected by cDNA molecules.

The NIH sought patent protection because the potential commercial rights for this technology would otherwise be forfeited when the research was published in the scientific literature. Lawyers also pointed out that publishing the cDNA sequences without patent protection might create a disincentive for a company to use the gene to develop a new therapy. They reasoned that a company might discover the entire sequence and function of one of Venter's genes but have its patent application denied because the gene had already been described by Venter and was therefore in the public domain.

The effort to patent the genes was opposed by James Watson and the joint NIH-DOE genome advisory panel. A strongly worded letter sharply criticizing the NIH was dispatched by the panel to federal officials, including the Secretary of Health and Human Services (supervisor of the NIH) and the Secretary of Energy. The letter deplored the potential impact of those patent applications. It held that the public is paying for decoding the genome and the public should be allowed to decide what to do with the information. To advance genome research, it added, scientists should share mapping and sequence data, as well as material resources being developed (Questionline 12.2).

The scientific community's response to the patent application was equally vociferous. British researchers announced their intention to file a patent claim on hundreds of genes isolated by techniques similar to those used at the NIH. Other scientists pointed out that the few genes previously patented were known to encode a useful and known protein, whereas the NIH claim was for thousands of genes of unknown use. Industrial scientists also indicated their unwillingness to spend money to study genes and gene products if the NIH "owns" the genes.

The debate heated up considerably in the fall of 1991 when the NIH applied for patents on 2000 more human genes. NIH director Bernadine Healy indicated that, as before, the applications were a defensive measure to prevent exploitation by the private sector. Opponents were quick to point out patenting human genes offends the spirit of scientific inquiry and that patenting a gene on the basis of a partial inventory of its components was less defensible because it would discourage anyone but the patent holder from conducting further research on the gene. Even Watson, the Center director, was livid. He publicly described the patenting as "sheer lunacy."

To win a patent, an invention must be novel, useful, and not obvious to someone working in the research field. There is debate as to whether the gene work at NIH meets that test. Because Venter's applications were for cDNA molecules prepared by reverse transcriptase, NIH lawyers contended that the federal government should be

QUESTIONLINE 12.2

1. **Q.** Is the human genome the only one being worked on?

   A. No. A number of genomes are being studied in different laboratories throughout the world, and many genomes have already been deciphered. The genome of *E. coli* is known, as are the genomes of several other bacteria and the yeast *Saccharomyces cerevisiae*. A nematode worm, the fruit fly, two plants, and the mouse are currently being studied. Each organism is of particular interest to various biologists in their unique fields of research.

2. **Q.** What is the patent controversy I sometimes read about?

   A. There is question as to whether the gene probes used to identify genes and the genes themselves can be patented. Patent supporters suggest that the genes can be used for private monetary gain and that discoverers should reap the rewards. Patent opponents point out that genes are public property whose sequences are being deciphered by public money and that all researchers should have free access to the sequences.

3. **Q.** Exactly how will the human genome be deciphered?

   A. The Human Genome Project will attempt to do two basic things: first, map the human genome, that is, find out what the genes are and where they are; and second, learn the exact nitrogenous base sequence of each gene, as well as all the intervening nonfunctional sequences. The total to be determined is roughly 3 billion base pairs.

able to commercialize the fruits of federally funded research and should protect itself against foreign patent rights.

In 1992, the U. S. Patent Office issued a decision that the gene fragments could not be patented without a known function. (By that time, Venter had left NIH to found a private corporation devoted to gene sequencing.) The NIH appealed the ruling, stating that it is preferable for the federal government to patent the human genome than for private companies to do so. But within two years, NIH dropped its appeal, effectively surrendering any profits that might accrue from its work on gene sequences.

But the issue was not resolved. By 1996 scientists called for immediate release of "raw" genomic sequence, in public databanks accessible to all. They maintained that immediate release helps various laboratories coordinate their efforts, provides unfinished sequences of value in genetic studies, and promotes use of the raw data. Indeed, the following year, research leaders from several countries added their voices and appealed for raw data release as quickly as robotic sequencing machines could generate them. By that time, the NIH had established **GenBank**, a public database accessed by anyone through the Internet. And the federal agency was stipulating that new grantees will be required to place all their data in the bank within a few days or weeks of discovery. Finally, it was encouraging grantees to refrain from taking out patents on raw sequence data.

The main issue of who owns genes continues, however. In 1998, two biotechnology companies claimed rights to *BRCA1*, the gene that increases a woman's predisposition to certain types of breast cancer. Both companies were producing gene-detection tests, and both had patent rights—one company to the gene itself and the other to linkage markers associated with the gene. The issue was still making its way through the court system at publication time.

GenBank

a database established by NIH and containing gene sequences available to the public.

▨ MAPPING THE HUMAN GENOME

As any road atlas will show, it is possible to drive across the United States while staying on Route 80. You begin at the George Washington Bridge in New York City and 3000 miles later, you end your journey in San Francisco. Along the way you can stop and absorb the local culture in such places as Youngstown, Ohio; Chicago, Illinois; Des Moines, Iowa; Omaha, Nebraska; Salt Lake City, Utah; and Reno, Nevada. The key is staying on Route 80 and finding a few helpful gas station attendants to help you along your way.

It is clear that to get from here to there, one needs a descriptive diagram or map. A primary goal of the Human Genome Project is to make a series of maps of each human chromosome. Figure 12.7 displays several types. The mapping process will take place in steps, each section of a map describing the order of genes or other markers and the spacing between them on a particular chromosome. As the process moves along, maps of finer detail will emerge, and the clarity of the map, or **map resolution**, will increase. Thus, one could name ten large cities on Route 80 between New York City and San Francisco (Youngstown, Chicago, Des Moines, Salt Lake City, etc.) or there could be 100 cities on the map (Youngstown, Akron, Chicago, Davenport, Des Moines, etc.) or 1000 cities (Youngstown, Akron, Norwalk, Fremont, Angola [all in Ohio]) or even 10,000 cities, villages, and hamlets. In each case the value of the map has increased; the first map has the lowest resolution, the last map the highest resolution.

The mapping process for the human genome occurs in two major steps: dividing the chromosomes into small fragments that can be cloned and characterized and

resolution

degree of clarity between two points.

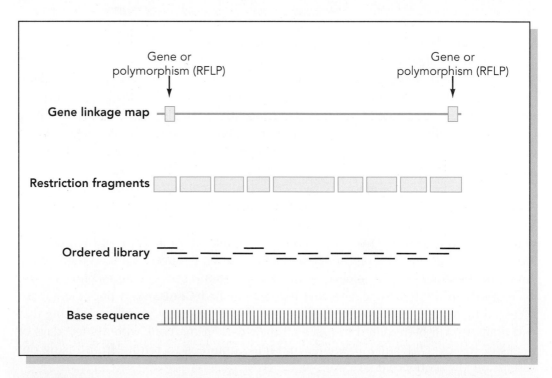

FIGURE 12.7

Various types of genome maps. The gene linkage map has the lowest resolution. It marks the location of genes by denoting the position of related RFLPs. The next map, the restriction fragment map, consists of random DNA fragments that have been sequenced. The ordered libraries of plasmids, cosmids, and artificially prepared chromosomes denote where DNA sequences are relative to one another. The base sequence map has the highest resolution and is the ultimate map. It denotes the location of every nitrogenous base in the human genome.

reconstructing the fragments to correspond to their locations on the chromosome. Once the mapping is completed, the next will be to determine the base sequence of the DNA fragments.

Two kinds of maps are under construction: gene linkage maps and physical maps. A **gene linkage map** is the coarsest type of map (i. e., has the lowest resolution). This map identifies the relative positions of genes and markers on the chromosome. A **physical map**, by contrast, describes the biochemical makeup of the chromosome itself. It is a high-resolution map that shows the location of genes and markers on the chromosome and number of nucleotides between them.

■ THE GENE LINKAGE MAP

The first and most traditional kind of genetic map is one based on linkage. The principle behind this map is that each individual has a unique package of genes, a package that arises during the formation of sperm and egg cells in the process of meiosis. In meiosis, chromosome pairs exchange pieces (geneticists call this process "crossing over"), and unique chromosomes result. The chromosomes then unite to form a unique individual.

The exchanges, however, are not all that random. This is because certain recognizable marker genes always seem to cross over with the genes. Large DNA fragments millions of nucleotides long move between chromosomes. Thus, two adjacent genes or markers are invariably inherited together, and the markers are said to be linked to the gene. Figure 12.8 illustrates how this occurs. Identifying the marker

gene linkage map
a chromosome map in which active genes are located by locating closely associated marker genes.

physical map
a chromosome map in which the location of the active genes and the number of bases between the active genes are known.

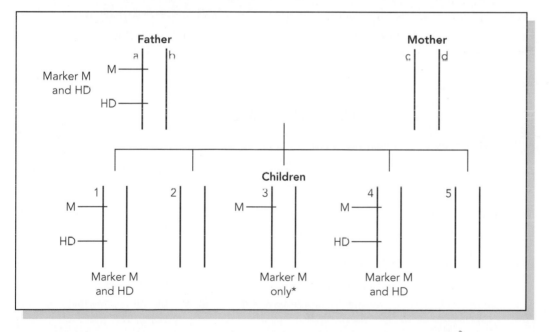

FIGURE 12.8

The gene linkage map is based on how close two genes are to one another. Chromosomal exchanges often occur during the production of egg cell and sperm cells, but if two genes lie close to one another, they will probably not separate when exchanges take place. In this diagram the father's chromosome contains the gene related to Huntington's disease (HD) and a marker gene (M). He and his wife have five children. Children 1 and 4 inherit chromosome a from the father, and children 2 and 5 inherit chromosome b. However, before child 3 was conceived, an exchange took place between chromosomes a and b during sperm cell production, and the child inherited chromosome a (because HD moved off to chromosome b). The fact that only one exchange took place in the chromosomes inherited by five children indicates that the HD and M genes probably lie close to one another. The frequency of gene exchanges and recombinations such as these help determine the distance between two genes on a gene linkage map.

at a chromosomal location identifies the gene at that same location. A **gene linkage map** thus describes the biochemical location of a gene on a chromosome.

The real advantage of gene linkage maps is narrowing the hunt for a gene down to a region of the chromosome only a few million base pairs long. To hunt for a gene, researchers can simply go to a freezer, secure the piece of DNA with the marker, and begin the work. Ferreting out the gene itself may be difficult, but with a gene linkage map the difference is like using a catalog to look for a book in a library compared with searching for the book without the catalog (Questionline 12.3).

The marker most commonly used in genetic analyses is the **restriction fragment length polymorphism** (RFLP, pronounced "rif-lip"). A RFLP is an apparently functionless stretch of DNA that can be detected because it resists restriction enzyme activity somewhere along the stretch. This resistance yields fragments of different lengths ("polymorphisms") that can be detected by electrophoresis. Chapter 7 explores RFLPs and their diagnostic use in more detail.

Another important marker is the **variable-number tandem repeats** (VNTRs, pronounced "vinters"). VNTRs are somewhat similar to RFLPs, except that VNTRs arise from the number of repeated base sequences between two points on the DNA molecule. Because there are many different VNTRs in a person's genome, the pattern of VNTRs is unique for an individual, and DNA fingerprinting is based on their presence (Chapter 9). More than 1500 VNTR and RFLP markers have been located thus far.

By studying the pattern of genetic disease in a family, DNA technologists can find the general location of the responsible gene on a chromosome by locating a marker occurring only in people with the disease. In a sense, this process works backward from the traditional mode of relating a disease to a gene. In previous decades, a disease would be related to an impaired function, then the gene would be pinpointed and its location on a chromosome noted. In contemporary times, a disease is studied in families, and marker analysis is used to pinpoint the responsible gene on a

RFLP

a stretch of DNA occurring in a chromosome and having different lengths in different individuals.

electrophoresis

e-lek″tro-fo-re′sis

VNTR

a stretch of DNA occurring in a chromosome and having repeated base sequences.

QUESTIONLINE 12.3

1. **Q.** What is a gene linkage map?

    A. A gene linkage map is one that indicates the approximate location of a particular gene by linking it to a marker base sequence. Gene markers such as RFLPs are used to "flag" the gene. By locating the RFLP, one locates the gene. The gene linkage map is very rough and is much like saying "look for the Arch and you are in St. Louis" or "find the Empire State Building and you are in New York City."

2. **Q.** What is a physical map?

    A. A physical map is one that indicates which genes follow which in a particular chromosome. It contains the number of bases between genes much like a road map tells the mileage between cities. As physical maps become more detailed, more genes are located in the spaces between known genes.

3. **Q.** How complex is the process of base sequencing?

    A. Sequencing the bases in the human genome is somewhat like identifying each brick in a road that circles the earth several times. It is an extraordinarily tedious process that requires high technology and advanced automation. Scientists break the DNA into mapped fragments about 1500 base pairs in length. The fragments are fed into an automated sequencing machine that tags the DNA with four different dyes, one for each of the four bases in DNA. The order in which the colors stream out of the machine represents the order of DNA's bases.

chromosome (even though no effects of the gene are observed). Then the gene is cloned and its functions are studied to locate a mutation that can account for its potential malfunction in the ill patient.

The contemporary approach to genetic disease is sometimes called **reverse genetics** because the disease gene is used to identify the metabolic defect. For the human genome project, the process can be used to identify a base sequence in a gene by working backwards: identify the mutated protein produced on the disease gene; learn the amino acid sequence of the correct protein occurring in normal individuals; and look up the genetic codes for the amino acids in the protein to learn the base sequence in the gene. Although the process appears simple and straightforward, the biochemistry is extraordinarily complex.

Using a **gene linkage map** to find one's way along a chromosome is somewhat similar to knowing landmark cities on a road map, but not the number of miles between them. Although the mileage between the cities is unknown (as they are on a physical map), nevertheless, a gene linkage map can be useful because the more markers (or "cities") that can be established, the more confident scientists can be of knowing where certain genes are. Gene linkage maps are already in use to establish the presence of genes for such diseases as Huntington's disease, neurofibromatosis, and other genetic disorders discussed in Chapter 7.

The closer a gene and marker are on a chromosome, the less likely any sort of change will occur between them. This means that if gene and marker are "**tightly linked**," a genetic recombination (such as a separation during crossing over) is unlikely to take place. Thus the frequency of recombination observed at a genetic site gives an estimate of how close the gene and marker are to one another. Figure 12.9 shows how a chromosome might look to a map constructor.

On a gene linkage map, the distances between genes and markers (or other genes)

reverse genetics
the concept of wrong gene analysis to identify a metabolic defection in an individual.

Huntington's disease
a genetic disease of the nervous system accompanied by thrashing movements.

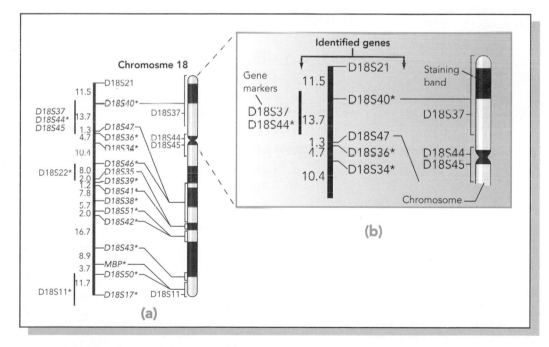

FIGURE 12.9

The gene linkage map. (a) Human chromosome number 18 as it might look to a biochemical geneticist after certain genes have been identified. (b) The top of the chromosome showing some of its details. The thick black bar next to the identified genes gives the distance between markers in centimorgans. The thin black bar delineates the region where they occur. The chromosome itself shows the characteristic staining bands seen under the microscope and the regions of high genetic activity.

centimorgan (cM)

a unit of molecular genetics equivalent to the space occupied by a million base pairs in a DNA molecule.

are measured in **centimorgans (cM)**, named for the geneticist Thomas Hunt Morgan. If a gene and marker have a recombination frequency of one percent (i. e., they are separate during one percent of all crossings ever observed), the gene and marker are said to be 1 cM apart. This unit is equated to one million base pairs. Hence 1 cM equals one million base pairs. Considering that the human genome has roughly three billion base pairs, a quick calculation shows that its length is about 3000 cM. At intervening distances of 50 cM or more, a gene and a marker are considered unlinked.

At this writing, the current resolutions of most human genome maps are about 10 cM (i. e., on an average, a gene and a marker are about ten million base pairs apart). A short-term goal of the project is to reduce the distance and increase the resolution of genetic linkage maps to 2 cM. It is hoped that a physical map will achieve a resolution of 0.1 cM. We shall see how that is accomplished next.

■ THE PHYSICAL MAP

The second type of chromosome map is the **physical map**. A physical map gives the actual distances in number of bases between genes on a chromosome. As such, it describes the structural characteristics of the chromosome itself. Physical maps are created by chemically snipping the chromosomes into pieces, cloning the pieces, and reordering them the way they originally occurred on the chromosome, as Figure 12.10 displays. Knowing where each piece comes from helps the DNA technologist know where a gene within that piece is located.

Assembling a physical map is somewhat analogous to solving a jigsaw puzzle. To be sure, a jigsaw puzzle with a few large pieces is much easier to work with than one with many smaller pieces, although less information will be learned about how the puzzle is constructed. For the human genome, a map with many large pieces is the **low-resolution physical map**. This map is based on the distinctive banding patterns of stained chromosomes as observed by light microscopy. The map, sometimes called a **cytogenetic map**, shows the locations of expressed DNA regions (the exons) on the chromosome.

cytogenetic map

a chromosome map in which the active genes respond to a chemical dye and display themselves as bands on the chromosome.

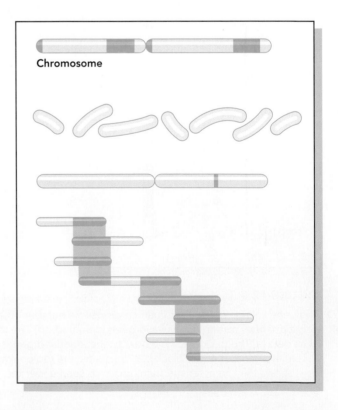

FIGURE 12.10

The basis for the physical map. The physical map is one in which chromosomes are cut into large fragments, then copied (or cloned) many millions of times. The copies are then analyzed with a goal of finding segments that overlap. Rearranging these overlapping segments reconstitutes the chromosome.

To create a cytogenetic map, messenger RNA for a particular exon is used in the laboratory to formulate a complementary DNA (cDNA) molecule. (These are the cDNAs for which patent rights are desired.) The cDNA molecule is then tagged with a fluorescent or radioactive label and used as a probe. When the tagged **cDNA probe** is combined with cells undergoing cell division, the probe will bind to (hybridize to) its complementary DNA molecule in an intact chromosome. Thus, the location of the exon that expressed the original mRNA molecule can be pinpointed.

A **typical band** seen on a chromosome by cytogenetic staining is about ten million base pairs long, although recent technical improvements have improved the length to about five million base pairs. (Figure 12.11 shows the bands in a cytogenic map.) Maps of this type have low resolution. Despite this drawback, the cytogenetic map is advantageous because it identifies expressed regions of gene activity (exons), which are regions of substantial genetic significance. When used with gene linkage maps, the cytogenetic map can suggest the approximate location of a disease gene and narrow the search for that gene considerably.

Although the cytogenetic map is useful, cloning the entire human genome requires identification of successively smaller pieces of base sequences, and more detailed

cDNA probe

a molecule of DNA formulated from the base code in mRNA and used to detect a complementary molecule of DNA.

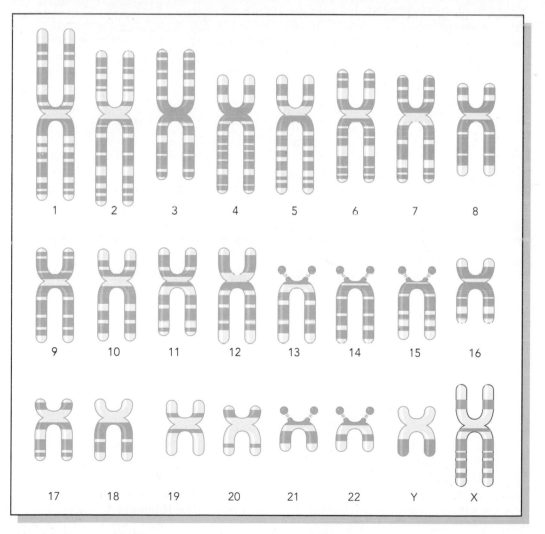

FIGURE 12.11

The cytogenetic map of human chromosomes. In this schematic drawing, bands can been seen where Giemsa stain has been attracted to the chromosome. Approximately 400 bands are detectable on a chromosome set such as this. In very rough numbers, each band represents about 7.5 million base pairs or almost twice that of the entire *E. coli* chromosome.

contig map

a chromosome map in which various DNA segments have been joined together to form a contiguous (continuous) DNA molecule.

maps with higher resolution will be developed. These **high-resolution physical maps** are prepared by aligning the sequences of adjacent DNA fragments to form a continuous map called a **contiguous** (or "**contig**") **map**. As contig maps of smaller and smaller DNA fragments are established, the resolution of the map will increase, and the highest resolution map will be one containing all the base pairs of each chromosome, that is, the complete human genome.

Contig maps consist of a linked library of small overlapping clones of DNA representing a complete chromosomal segment. Early in the genome project scientists began maintaining a central bank of DNA clones and distributing the clones to laboratories all over the world for mapping. It appeared that large DNA segments (contigs) from 10,000 to a million base pairs long would have to be used. However, a set of alternative DNA markers called sequence tag sites were soon found. The markers are now in use.

sequence tag site (STS)

a DNA segment that is attached to a selected location on a DNA molecule to mark the location previous to analysis.

Sequence tag sites, or **STSs**, have removed the need to centralize a bank of DNA clones. The STSs are short sequences of DNA about 200 to 500 base pairs long. They can be used to mark a spot on a chromosome and locate adjacent segments of a long DNA molecule. The STS markers are attached to the DNA molecule and the molecule is then split into fragments. Now the fragments are analyzed to detect the STSs (using the polymerase chain reaction to amplify the DNA). The underlying premise is that if two fragments share an STS, they must overlap. Figure 12.12 illustrates the procedure.

An STS can be synthesized in the laboratory by chemically stitching together a set of known nucleotides to form a single-stranded DNA probe. The STS is used to bind to and identify a particular segment of DNA (as explained previously) on which the researcher wishes to work. A long-range goal is to synthesize approximately 30,000 STSs that would span the human genome at intervals of about 100 base pairs.

The use of STSs enables a researcher to isolate and work on a particular stretch of DNA while letting other researchers know what DNA segment is being investigated. Thus, it provides a common language among DNA technologists. Also, and very importantly, it eliminates the need to obtain a particular DNA segment from a central bank of DNA clones. DNA technologists have proposed that all regions of the chromosome, introns as well as exons, be marked by this international system. In addition, the polymerase chain reaction has made it possible to amplify STS exchanges rather than cloned DNA exchanges.

Having identified the DNA segment to be studied, the DNA technologist must clone the segment in a useful vector organism. In early DNA cloning efforts, no more than 40,000 bases could be cloned in a single experimental organism (e. g., *E. coli*). The implication was that tens of thousands of clones would have to be analyzed to map a single human chromosome. A search was therefore instituted to locate a vector where much larger DNA fragments could be cloned. The result was the yeast artificial chromosome.

yeast artificial chromosome (YAC)

a segment of DNA derived from a yeast cell and used as a base for attachment of other DNA segments.

Yeast artificial chromosomes (YACs) were introduced in 1987 by **Maynard Olson** and his colleagues at Washington University. A YAC is a fragment of yeast DNA that will accept a stretch of foreign DNA between 250,000 and 500,000 base pairs in length, or about fifteen times that handled in other cloning systems. Such a foreign DNA fragment is stitched to a yeast DNA fragment representing less than one percent of the total DNA in the yeast. The yeast fragment is valuable because it contains three essential genes needed for replication: the genes for the telomere, the genes for the centromere, and the gene for the origin of replication. The **telomere** is a region at the end of the yeast chromosome that protects the chromosome from nuclease degradation; the **centromere** is the region where spindle fibers form during mitosis and attach to the chromosome; and the **origin of replication** is the site where DNA polymerase initiates DNA duplication during mitosis. Figure 12.13 illustrates the YAC.

As long as the YAC contains the three essential genes, it will replicate, and if long stretches of human DNA (called megabase fragments) are attached to it, they will

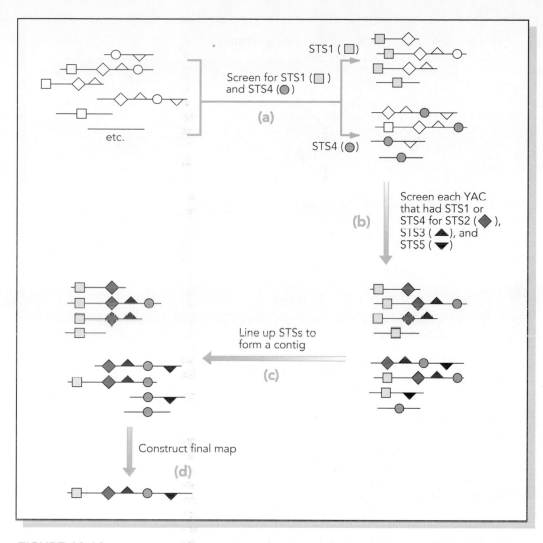

FIGURE 12.12

Making a physical map (a contig map) using STSs. (a) At the top left, a long DNA molecule has been fragmented into random segments, and several DNA fragments are shown. Different symbols on the fragments represent different STSs that have been placed at specific intervals. In step a, the DNA technologist screens for STS1 and STS4. At the top right, all the fragments with STS1 and STS4 are shown dark. (b) Now each of the fragments is screened for STS2, STS3, and STS5. At the bottom right, the STSs detected in each fragment are shown. (c) Next, the fragments are lined up to form a partial contig map using the overlapping STSs as a guide. (d) The final contig map locates adjacent segments on the original long DNA molecule. The length of the fragments can be measured by electrophoresis.

replicate as well. Thus it becomes possible to clone large segments of genomic DNA using YAC methods and to drastically reduce the number of clones that must be analyzed. Already, many YACs have been produced that span an entire human gene.

To obtain a physical map of contigs, the DNA technologist obtains a YAC and cuts it into smaller fragments for analysis. The actual analysis is performed on the ends of the YAC fragments and computers search for ends that can overlap. Once **overlapping ends** have been located, the fragments can be coalesced to form a single long stretch of DNA, the contig map. If one or more STSs can be located in the fragments, then additional evidence is presented that the fragments overlap. Also, the physical and genetic linkage maps can be prepared.

To obtain fragments from the YAC, yeast cells containing foreign DNA are digested within small blocks of agarose gel. Restriction enzymes are then added to the cell

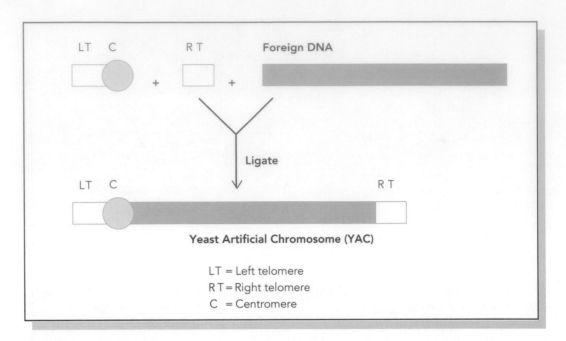

FIGURE 12.13

Construction of the yeast artificial chromosome (YAC). The construction involves two opposite ends of a yeast chromosome, each with a telomere (telomeres protect against DNA degradation). These ends are represented as LT (left telomere) and RT (right telomere). The LT is attached to a centromere (C), which is necessary for DNA replication. A large segment of foreign DNA is added, and the three segments are ligated. What results is the YAC. The large DNA segment can now be cloned easily.

debris within the blocks to produce the fragments. The fragments are then analyzed by Southern blotting. Named for Edward Southern, this technique is based on electrophoresis and gene probes. Also used for analyses in DNA fingerprinting (Chapter 9), Southern blotting produces a "fingerprint" of the YAC fragments.

The techniques for physical mapping are tedious and complex, but they can be accelerated, although with cruder results. In 1992, French researcher **Daniel Cohen** announced a process that forces yeast cells to accept pieces of human DNA a million or more bases long (double the previous size). The YAC thus formed came to be called the **megaYAC**. To put together a physical map, Cohen stitched together half the previous number of DNA pieces. Moreover, his laboratory found a way to automate the Southern blotting procedure using robots and five technicians instead of hundreds of skilled workers. The process permitted him to prepare a physical map of human chromosome 21.

But megaYACs were not the panacea that genome researchers hoped they would be. It often happens that the yeast enzyme stitches the foreign DNA fragments to one another instead of stitching the fragments to the yeast chromosome. The result is a chimera consisting of various foreign DNA segments. In the megaYAC, there may be deletions of base segments in the foreign DNA, and in some cases, regions of genomic DNA may be toxic to yeast or too fragile to be cloned without breaking.

Despite its limitations, the megaYAC has been used by Cohen to develop a high-resolution physical map for **human chromosome 21**. In their seminal article published in 1992 in *Nature*, Cohen and thirty-five coauthors acknowledged that they could not pinpoint seven genome markers known to be on chromosome 21 and that deletions were present on a significant percentage of the megaYACs they used. Nevertheless, his group assembled a set of overlapping DNA segments ostensibly in the correct order.

Cohen's group used **sequence tag sites** to obtain 198 equally spaced biochemical "landmarks" on chromosome 21, and they split the chromosome of about forty-three

MegaYAC

an unusually large YAC.

million bases into manageable chunks. In completing the map, the researchers skipped portions of the short arm of the chromosome, an arm composed of repeated sequences that are biologically less interesting because they do not appear to encode proteins. As noted, they also could not establish sites for seven markers, although they found the approximate regions where they occur. The genes for amyotrophic lateral sclerosis (Lou Gehrig's disease), some forms of epilepsy, and Alzheimer's disease lie on chromosome 21.

Almost simultaneously with Cohen's announcement came a report by **David Page** and his group (Figure 12.14) at the Whitehead Institute for Biomedical Research. The Page group reported that it had prepared a high-resolution physical map of human **chromosome Y**, the chromosome of human maleness. Their article, published in *Science*, described their use of DNA from a man with three extra Y chromosomes in his cells. The group used the DNA to form megaYACs having 650,000 bases, as Cohen's group had done. Then they generated nearly 200 STS markers and used them to screen the megaYAC clones and assemble the DNA segments in order. As mentioned previously, if two segments share one or more STS markers, they must overlap. The result was a physical map containing twenty-eight million bases with no apparent gaps and STS markers spaced roughly every 220,000 bases.

The mapping of two human chromosomes was hailed as a milestone in genetics. For the first time, researchers successfully mapped tens of millions of base pairs in a continuous set of fragments. The two chromosomes mapped make up only two percent of the total human genome, but the methods used to map them can conceivably be used on a grand scale to decipher the entire human genome. To be sure, the map would be a low-resolution map with widely space STS markers, but in the view of some researchers, it is better to obtain the full map before concentrating on the sequence of base pairs.

Having a complete physical map will revolutionize the process of locating a disease-related gene. As one researcher noted, if the disease gene is linked to RFLP or another marker, the researcher will only have to screen a library of megaYACs for the one that contains the marker. Then the researcher can "parachute" to within a million bases of the target gene. Once they are there, however, they will need a finer map, a **sequence map**. Such a map contains every single base, or as another scientist has

sequence map

a chromosome map in which the nature and position of each nitrogenous base is known.

(a)   (b)

FIGURE 12.14

Mappers of the human genome. (a) David Cohen, the French investigator who used megaYACs to develop a high-resolution map for human chromosome 21. (b) David Page (far left) at his research group at the Whitehead Institute in Cambridge, Massachusetts. Page's group mapped the Y chromosome of human males.

noted, every single brick on the highway from New York to San Francisco. That sequence map will be determined by techniques we will examine next.

■ SEQUENCING THE GENOME

When the physical map of the human genome has been completed, the name and position of each nitrogenous base will be known. Having such a physical map will be significant because the effort spent locating genes then can be shifted to determining the functions of the genes.

One of the major methods for sequencing a DNA fragment is the **Sanger method**, named for Frederick Sanger, the biochemist who won separate Nobel Prizes for explaining the structure of insulin (1958) and the sequence of bases of an RNA virus (1980). The Sanger method is also known as the **chain-termination method** for how it works and the **dideoxy method** for the biochemicals it uses. The method uses an enzymatic procedure to synthesize DNA chains of varying lengths, stopping the DNA replication at one of the four bases. Each of the resulting fragment lengths is then analyzed by electrophoresis, and a base sequence can be elucidated from the results.

The Sanger method works this way: The technologist begins with a single-stranded DNA fragment of unknown base sequence. To this is attached a primer sequence complementary to one end (the origin) of the fragment. This primer provides a starting point for DNA synthesis. DNA polymerase is then mixed in to initiate DNA synthesis. Next, the reactants are placed into four separate tubes. A mixture of the four different nucleotides (A,T,C, and G) is added to each tube, and DNA synthesis by primer extension is encouraged, as Figure 12.15 shows.

Each of the four tubes now receives a different radioactively labeled nucleotide and a different synthesis-stopping dideoxynucleotide (dd-nucleotide). The first tube, for example, receives ddA (dideoxyadenosine). In this tube, DNA synthesis will stop whenever the growing chain incorporates the ddA. Thus, the tube will contain a series of different-length DNA fragments. These fragments correspond to the various lengths of DNA made before ddA is encountered. And the fragments all carry a radioactive label because the primer molecule was labeled.

To see , consider what will happen in the sample tube containing dideoxyadenosine. Let us assume that adenine is the fourth and sixth base in the unknown DNA fragment whose base sequence we are attempting to learn (G-G-C-A-T-A-G-C-T). When the reaction occurs in the tube, DNA fragments containing ddA will form after four bases are reached (G-G-C-ddA) and after six bases are reached (G-G-C-A-T-ddA). But recall that the primer was labeled with radioactivity, so now we have a radioactive four-base fragment (*G-G-C-ddA) and a radioactive six-base fragment (*G-G-C-A-T-ddA). The two fragments can then be separated on an electrophoresis apparatus, and bands will show up where the adenine was placed in the DNA chain. Working backwards from the base sequence of the extended primer, we can deduce the base sequence of the original fragment (Figure 12.16).

Let us take another example. In tube two, we place ddC (dideoxycytidine). In this tube DNA synthesis ceases wherever cytosine is encountered because the ddC halts the reaction. Now consider the base sequence above (i. e., G-G-C-A-T-A-G-C-T). Cytosine is the third and eighth bases. Thus, a labeled fragment will form that is three bases long (*G-G-ddC) and another that is eight bases long (*G-G-C-A-T-A-G-ddC). When separated according to size by electrophoresis, an X-ray band shows up in the gel for three-base lengths and eight-base lengths in the cytosine area.

If the Sanger method is followed and four tubes are used, the mixtures resulting from the reaction will contain one-base DNA fragments, two-base DNA fragments,

chain-termination method

a method of base sequencing in which the synthesis of DNA during replication is halted after insertion of an unusual nucleotide.

dideoxynucleotide

di″de-ox′e-ri″bo-new′cle-o-tide

dideoxyadenosine

di″de-ox′e-a-den″o-sen

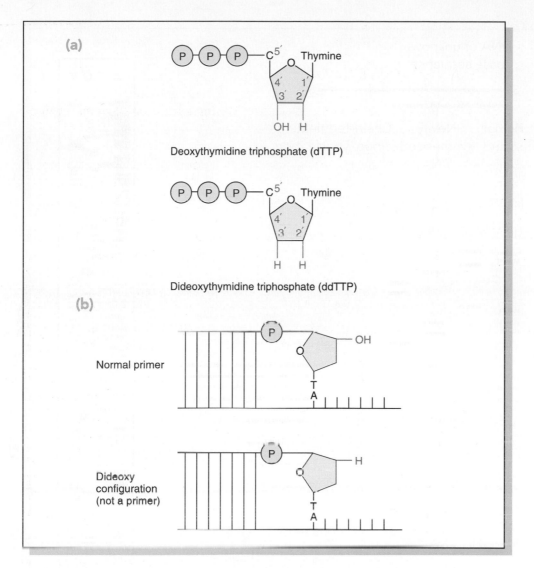

(a)

Deoxythymidine triphosphate (dTTP)

Dideoxythymidine triphosphate (ddTTP)

(b)

Normal primer

Dideoxy configuration (not a primer)

FIGURE 12.15

How dideoxynucleotides interrupt DNA synthesis by interrupting chain elongation. (a) Note the difference at position 3′ between normal deoxythymidine triphosphate (dTTP) and dideooxythymidine triphosphate (ddTTP). The absence of an –OH group prevents binding of the next nucleotide. (b) The normal primer containing dTTP and the abnormal molecule containing ddTTP are shown. The abnormal molecule cannot be elongated any further.

three-base DNA fragments (ending in C), four-base DNA fragments (ending in A), five-base DNA fragments, six-base DNA fragments (A), seven-base DNA fragments, eight-base DNA fragments (C), nine-base DNA fragments, and continuing numbers of different-base DNA fragments, each containing a terminal labeled base (note the ones we have determined). By simply reading the bands on the X-ray film, the DNA technologist can determine the base sequence of the unknown fragment.

In 1977 Frederick Sanger used his method to determine the entire genome sequence of a bacteriophage (bacterial virus). Unfortunately, the available gel-based sequencing technologies are inadequate for sequencing the entire human genome. The major problem areas are time and expense. As of 1993, for example, the largest continuous DNA molecule sequenced was approximately 350,000 base pairs in length. By comparison the smallest human chromosome contains 50,000,000 base pairs. In addition, the available equipment (as of 1997) can sequence only 50,000 to 100,000

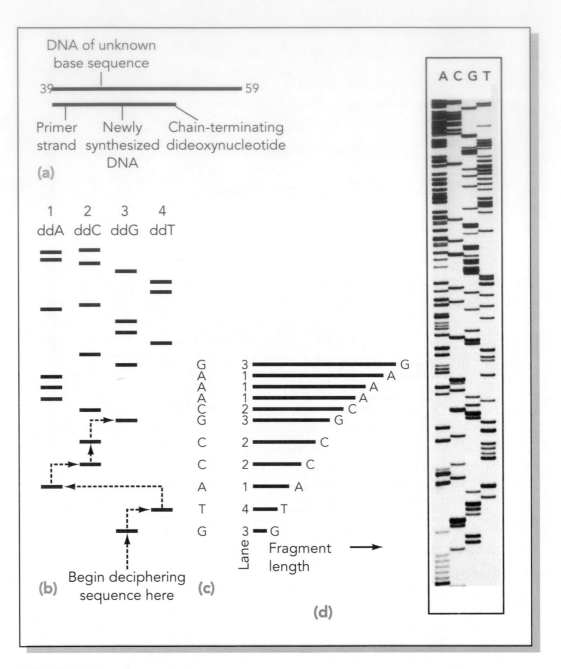

FIGURE 12.16

The Sanger method for DNA sequencing. (a) The process begins with a fragment of DNA of unknown base sequence. A primer DNA fragment is chosen so that its base sequence is complementary to one point in the unknown DNA molecule. Also added is a reaction mixture containing all four radioactive nucleotides and small amounts of a single dideoxynucleotide (e. g., ddA, ddC, ddG, or ddT). The dideoxynucleotide serves as a chain terminator. (b) DNA polymerase catalyzes the synthesis of DNA complementary to the unknown DNA, and the products of synthesis are then separated by electrophoresis. For a given dideoxynucleotide, all fragments having that base give rise to radioactive bands on the gel. Reading the lanes in the gel, it is clear that the smallest labeled fragment (the fragment with radioactive G) has moved the fastest and appears at the bottom of the gel. The fragment with ddT moves next fastest, and we can see evidence for its presence in lane 4; then ddA in lane 1; ddC in lane 2; ddC in lane 2, and so on). The sequence of the original unknown DNA would be the complement of this sequence, or C-A-T-G-G-etc. (d) A photograph of a typical sequencing film. Beginning at the bottom, the sequence reads C-A-A-A-A-A-A-C-G-G-etc.

base pairs per year at an approximate cost of $1 to $2 per pair. It has been estimated that to sequence the entire human genome at this rate, a total of 30,000 work-years would be consumed at a cost of $3 billion for the sequencing process alone.

What are the alternatives? One possibility is a procedure called **Maxam-Gilbert sequencing** or the **chemical degradation method**. This procedure was worked out by Allan Maxam and Walter Gilbert (Nobel Prize winner with Sanger in 1980). It uses "rare-cutter" enzymes to cleave DNA at specific bases, thereby resulting in fragments of different lengths. Investigators can then use Southern blotting and pulse-field electrophoresis to identify up to forty different fragments of a single sequencing gel. This method, however, suffers from the same cost drawbacks and tedium as the Sanger methods. The two methods are now used to sequence small regions of interest in the human genome, but they would be impractical for large-scale sequencing projects.

To resolve the sequencing dilemma, a major focus of the Human Genome Project is developing an **automated sequencer** that can accurately sequence more than 100,000 bases per day at an approximate cost of $0.50 per base as of 1998 and $0.20 per base by 2000. An important word is "accurately" because a single nucleotide mistake can have disastrous consequences. Inserting or deleting a single base, for example, can shift the reading frames and predict an incorrect amino acid sequence in the protein encoded. Until 1997, the error rate was about 0.1 percent (a 99.9 percent accuracy rate, or one error per 1000 bases). If one exon is roughly 200 nucleotides in length, this rate implies that one error may occur in every five exons. This error is of significant magnitude, and it remains a point of considerable controversy. By 1998, Francis Collins of the National Human Genome Research Institute was pushing to lower the error rate to 0.01 percent (a 99.99 percent accuracy rate, or one error per 10,000 bases).

Once a base sequence of a particular chromosome area has been completed, it is possible to increase the length of the strand by a process euphemistically called **chromosome walking**. Chromosome walking fills the gap between two segments of DNA whose base sequences are know by essentially extending one tip out to reach the other. It is a method of systematically moving along a chromosome from a known location until the opposite side of the gap is reached.

The biochemistry of chromosome walking begins by obtaining the DNA from the gap to be sequenced. The DNA then is broken into segments approximately 40,000 base pairs long. A marker is now used to identify an appropriate gene probe (from a probe collection) that has a section of base pairs identical to the base pairs of a segment. The probe is then reproduced and sequenced. The end of this probe protrudes further into the gap than before and it therefore extends the DNA chain while narrowing the gap.

Now the process is repeated using an adjacent segment from the gap to be sequenced. The first probe is taken and a second probe is made from its far end. This probe is used to reprobe the adjacent segment. When a match is made between the base pairs of the second probe and segment, a new probe with that base sequence is obtained from a genomic library and its base sequence is determined. As before, this sequencing extends the DNA chain and narrows the gap still further. By these repeated cycles of walking, the gap is gradually closed, and eventually the two ends of the DNA are joined. The sequencing of that section is now complete. Figure 12.17 depicts the process.

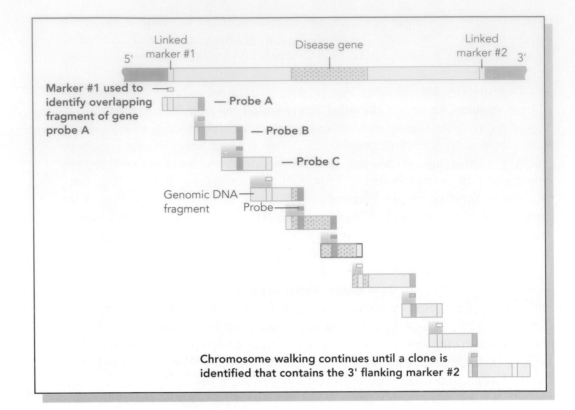

FIGURE 12.17

Chromosome walking to identify the flanking base sequences around a gene of interest. In this example, a disease gene has been identified within a gap flanked by linked markers #1 and #2. To begin the "walk" marker #1 is used to identify a gene probe (A) from a probe collection with a section of bases identical to those of the marker. This probe has ends that protrude out into the gap. The probe is cloned and is then used to identify another gene probe (B) from a probe collection that has an overlapping section of bases. Repeating the process, the far end of probe B is cloned and used to identify a gene probe (C) with yet another set of overlapping bases. Each time this is accomplished, the genomic DNA fragment is extended into the gap. Eventually a gene probe will be identified with a set of bases identical to the marker #2, and the gap will be closed.

chromosome walking

a base sequencing method in which a chromosome is analyzed by extending the tip out a bit at a time and determining the base sequence of the tip.

A substantial problem in chromosome walking is the time and expense needed to make the probes used for successive matchings. Traditionally, this has involved inserting the probes into closed-loop strands of DNA from a bacterial virus (bacteriophage). The fragments are then sequenced by adding a short primer of fifteen or more bases that bind to a known sequence on the vector phage and serve as a starting point for producing a series of strands complementary to the unknown DNA. These strands are then analyzed by Southern blotting and electrophoresis to read the sequence. Then the technician must determine where each fragment is fitted on the original continuous DNA by comparing the overlapping sequences. The process has been called the **"shotgun" technique** because it uses a random selection of DNA segments that must be pieced together.

Simplification and automation are the two key concepts that will attend the sequencing techniques as they continue to evolve. In 1993, for example, two groups published papers on **primer walking**. In this process, a single long segment of DNA can be inserted as a primer to a vector. Then, after a sequencing run, a new DNA primer is bound to the first primer and a new sequencing is performed. The process is repeated over and over again until the whole DNA segment has been sequenced. Six-base sequences called hexamers are being researched as ways of binding the

(a)

(b)

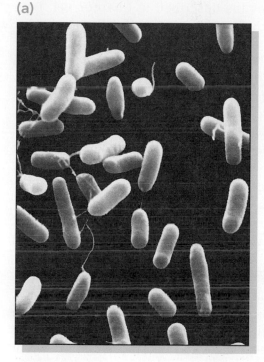

FIGURE 12.18

The first completed genome. (a) *Haemophilus influenzae*, the bacterial rod whose genome was the first to be deciphered. (b) J. Craig Venter, one of the researchers to complete the sequencing.

successive DNA primers to the former ones, a step necessary for successful automation of the procedure. Eventually, one researcher believes, a sequencing job of several years (such as for *E. coli*) might be reduced to a few weeks.

■ COMPLETED GENOMES

History was made in May 1995, when the race to provide the first complete genome of a free-living organism was won by **J. Craig Venter** (Figure 12.18) and his associates at The Institute for Genomic Research (TIGR). Scientists at TIGR, a private company, worked with Nobel laureate **Hamilton Smith** and his coworkers at Johns Hopkins University. Jointly, the two groups unveiled the sequence of 1.8 million base pairs in the genome of the bacterium *Haemophilus influenzae*.

Haemophilus influenzae is a possible cause of secondary infection in patients with pneumonia, and strain b of *H. influenzae* induces bacterial meningitis in young children. The bacterium's genome consists of 1749 genes and contains the entire information needed to sustain life (compared with viruses, which rely on genetic information from their host cells). The Venter-Smith team fragmented the bacterial genome with ultrasonic (high-pitch) vibrations, then sequenced the fragments, and without a map, arranged them in order using innovative computer software and sequence overlap information. By increasing the computer's efficiency to process more information, they were able to deduce in a few short months the genome for a second organism for *Mycoplasma genitalium*. This reproductive tract pathogen is among the smallest bacteria known and has one of the smallest genomes (580,000 base pairs).

Among the noteworthy points of the *H. influenzae* genome were its relative absence of noncoding (intron) DNA. This finding was not unexpected because most bacteria contain few introns. However, the scientists noted that more than forty percent of the genes (736 of 1743) appear to have no counterparts among genes found in other

Haemophilus
he-mof´i-lus

Mycoplasma
my-co-plas´mah

genitalium
gen´´i-tal´-i-um

Krebs cycle

a cyclic series of enzyme-catalyzed biochemical reactions in which energy-rich compounds are formed from the breakdown of carbohydrates.

Methanococcus

meth″an-o-kok′us

jannaschii

jan′ah-shi

Archaea

a recently identified domain of living things consisting primarily of bacteria living under extremely harsh environmental conditions.

Staphylococcus

staph″i-lo-coc′cus

neurofibromatosis

nu″ro-fi″bro-mah-to′sis

organisms. By "knocking out" these genes (as described in Chapter 11), DNA technologists can determine the proteins they encode and establish their unique functions in the bacterium. In addition, the genome did not contain genes for three enzymes that function in the Krebs cycle. This chemical process is widely found among living things, and the absence of three enzymes may shed light on an alternative mechanism in bacteria. Other features of the bacterium's genome are listed in Table 12.2.

In his 1995 report, Venter and his coworkers suggested that the "shotgun" sequencing method would hasten the discovery of other genomes. Within a year, they fulfilled their promise with the complete sequencing of the bacterium ***Methanococcus jannaschii***. This organism represents the Archaea, a domain of living things distinct from the traditional bacteria (prokaryotes) and the more complex eukaryotes (e. g., fungi, protozoa, plants, and animals). First proposed by **Carle Woese** and his coworkers in 1977, the existence of the **Archaea** domain was viewed as a threat to an essential tenet of biology that all living things are either prokaryotes or eukaryotes. But Venter's research team solidified Woese's proposal by showing that the genes of *M. jannaschii* are two-thirds different from those of traditional bacteria (such as *E. coli*). For instance, the proteins that replicate the DNA in *M. jannaschii* have no counterpart in traditional bacteria. Moreover, its protein structures more closely resemble those of eukaryotes than prokaryotes. And the proteins of RNA synthesis are quite different than those of *E. coli* and other bacteria. With evidence supporting the third domain, Archaea, a new view of life's organization on Earth continued to unfold.

That same year, the United States government provided budgeting support to begin the sequencing of the small flowering plant, ***Arabidopsis thaliana***. Its initial investment of $12.7 million ended an apparent omission of interest in complex plants and pleased botanists who see *A. thaliana* as the plant's equivalent of *E. coli*, the workhorse of microbiology. Plant scientist **David Meinke** of Oklahoma State University was appointed to head the plant genome project. Eighteen other laboratories from Europe and Asia are also involved in the international effort.

The genome sequences for another bacterium and a virus were also announced in 1996. In March, scientists from Human Genome Sciences published the sequence of 2.8 million base pairs in the 2400 genes of ***Staphylococcus aureus***. Strains of this organism cause food poisoning, skin infections, and toxic shock syndrome. The virus sequenced was the **molluscum contagiosum virus**, a cause of benign skin tumors. Its relatively small genome contains 190,000 base pairs organized to 163 genes.

But the big news of 1996 was completion of the genome for the baker's yeast ***Saccharomyces cerevisiae***. In a field already littered with milestones, the sequencing of *S. cerevisiae* marked the first insight into the eukaryotic genome. The project encompassed seven years, required $30 million, and needed the resources of thirty-seven laboratories and 100 research groups coordinated by **Andre Goffeau** of Belgium's Catholic University of Louvain. Sixteen chromosomes were analyzed, twelve million bases were sequenced, and 6000 genes were identified. The genome was seven times larger than that of *H. influenzae*.

The unveiling of the **yeast genome** marked the first time scientists have sequenced all the genes of an organism remotely similar to a human being. Immediately a practical application emerged: Two newly identified yeast genes encode a protein that inhibits the *ras* gene, which regulates cell division. In human neurofibromatosis, a disease accompanied by brain tumors, scientists have found a defective gene at fault. The gene resembles the yeast genes in structure and function, thereby implying that it may be unable to inhibit the *ras* gene in a human cell. Without

TABLE 12.2
Features of the Genome of Haemophilus Influenzae.

DESCRIPTION	NUMBER
Double-stranded templates	19,687
Forward-sequencing reactions (M13-21 primer)	19,346
Successful (%)	16,240 (84)
Average edited read length (bp)	485
Reverse sequencing reactions (M13RP1 primer)	9,297
Successful (%)	7,744 (83)
Average edited read length (bp)	444
Sequence fragments in random assembly	24,304
Total base pairs	11,631,485
Contigs	140
Physical gap closure	42
PCR	37
Southern analysis	15
8 Clones	23
Peptide links	2
Terminator sequencing reactions*	3,530
Successful (%)	2,404 (68)
Average edited read length (bp)	375
Genome size (bp)	1,830,137
G + C content (%)	38
rRNA operons	6
rrnA, rrnC, rrnD (spacer region) (bp)	723
rrnB, rrnE, rrnF (spacer region (bp)	478
tRNA genes identified	54
Number of predicted coding regions	1,743
Unassigned role (%)	736 (42)
No database match	389
Match hypothetical proteins	347
Assigned role (%)	1,007 (58)
Amino acid metabolism	68 (6.8)
Biosynthesis of cofactors, prosthetic groups, and carriers	54 (5.4)
Cell envelope	84 (8.3)
Cellular processes	53 (5.3)
Central intermediary metabolism	30 (3.0)
Energy metabolism	105 (10.4)
Fatty acid and phospholipid metabolism	25 (2.5)
Purines, pyrimidines, nucleosides, and nucleotides	53 (5.3)
Regulatory functions	64 (6.3)
Replication	87 (8.6)
Transcription	27 (2.7)
Translation	141 (14.0)
Transport and binding proteins	123 (12.2)
Other	93 (9.2)

*Includes gap closure, walks on rRNA repeats, random end-sequencing of 8 clones for assembly confirmation, and alternative reactions for ambiguity resolution.

the inhibition, the human cell may divide without control and develop to a tumor.

On a larger scale, the sequencing of the yeast genome revealed many genes wholly new to biology (which demonstrates how little biologists know about the simplest cell). It also showed a high degree of **redundancy** in the genome; that is, many genes are repeated over and over again, the significance of which remains unexplained. There is also the opportunity to study the genes associated with regulating the fermentation process, a chemical sequence not often found elsewhere. And the sequence data should boost evolutionary studies of yeasts and eukaryotes in general.

But knowing the genes' compositions does not necessarily give information on what the gene product does. Thus, in 1997, **Ronald W. Davies** of Stanford University launched a project to determine the role for each gene of a yeast. With the cooperation of an international consortium of universities, Davies' group set out to prepare 6000 strains of yeast, each strain missing one of the 6000 yeast genes. An identifying tag of DNA is substituted for the missing gene and various conditions are established (e. g., drought, high salt, high sugar) to call different genes into play. The ambitious project must take into account the duplicate alleles for each gene and rely on many technological advances not yet available.

In 1997, the science of whole genome sequencing took off. Among the bacterial genomes sequenced were those of *Helicobacter pylori*, a cause of gastric ulcers; *Borrelia burgdorferi*, the agent of Lyme disease; and *Streptococcus pneumoniae*, among the most serious causes of bacterial pneumonia. The successes give biologists new starting points from which to develop diagnostic tests and vaccines.

That same year, European and Japanese biochemists determined the sequence of bases in the genome of *Bacillus subtilis*. This organism is familiar to most students of microbiology because it is widely used as a laboratory test organism. It is also familiar to industrial microbiologists because of its value as a source of enzymes for the production of vitamins, detergents, and food products. Indeed, it is hoped that knowledge of the genome will help industrial microbiologists boost enzyme production and lower prices. Another genome completed in 1997 was that of the Archaea *Archaeglobus fulgidus*. Information from this genome verified the uniqueness of the Archaea, as first indicated with *M. jannaschii* a year earlier.

But probably the outstanding achievement of 1997 was the complete sequencing of *Escherichia coli*. The long-awaited announcement was made by **Frederick R. Blattner** of the University of Wisconsin–Madison. Studied for decades and a regular on the pages of this book, *E. coli* is the organism of choice for studying how bacteria work. Because there is enormous biological literature on the organism, a gene and its gene product can be fit into the vast understanding of its biology. Moreover, the *E. coli* genome stands as a reference point for understanding the biology of other organisms as their genomes are uncovered.

Blattner's research group identified 4,638,858 base pairs in the **4288 genes** of the *E. coli* genome. Over half bear no strong resemblance to any known genes, but 1827 genes were previously characterized. The project took five years to complete, and Blattner's group deposited the last few sequences of the genome into public databases just days before they were posted by a Japanese research team.

The Institute for Genomic Research was back in the news in 1997, having participated in the sequencing for the genome of *Treponema pallidum*, the cause of syphilis. Analysis of the genome revealed that the organism cannot encode various enzymes and complex molecules normally required for independent living (e. g., it cannot synthesize numerous fatty acids or nucleotides). It thus has a strong reliance on its host organism and is extraordinarily difficult to cultivate in the laboratory.

redundancy
repeating.

Helicobacter
hel″i-co-bak′ter

pylori
py-lor′i

Borrelia
bor-rel′i-ah

burgdorferi
burg-dorf′er-i

Archaeglobus
ark″a-e-glo′bus

fulgidus
ful′gi-dus

Another pathogen, **Mycobacterium tuberculosis**, was also sequenced in 1997, the work performed by investigators from France and England. The agent of tuberculosis, *M. tuberculosis*, has a complex cell wall that protects it against body defenses and antibiotics. The newly identified genes will help scientists understand and possibly overcome this defense.

By 1998 and beyond, whole genome sequencing of microorganisms was reaching unprecedented heights. Venter's group at TIGR was predicting that new genomes could be completed every two months, and research groups from around the world were participating in well-coordinated and well-financed projects. The typhus fever microbe **Rickettsia prowazekii** was sequenced that year; *Science* magazine was publishing updates on the effort to sequence the worm **Caenorhabditis elegans** (all 97 million bases were finally decoded by December 1998); and botanists were eagerly awaiting genomic information on the simple plant **Arabidopsis thaliana** while beginning the effort to learn the sequence of bases in corn (maize) cells.

THE STUDY OF GENOMES

The publication of prokaryotic and eukaryotic genomes presented scientists with great volumes of data and, for the first time in history, made available the entire genetic content of a living organism. But it also created the dilemma of what to do with the data; as studies continued and years passed, there emerged the new science of **genomics**, referring to the mapping, sequencing, and study of genomes. The term "genomics" was coined in 1986 by Thomas Roderick as part of the naming of a new journal *Genomics*. Allied to genomics is the growing field of **bioinformatics** (computational biology), in which the massive amounts of sequence and biological information are documented, sorted, and analyzed.

Genomics has gradually subdivided into a number of branches. Among these is **structural genomics**, which has as its purpose the construction of high-resolution genetic, physical, and transcript maps of an organism. Another branch is **functional genomics**, in which scientists attempt to translate base sequences into cellular activities and systems and to use the vast resource of information from structural genomics to understand fundamental life principles. (One practical possibility is producing drugs on the basis of the activities of a gene.) Other scientists have developed the field of **nutritional genomics**, where they use gene manipulations to increase the quality and quantity of the world's food supply.

Still another branch is **comparative genomics**, in which genomes from individual organisms are studied to discover similarities and differences that help us understand all living things. Comparative genomics will continue to attract interest in the future for several reasons: Studying the genomes of diverse organisms helps scientists develop a minimum inventory of genes needed by a free-living organism. They may be able to deduce which functions are common to disease organisms and reconstruct a hypothetical minimal cell akin to the first cells and without genetic luxuries. Comparative genomics may also give a clue to how genetic information arose by evaluating DNA's transformation to a chromosomal entity. Finally, it may be possible to understand how a cell regulates its genes to use its resources efficiently.

For the microbiologist, the field of **medical genomics** offers fresh insights into the disease process. From genome analysis, for example, scientists can deduce the importance of a specific function. Thus, *M. genitalium* uses a full five percent of its gene content to encode a gene, allowing it to adhere to the reproductive tissue while evading the host's immune system. Research directed toward altering this gene would be productive. Furthermore, *H. influenzae* lacks the genes for three metabolic enzymes, a

---

Mycobacterium

my″co-bac-ter′i-um

tuberculosis

an airborne respiratory tract disease accompanied by spreading tubercles on the lung surface.

genomics

je-no′mics

the study of the genomes of organisms to discover functional, nutritional, comparative, or other qualities.

comparative genomics

the discipline in which genomes of organisms are compared to locate significant similarities and differences.

virulent

the quality of an organism that denotes its ability to cause disease.

fact that may help microbiologists understand its ability to thrive in the human body. And by comparing the genomes of virulent and avirulent strains, a window should open to the pathogenic process.

Oncologists are eagerly awaiting the results of the **Cancer Genome Anatomy Project** begun in 1997. This ambitious effort aims for a catalog of all the genes expressed in cancer cells and will include the cellular origin of the gene and the degree of the cell's malignancy. It is hoped that with the click of a mouse, researchers can determine how gene expression changes as a cancer progresses and understand how tumors develop. Because the effort is funded by the federal National Cancer Institute, the information can be obtained without charge.

Other examples of the value of genomics are sprinkled throughout this chapter. (For example, the genome of *M. jannaschii* confirmed its classification with Archaea.) In the industrial setting, pharmaceutical scientists will sift through genomes, panning for potential drugs and therapies. Others will look for genes useful for diagnostic and identification purposes. Still others will watch for genes that encode enzymes of industrial value and not heretofore known. In these examples, the science of genomics provides a powerful tool to biologists, while justifying the high cost and enormous technical challenge of sequencing complete genomes.

Genomics is also the basis for an ongoing study announced in 1997. Spurred by the observation that some individuals are more susceptible to toxic chemicals than others, scientists at the National Institutes of Health (NIH) put together an effort to sequence DNA from about 1000 individuals to identify a link between genes and toxicity. The **Environmental Gene Project**, as the study is known, is attempting to pinpoint "environmental risk genes," such as those that govern susceptibility to the nerve gas sarin. In resistant individuals, a gene-encoded enzyme apparently converts the gas to a less toxic chemical about ten times more quickly than in susceptible individuals. Susceptible individuals lacking the gene could be advised to avoid certain chemicals.

Environmental Gene Project

the study of human genomes to locate genes that place individuals at risk from environmental substances.

Directors of the project hope to identify about 200 candidate genes involved in such environmental toxins as ozone, beryllium, and lead, then look for versions shared by more than one percent of the population of 1000 individuals. Population studies would then determine how important the genes are to the toxicology of the chemical. This latter step is complicated by the unusually large variation of alleles.

A more ambitious project called the **Human Genome Diversity Project (HGDP)** was also announced in 1997 (Figure 12.19). The work is designed to give a better understanding of the origin and evolution of humans. First suggested by geneticist **Luca Cavalli-Sforza** in 1991, the initiative is an offshoot of the Human Genome Project and is headed by its director Francis Collins.

The HGDP is intended to survey DNA variations in populations to study cultural and genetic diversity among humans around the world. Researchers will sequence snippets of DNA from hundreds of individuals of different racial backgrounds. The snippets are small stretches of DNA differing in a single nucleotide base. They are called **single nucleotide polymorphisms** (**SNPs**, or "snips"). The project seeks to collect 100,000 SNPs in human DNA donated by individuals in four groups: Africans, Asians, Europeans, and Native Americans. Between 4000 and 8000 individuals will supply samples of blood, hair, and saliva.

The diversity project is estimated to take five to ten years to complete. As samples are collected, they will be analyzed and results will be placed in computer databases for researchers worldwide. Among the concerns of opponents are issues of informed consent and confidentiality, as well as sampling techniques that could violate cultural

FIGURE 12.19
The Human Diversity Project will lead to a better understanding of the origin and evolution of humans. Among the concepts to be studied are human migration patterns and customs nfluencing patterns of reproduction.

or religious beliefs. Project organizers are also aware that they may be racing with industrial biochemists to patent the newly discovered SNPs. A new competition for gene discovery may be in the offing.

THE FUTURE OF GENOME RESEARCH

The linkage and physical maps and the base sequences generated by genome research encompass a vast amount of data for storage and interpretation. It is important, therefore, that scientists develop database hardware and software for accessing and manipulating the reams of information generated. For example, databases will have to represent map information (linkage maps, STSs, physical maps, and disease genes) and base sequences (genomic sequences, cDNA sequences, and proteins encoded by those sequences). The map information must link to the base sequence information in the database, and the two must link with information already published in the literature.

It will also be important to provide computer networks that analyze data from maps and sequences. Where, for instance, do genes begin and end, and what are the exons, introns, and regulatory sequences? How do the sequences from one species relate to those from another, and do similarities exist? Will understanding the base sequence of a yeast chromosome help us understand a similar sequence in a human chromosome?

All these questions will be resolved as the years unfold, and the fruits of DNA technology will continue to come to the fore. To some researchers, the human genome is not a library but a pharmacy. They see the genes as mechanisms for diagnosing disease, and they envision DNA as molecular snippets they can use to send patients home cured of their illness. In that sense, all the DNA knowledge uncovered in the past half-century is merely a preamble to the startling discoveries waiting to be made in the decades ahead. For many physicians, DNA will be a pharmacological substance of extraordinary potency that can treat disease and its symptoms, while correcting the imperfections that make patients susceptible to disease.

Before this new world arrives, however, society must work out several dilemmas. A recent *Time* magazine poll, for example, found that respondents were ambivalent about DNA research and its implications. Asked whether they would take a genetic test to uncover diseases they would one day contract, almost half the respondents said they

ambivalent
undecided

would prefer to remain ignorant. Respondents were also worried that the secrets of their DNA might be used against them: For instance, would they be able to purchase life insurance if they were found to carry genes for a debilitating disease? And will genetic privacy outweigh the public good? Or will the reverse prevail?

While issues as these continue to be sorted out, the research in DNA technology moves ahead steadfastly: As of 1999, human genes were being located at a rate of almost one per day; more than 100 trials of gene therapy were under way at hospitals around the world; diabetic individuals were receiving genetically engineered insulin; DNA fingerprinting was an accepted method of forensic medicine: transgenic pigs were producing human hemoglobin; the **knockout mouse** was a staple of laboratory research; and newspaper headlines that once trumpeted the race for the moon now screamed about the discoveries and advances in DNA technology.

knockout mouse

a laboratory mouse in which a gene for one or more proteins can be eliminated at the whim of the researcher.

In addition, the Human Genome Project was progressing steadily. As of 1999, the genomes of seventeen microorganisms were completely sequenced; the genome of the nematode *C. elegans* was seventy-one percent sequenced, the fruit fly's genome was six percent sequenced; and the mouse's genome 0.2 percent. In the human genome, sixty million base pairs were sequenced, representing about two percent of the total genome. More importantly, the technology was in place to begin high-speed sequencing, and researchers had completed a framework physical map for most of the human genome. Some optimistic researchers were even predicting that the 2005 completion date might be beaten (and the most optimistic were hoping to complete the project by 2002).

Although the Human Genome Project has been likened to the search for the Holy Grail or the quest to land on the moon, geneticist **Eric Lander** has written that it is better understood in terms of the **Periodic Table**. In chemistry, the Periodic Table is a systematized chart of the hundred-plus elements arranged with one another to highlight their differences and similarities. With its development by Dmitri Mendeleev in 1869, chemistry became an organized science.

Periodic Table

a chart that lists all the known and synthesized chemical elements and groups them according to similar and dissimilar properties.

The Human Genome Project, says Lander, will chart the hundred thousand genes as they reflect functional relationships in humans. Just as chemists can see how atoms work together to make molecules, biologists will now be able to see how genes work together to make organisms. A global perspective of the cell will emerge from these data. But broad-based goals should be expressed in narrower, more immediate objective. Therefore, in an article in 1996 in *Science*, Lander set out a number of more precise projects once the human genome has been deciphered:

1. Routine resequencing of multimegabase regions of DNA;
2. Systematic identification of all common variants in human genes;
3. Rapid de novo sequencing from other organisms;
4. Simultaneous monitoring of the expression of all genes;
5. [Development of] genetic tools for manipulating cell circuitry;
6. Monitoring the level and modification state of all proteins;
7. Systematic catalogs of protein interactions;
8. Identification of all essential protein shapes;
9. Increased attention to ethical, legal, and social issues;
10. Public education.

In Lander's view, the wheels must be set into motion to implement the program as soon as feasible. Furthermore, a major commitment must be made to train the next generation of scientists. Last, the new technologies must be disseminated as widely as possible to ensure global involvement in using the emerging periodic table of life.

For many decades humans stood by and watched the "game of life." They marveled at the wonders of nature; they searched out and cataloged the plants, animals, and microorganisms; and they spent exhaustive hours trying to understand how these organisms fit into the scheme of things.

Then the age of DNA technology dawned. Now a great number of observers became manipulators. These researchers learned how to change the character of organisms at its most fundamental level. They isolated the organism's DNA, changed its chemistry, and inserted the new DNA into recipient cells to see what would happen. Bacteria began producing human hormones and antiviral proteins, and human cells deficient in a certain protein learned how to produce the protein. Researchers even brought new species into existence—in 1987 they produced an animal with the body of a goat and the head of a sheep.

We stand at the brink of an adventure that will carry us through the twenty-first century and beyond. The implications of DNA technology are so colossal that the human mind has yet to imagine all its possibilities. A colleague of mine once said, "If you dip a tennis ball into the ocean, the water dripping from the tennis ball represents all that is known. The ocean represents all that is waiting to be learned."

DNA technology will continue to make an impact on our lives for many generations to come, and we can proudly tell our grandchildren that we were there at the beginning.

SUMMARY

The effort to map and learn the sequence of bases in the human genome is one of epic proportions encompassing a planned fifteen years of work at a cost of $3 billion. Knowing the content of the human genome will help researchers devise improved diagnoses and treatments for genetic diseases while helping developmental biologists learn how humans progress from single fertilized egg cells to adults. It will be of value to evolutionary and comparative biologists in ways not yet imagined while spurring the development of technology.

The Human Genome Project is being funded by the federal government through the National Human Genome Research Institute, a hybrid group overseen by the National Institutes of Health and U. S. Department of Energy. In addition to the learning the genome of humans, the project will support efforts to decipher the base sequences of the nematode *C. elegans*, the fruit fly *Drosophila*, two plants, the yeast, and the mouse *M. musculus*. Much controversy has been generated by efforts to obtain patents on the gene probes developed in connection with the project, and the debate remains unresolved.

An important goal of the Human Genome Project is to prepare gene linkage and physical maps of the chromosomes. A gene linkage map is a low-resolution scheme of a chromosome that pinpoints genes on the basis of their connections to markers such as RFLPs and VNTRs. The process is somewhat similar to knowing that cities are along a certain highway but not knowing the mileage between them. A physical map, by comparison, gives the actual number of bases between genes on a chromosome, and thus, the map resolution is higher.

To form a physical map, messenger RNA molecules are used to form cDNA probes of known composition, which are then used to bind to and identify certain gene segments. Overlappings are then joined to form a continuous map. Sequence tag sites and yeast artificial chromosomes have been used to assist the physical map development. The actual sequencing of bases is a tedious process performed by the

Sanger method by the use of labeled dideoxynucleotides to mark the locations of bases in the genome. Automation and computerization are essential to this phase of the project. "Chromosome walking" is used to fill gaps between identified gene sequences.

Through 1998, scientists had sequenced a number of organisms including the bacterium *Haemophilus influenzae*, the first organism to have its genome deciphered; the yeast *Saccharomyces cerevisiae*, the first eukaryote to be sequenced; *Escherichia coli*, the widely used test organism in DNA technology; and the pathogens *Borrelia burgdorferi*, the agent of Lyme disease, *Mycobacterium tuberculosis*, the cause of tuberculosis, and *Staphylococcus aureus*, a cause of food poisoning. Scientists were also developing the sciences of functional, nutritional, and comparative genomics in which knowledge of the genome is put to practical and esoteric use.

The Human Genome Project is one of colossal magnitude that will have an impact on the myriad branches of science for decades to come. Although many issues remain to be resolved, the project remains as the crowning achievement of DNA research in the twentieth century and the bedrock for research in the twenty-first century.

REVIEW

The Human Genome Project stands as one of the great adventures undertaken by scientists. To test your knowledge of the project, answer the following review questions.

1. About how many base pairs make up the human genome and how does this number compare with the genomes of other selected organisms.

2. How will deciphering the human genome be of benefit to geneticists, physicians, developmental biologists, comparative biologists, evolutionary biologists, and technology development?

3. Describe the origin, development, and current supervisory offices that oversee the Human Genome Project. Who is funding the project?

4. The human genome is but one of several genomes currently being deciphered. What other organisms are involved and why is each involved?

5. What controversy has arisen regarding patent rights for genes, and what are some of the opposing views in the controversy?

6. What is meant by a gene linkage map and how is it developed for the human genome? What is meant by a physical map, and how is it developed?

7. Describe the roles of contigs, STSs, YACs, and pulse-field electrophoresis in the development of a physical map.

8. How is the Sanger method used to discover the sequence of bases in a specified segment of DNA in the genome? Describe the process of chromosome walking.

9. Name several organisms whose genomes have been successfully sequenced and indicate the significance of knowing each genome.

10. Describe the science of genomics, indicate its various branches, and specify some advantages to its study.

⧜ FOR ADDITIONAL READING

Bussey, H. "Chain of Being." *The Sciences* (March/April, 1996).

Charlebois, R. L. "The Modern Science of Bacterial Genomics." *ASM News* 62(5) (1996): 255–260.

Marshall, E. "Whose Genome Is It, Anyway?" *Science* 273 (1996): 1788–1791.

Marshall, E. "A Catalog of Cancer Genes at the Click of a Mouse." *Science* 276 (1997): 1023–1027.

Marshall, E., and Pennisi, E. "NIH Launches the Final Push to Sequence the Genome." *Science* 272 (1997): 188–191.

Nowak, R. "Bacterial Genome Sequence Bagged." *Science* 269 (1995): 468 472.

Pennisi, E. "DNA Sequences' Trial by Fire." *Science* 280 (1998): 814–818.

Rowen, L., et al. "Sequencing the Human Genome." *Science* 278 (1997): 605–610.

Shreeve, J. "The Code Breaker." *Discover* (May, 1998).

Strauss, E., and Falkow, S. "Microbial Pathogenesis: Genomics and Beyond." *Science* 276 (1997): 707–710.

Syvanen, M., and Young, R. "The Human Genome Project: An Alternate View." *ASM News* 57(9) (1991): 444–448.

Travis, J. "Long-awaited Bacterial Genome Debuts." *Science News* 151 (1997): 84.

Glossary

A

activator protein   a protein that reacts with a regulatory site on a DNA molecule and encourages gene expression.

adenosine   a nucleotide containing ribose or deoxyribose, phosphate, and adenine.

adenovirus   a common cold virus composed of a DNA genome and an icosahedral capsid; used as a vector in gene therapy.

agarose   a gelatinous material through which biochemicals migrate in electrophoresis techniques.

Agrobacterium tumefaciens   a rod-shaped bacterium that causes crown gall in plants by inserting its DNA into plant cells and transforming the cells to form tumors; tumor-inducing plasmids from *A. tumefaciens* are used to insert genes to plant cells.

AIDS   a viral disease characterized by loss of the immune system's T-lymphocytes and often accompanied by disease caused by opportunistic microorganisms.

amino acid   a chemical compound that contains at least one amino group and one organic acid group; amino acids are the units of a protein molecule.

anticodon   a three-base sequence on a tRNA molecule that complements the codon on an mRNA molecule and places a particular amino acid at a particular location in a protein molecule.

antisense molecule   an RNA molecule that reacts with and neutralizes the mRNA molecule used for synthesis of a particular protein; an antisense molecule effectively halts the production of that protein.

autosomes   the 44 chromosomes of the human genome that are not involved in sex determination.

B

Bacillus thuringiensis   a bacterial rod whose toxic crystals act as an insecticide against certain species of arthropods; source of genes for inducing arthropod resistance in plants.

bacteriophage   a virus that replicates within a bacterium or integrates its genetic material into the chromosome of the bacterium; used as a vector in DNA technology.

beta-galactosidase   a lactose-digesting enzyme encoded by a well-known operon found in a bacterial cell.

biolistic   a cylinder containing a cartridge with a nylon projectile, which carries spheres that insert genes into plant cells.

biotechnology   the subdiscipline of biology in which biochemically based processes such as DNA technology are used for practical purposes.

bovine growth hormone (BGH)   a hormone produced by genetic engineering techniques; BGH encourages increased milk production in dairy cattle; also called somatotropin.

C

callus   a random mass of plant cells developing from a single cell and capable of forming a mature adult plant.

"cap"   a nucleotide containing 7-methyl guanosine that is added to the lead end of an mRNA molecule during its synthesis.

cDNA   the DNA molecule that results from reverse transcriptase activity on RNA; the nitrogenous based sequence in cDNA complements the base sequence in RNA.

cDNA library   a mass of cells that have incorporated into their plasmids a specific set of genes in such a way that the genes can be obtained in a relatively easy manner.

centiMorgan (cM)   a unit of molecular genetics equivalent to the space occupied by a million base pairs in a DNA molecule.

chimera   a plasmid or other vector that contains DNA not normally found in the plasmid.

chimeric plasmid   a plasmid that has been biochemically reengineered to contain one or more foreign genes.

chromosome   a structural element of protein and DNA that serves as the repository of hereditary information in the cell; segments of the chromosome are genes.

chromosome walking   a base sequencing method in which a chromosome is analyzed by extending the tip out a bit at a time and determining the base sequence of the tip.

clone   a group of organisms, cells, molecules, or other objects all originating from a single individual; synonymous with colony.

codon   a three-base sequence on an mRNA molecule that works with the anticodon on a tRNA molecule to specify where a particular amino acid is placed in a protein.

complementary   being the opposite member of a well-recognized pair; also an adjective describing the bases in the double-stranded DNA molecule; specifically, adenine is complementary to thymine, and cytosine is complementary to guanine.

conjugation   the genetic recombination process in which one live bacterium acquires the fragments of DNA from another live bacterium and expresses the proteins encoded by the acquired DNA.

contig map   a chromosome map in which various DNA segments have been joined together to form a contiguous (continuous) DNA molecule.

cosmid   a linear stretch of DNA whose ends can be biochemically stitched together to form a loop; a cosmid can be used as a vector for genes in DNA technology.

cytogenetic map   a chromosome map in which the active genes respond to a chemical dye and display themselves as bands on the chromosome.

D

dideoxynucleotide   a nucleotide that lacks two oxygen atoms in the carbohydrate portion.

DNA   an acronym for deoxyribonucleic acid, the organic substance of heredity and the material of which genes are composed.

DNA fingerprinting   a DNA matching technique in which certain base sequences in the DNA of a cell sample are determined and compared to the DNA base sequences of a second cell sample to learn whether they are identical within established confidence levels.

DNA ligase   an enzyme that joins fragments of DNA by linking the phosphate group of one nucleotide to the deoxyribose molecule of the next nucleotide.

DNA polymerase   an enzyme that functions in the synthesis of DNA; the enzyme utilizes a template molecule, a series of nucleotides, and other essential biochemicals.

DNA probe   a synthetic, small, single-stranded DNA molecule that binds to a target, complementary DNA strand in a mixture of biological compounds.

double helix a double-stranded molecule wound to form a helix; the form taken by the DNA molecule in a chromosome.

Drosophila melanogaster the scientific name for the fruit fly used in DNA technology experiments.

E

*Eco*RI a restriction enzyme derived from strain R of *Escherichia coli*.

electrophoresis a biochemical laboratory technique in which molecules move through a gel according to their size and chemical charges within an electrical field.

electroporation a method of inserting genes to plant cells whereby electric currents force gold beads carrying DNA into the cells.

endonuclease a cellular enzyme that cleaves a DNA molecule during a metabolic reaction.

enzyme a protein that catalyzes a chemical reaction of metabolism while itself remaining unchanged.

Escherichia coli a species of bacterium normally found in the human intestine and used in DNA technology as a recipient organism for gene sequences.

eukaryote an organism whose cells have eukaryotic properties (examples are fungi, protozoa, plants, and animals).

eukaryotic pertaining to a complex organism whose cells have a nucleus and organelles, multiply by mitosis, and have other features that separate them from simpler prokaryotic cells; plant, animal, and human cells are eukaryotic.

exon a section of mRNA that specifies an amino acid sequence; an exon is retained during the production of the final mRNA molecule and is expressed in the amino acid sequence of the protein.

F

forensic having to do with laws and legal proceedings.

G

gene a segment of a DNA molecule that provides chemical information for the synthesis of protein in a cell; a collection of genes and intervening DNA sequences compose a chromosome.

gene bank a repository for storing an individual's cells until such time as a genetic analysis can be performed.

gene library a series of cells or unicellular organisms that are used to store genes from foreign cells; the genes obtained from the cells of a gene library are used in DNA technology experiments.

gene linkage map a chromosome map in which active genes are located by locating closely associated marker genes.

gene probe a molecule of DNA or RNA that unites with its complementary stretch of DNA or RNA molecule when introduced to a mass of nucleic acid molecules.

genetic code the sequence of nitrogenous bases in a DNA molecule that specifies a sequence of amino acids in a protein.

genome the total nuclear DNA component of all the genes of a cell or single-celled microorganism.

genomic library a colony of cells that have incorporated into their plasmids all the genes derived from another cell in such a way that the genes can be obtained in a relatively easy manner.

glycoprotein a protein that contains one or more carbohydrate molecules bonded to one or more of the amino acids in the protein.

glycosylation the biochemical process in which molecules of carbohydrate are attached at various points to the amino acids of a protein.

glyphosphate an herbicide used to destroy unwanted plants; genes for glyphosphate resistance have been inserted to agricultural plants.

Golgi apparatus a series of flattened membranes in the eukaryotic cell along which proteins are processed before synthesis is complete; also called the Golgi body.

gp 120 a glycoprotein located in the envelope of the human immunodeficiency virus (HIV); an essential substance for the binding of HIV to its host cell in the replication process; can be synthetically produced by DNA technology for use in a vaccine.

H

helix a coil resembling a telephone cord; the form taken by the DNA molecule in a chromosome.

hemophilia B a form of hemophilia in which the patient fails to produce an adequate blood clot due to the absence of an essential clotting factor; synthetic clotting factor can alleviate the symptoms.

hepatitis B a viral disease of the liver caused by a fragile virus that passes from blood to blood or semen to blood; a vaccine against hepatitis B is produced by DNA technology.

human genome the repertoire of the approximate 100,000 genes of a human cell.

"human mouse" a mouse that has been anatomically changed to harbor functional cells of the human immune system.

Huntington's disease a genetic disease characterized by progressive deterioration of the nervous system and accompanied by thrashing movements.

hypercholesterolemia a genetic disease accompanied by a higher than normal level of cholesterol in the blood owing to the absence of cholesterol receptors.

I

ice-minus bacterium a genetically altered bacterium that produces proteins, that induce ice crystal formation at lower than normal temperatures.

insulin a pancreatic hormone composed of 51 amino acids in two interconnected chains; insulin facilitates the passage of glucose from the bloodstream into the body cells; insulin is also produced by genetically altered bacteria.

interferon proteins produced by human cells in response to viruses; interferon stimulates the production of biochemical substances that yield protection against viral penetration to cells; also produced by genetically altered bacterial cells.

intron an intervening section of the preliminary messenger RNA (mRNA) molecule that is removed before the production of the final mRNA molecule; the base sequence in an intron is not expressed in the final protein.

K

kilobase a unit consisting of a thousand nitrogenous bases in a DNA or RNA molecule.

"knockout mouse" a mouse that has been genetically altered to lack the genes for an entire organ or organ system.

L

lactoferrin   a protein found in animal milk that has the ability to bind iron molecules; may be inhibitory to bacteria; lactoferrin is produced in genetically altered animals.

M

mammalian   applying to warm-blooded vertebrates that have mammary glands and hair; examples of mammals are rodents, barnyard animals, primates, and humans.

megaYAC   an unusually large yeast artificial chromosome (YAC).

messenger RNA (mRNA)   a molecule of RNA synthesized with a nitrogenous base code complementary to that of DNA; mRNA carries the genetic "message" to the cell cytoplasm for the synthesis of protein.

microinjection   a biochemical technique in which a microscopic syringe is used to penetrate the cell and propel DNA fragments into the nucleus of the cell.

Mus musculus   the species of mouse commonly used in DNA technology.

mutation   a change in the characteristics of an organism arising from a change in the genes of the organism.

N

nanometer   a unit of measurement equivalent to a billionth of a meter.

nucleic acid   an organic substance composed of nucleotides joined to one another by phosphate bonds.

nucleotide   a building block unit of a nucleic acid, consisting of a five-carbon carbohydrate molecule (such as ribose or deoxyribose), a phosphate group, and one of five available nitrogenous bases (adenine, thymine, cytosine, guanine, or uracil).

O

oligonucleotides   small, synthetic stretches of RNA or DNA; a type of synthetic gene; also called oligos; also, small segments of DNA that can act as antisense molecules.

oncogenes   genes associated with tumors and cancers.

"oncomouse"   a mouse that has been genetically altered to be highly prone to breast cancer.

operator   a series of nitrogenous bases in the DNA of an operon where a repressor or activator protein is able to regulate gene expression.

operon   the complex of structural genes and regulatory genes in a chromosome that function together to bring about expression of the structural genes in a protein molecule.

origin of replication   a sequence of nitrogenous bases on a DNA molecule that signals the beginning point for replication of the DNA molecule.

P

peptide   a small protein.

phage   abbreviation for bacteriophage; a virus that replicates within bacteria or incorporates its nucleic acid to the bacterial genome; useful as a DNA vector.

"pharm" animal   an animal genetically altered to produce a certain pharmaceutical product.

phosphate group   a molecule of phosphorus, oxygen, and hydrogen atoms derived from phosphoric acid and present in both DNA and RNA; one or more hydrogen atoms are lost from the phosphoric acid molecule in the formation of a phosphate group.

phosphodiester bond   a chemical bond forming between the free 5'-phosphate of one nucleotide and the deoxyribose molecule of the second nucleotide.

physical map   a chromosome map in which the location of the active genes has been pinpointed and the number of bases between the active genes is known.

plasmid   a closed loop of DNA containing about a dozen genes and existing in multiple copies in the cytoplasm of bacterial cells; a possible vector for genes in DNA technology.

poly-A tail   a sequence of adenine-containing nucleotides at the end of an mRNA molecule that does not have a known function but can be used in DNA technology to identify an mRNA molecule.

polygalacturonase   an enzyme in certain plants that digests the pectin to encourage rotting; inhibiting the activity of this enzyme by proteins encoded by inserted genes forestalls rotting.

polymer   a chemical molecule consisting of repeating units of a particular substance.

polymerase chain reaction   the biochemical process in which a segment of DNA is multiplied millions of times over through the activity of the enzyme DNA polymerase.

primer DNA   a segment of DNA used to initiate DNA synthesis, such as in the polymerase chain reaction.

prokaryote   an organism whose cells have prokaryotic properties (an example is a bacterium).

prokaryotic   pertaining to a simple organism whose cells lack a nucleus or organelles, multiply by simple fission, and have other features that separate them from more complex eukaryotic cells; bacterial cells are prokaryotic.

promoter site   a sequence of nucleotides that initiates the transcription of the genetic code in DNA to messenger RNA.

protease   a protein-digesting enzyme; in bacteria, proteases commonly digest foreign proteins.

Pseudomonas syringae   a bacterial rod whose proteins induce ice crystal formation in nature; the bacterium can be genetically altered to induce crystal formation at a lower temperature.

purine   a nitrogenous base having two carbon-rich rings of atoms in its molecule; examples are adenine and guanine.

Q

Q-beta replicase   an enzyme that catalyzes RNA synthesis using RNA as a template; used in signal amplification in the polymerase chain reaction.

R

RAC   Recombinant DNA Advisory Committee; a supervisory group within the U.S. National Institutes of Health that lends approval or disapproval to proposed experiments involving gene therapy.

radioactive   able to emit detectable radiations.

recognition sequence   a sequence of nitrogenous bases on a DNA molecule that is chemically recognized by a restriction enzyme as its location of activity.

regulatory site   a sequence of bases on a DNA molecule where gene expression can be controlled by reactions with repressor or activator proteins.

repressor protein   a protein that reacts with a regulatory site on a DNA molecule and restricts expression of the gene by inhibiting transcription.

resolution   degree of clarity between two points.

restriction enzyme   a digestive enzyme that biochemically cuts a DNA molecule at a restricted site.

restriction fragment length polymorphism   (see RFLP)

retrovirus   a virus that penetrates a host cell and uses its reverse transcriptase to encode a DNA molecule complementary to the retroviral RNA; useful as a DNA vector.

reverse transcriptase   an enzyme that utilizes the base sequence in an RNA molecule as a model for synthesizing a complementary DNA molecule.

RFLP (restriction fragment length polymorphism)   a DNA segment isolated by restriction enzymes and existing in any of various lengths; RFLPs are used in DNA diagnoses and fingerprinting but do not have a known function.

Rhizobium   a rod-shaped bacterium that lives symbiotically with leguminous plants and fixes nitrogen into organic compounds useful to the plants; a source of genes for transgenic plant experiments.

ribose   a five-carbon carbohydrate molecule and one of the components of RNA; similar to deoxyribose, except with one extra oxygen atom.

ribosome binding site   a series of nitrogenous bases in the DNA of an operon where the base code in mRNA is specified to permit the mRNA to bind to the ribosome.

RNA polymerase   the enzyme that functions in transcription and synthesizes an RNA molecule with bases complementary to those in DNA.

S

sequence map   a chromosome map in which the nature and position of each nitrogenous base is known.

sequence tag site (STS)   a DNA segment that is attached to a selected location on a DNA molecule to mark the location prior to analysis.

signal peptide   a small protein that facilitates the movement of a manufactured protein out of the cell into the extracellular environment.

Southern blotting   a laboratory technique in which molecules are separated by electrophoresis and are then blotted off the gel for identification purposes.

"sticky ends"   a single-stranded string of nucleotides that extends from the end of a fragment of DNA.

structural genes   a group of genes in an operon that encodes a protein and whose expression is controlled by a number of regulatory genes.

subunits   bacterial molecules or fragments of viruses that can be utilized in a vaccine; produced by DNA technology.

T

Taq **polymerase**   a heat-tolerant enzyme derived from the bacterium *Thermus aquaticus*; the enzyme is a DNA polymerase used in the polymerase chain reaction.

telomere   a stretch of DNA at the end of a yeast chromosome that protects the chromosome against degradation by nuclease.

termination site   a series of nucleotides that signals the end of transcription of DNA to messenger RNA.

thymidine   a nucleotide containing deoxyribose, phosphate, and thymine.

thymidine kinase (TK)   an enzyme that functions in the synthesis of DNA by releasing the thymidine from previously used DNA and making it available for new DNA formation; used as a marker to ascertain gene insertion.

Ti plasmid   a plasmid of *Agrobacterium tumefaciens* that carries genes to induce tumors in an infected plant.

tissue plasminogen activator (TPA)   a protein that acts as a protease and destroys blood clots; TPA can be produced by DNA technology.

transcription   The process wherein enzymes synthesize an RNA molecule using a strand of DNA as a template; the bases in the resulting RNA complement those in the DNA.

transduction   the genetic recombination process in which bacterial viruses acquire fragments of bacterial DNA during viral replication and transport those fragments into another live bacterium where the DNA is expressed.

transfer RNA (tRNA)   a molecule of RNA that binds to specific amino acids and transports them to the ribosomes where protein synthesis takes place; different tRNA molecules exist for different amino acids.

transformation   The genetic recombination process in which bacteria acquire fragments of DNA from the external environment and express the proteins encoded by the genes in those fragments.

transgenic plant   a plant whose cells have been biochemically altered to contain one or more foreign genes.

translation   the biochemical process involving enzymes, ribosomes, and other biochemical components in which an mRNA molecule provides a nitrogenous base code for the placement of amino acids in the synthesis of a protein.

tumor-infiltrating lymphocytes (TILs)   certain lymphocytes of the immune system that attack and destroy tumor cells.

tumor necrosis factor (TNF)   a protein produced by human macrophages that encourages destruction of tumor cells; TNF can be produced by the methods of DNA technology.

U

ultramicroscopic   below the ability of the light microscope to visualize.

V

variable-number tandem repeat   (see VNTR)

vector   an entity that transports foreign DNA into a cell of an organism where the DNA is expressed; a plasmid, cosmid, or virus can serve as a vector.

vector incompetence   when an organism is unable to establish itself in a host or vector that normally carries the organism.

virus   a segment of nucleic acid surrounded by a protein coating and, in some cases, an envelope; capable of replicating in host cells and causing disease in some instances; in other instances, the viral nucleic acid integrates to the cellular chromosome; viruses are often used as gene vectors.

VNTR (variable-number tandem repeat)   a segment of DNA that contains a specified number of repeating nitrogenous base sequences; VNTRs are utilized in DNA fingerprinting techniques.

Y

yeast artificial chromosome (YAC)   a segment of DNA derived from a yeast cell and used as a foundation molecule for attachment of other DNA segments.

Credits

LINE ART

CHAPTER 1

Figures 1.2, 1.6, 1.16: From Linda R. Maxson and Charles H. Daugherty, *Genetics*, 3rd edition. Copyright © 1992 Wm. C. Brown Communications, Inc. Reprinted by permission of Times Mirror Higher Education Group, Inc., Dubuque, Iowa. All Rights Reserved.

Figure 1.5b: From Stephen A. Miller and John P. Harley, *Zoology*, 2nd edition. Copyright © 1994 Wm. C. Brown Communications, Inc. Reprinted by permission of Times Mirror Higher Education Group, Inc., Dubuque, Iowa. All Rights Reserved.

Figure 1.9: From Randy Moore and W. Dennis Clark, *Botany.* Copyright © 1995 Wm. C. Brown Communications, Inc. Reprinted by permission of Times Mirror Higher Education Group, Inc., Dubuque, Iowa. All Rights Reserved.

CHAPTER 2

Figures 2.1, 2.8, 2.11: From Robert H. Tamarin, *Principles of Genetics*, 4th edition. Copyright © 1993 Wm. C. Brown Communications, Inc. Reprinted by permission of Times Mirror Higher Education Group, Inc., Dubuque, Iowa. All Rights Reserved.

Figures 2.3, 2.4: From Geoffrey Zubay, *Biochemistry*, 3rd edition. Copyright © 1993 Wm. C. Brown Communications, Inc. Reprinted by permission of Times Mirror Higher Education Group, Inc., Dubuque, Iowa. All Rights Reserved.

Figure 2.9: From Robert F. Weaver and Philip W. Hedrick, *Genetics*, 2nd edition. Copyright © 1992 Wm. C. Brown Communications, Inc. Reprinted by permission of Times Mirror Higher Education Group, Inc., Dubuque, Iowa. All Rights Reserved.

Figures 2.9b, 2.15, 2.16: From Thomas J. Gagliano, Gagliano Graphics, Albuquerque, NM

Figures 2.10, 2.13, 2.14: From Sylvia S. Mader, *Biology*, 4th edition.

Copyright © 1993 Wm. C. Brown Communications, Inc. Reprinted by permission of Times Mirror Higher Education Group, Inc., Dubuque, Iowa. All Rights Reserved.

Figure 2.12 publ: Reprinted by permission from *Nature*, Vol. 171 No. 4356, pp. 737-738. Copyright © 1953 Macmillan Magazines Ltd., London, England.

CHAPTER 3

Figures 3.2, 3.5bc, 3.6, 3.11, 3.12a, 3.13, 3.14, 3.20: From Sylvia S. Mader, *Biology*, 4th edition. Copyright © 1993 Wm. C. Brown Communications, Inc. Reprinted by permission of Times Mirror Higher Education Group, Inc., Dubuque, Iowa. All Rights Reserved.

Figures 3.4, 3.15, 3.20: From Thomas J. Gagliano, Gagliano Graphics, Albuquerque, NM

Figure 3.17: From Eldon D. Enger, et al., *Concepts in Biology*, 7th edition. Copyright © 1994 Wm. C. Brown Communications, Inc. Reprinted by permission of Times Mirror Higher Education Group, Inc., Dubuque, Iowa. All Rights Reserved.

Figure 3.16: From Geoffrey Zubay, *Biochemistry*, 3rd edition Copyright © 1993 Wm. C. Brown Communications, Inc. Reprinted by permission of Times Mirror Higher Education Group, Inc., Dubuque, Iowa. All Rights Reserved.

CHAPTER 4

Figures 4.3bc, 4.11: From Lansing M. Prescott, et al., *Microbiology*, 2nd edition. Copyright © 1993 Wm. C. Brown Communications, Inc. Reprinted by permission of Times Mirror Higher Education Group, Inc., Dubuque, Iowa. All Rights Reserved.

Figure 4.4: From Robert F. Weaver and Philip W. Hedrick, *Genetics*, 2nd edition. Copyright © 1992 Wm. C. Brown Communications, Inc. Reprinted by permission of Times Mirror Higher Education Group, Inc., Dubuque, Iowa. All Rights Reserved.

Figure 4.9: From Robert H. Tamarin, *Principles of Genetics*, 4th edition. Copyright © 1993 Wm. C. Brown Communications, Inc. Reprinted by permission of Times Mirror Higher Education Group, Inc., Dubuque, Iowa. All Rights Reserved.

CHAPTER 5

Figures 5.4, 5.10: From Thomas J. Gagliano, Gagliano Graphics, Albuquerque, NM

Figure 5.13: From Robert F. Weaver and Philip W. Hedrick, *Genetics*, 2nd edition. Copyright © 1992 Wm. C. Brown Communications, Inc. Reprinted by permission of Times Mirror Higher Education Group, Inc., Dubuque, Iowa. All Rights Reserved.

CHAPTER 6

Figure 6.2: From Eldon D. Enger, et al., *Concepts in Biology*, 7th edition Copyright © 1994 Wm. C. Brown Communications, Inc. Reprinted by permission of Times Mirror Higher Education Group, Inc., Dubuque, Iowa. All Rights Reserved.

Figure 6.4: From Linda R. Maxson and Charles H. Daugherty, *Genetics*, 3rd edition. Copyright © 1992 Wm. C. Brown Communications, Inc. Reprinted by permission of Times Mirror Higher Education Group, Inc., Dubuque, Iowa. All Rights Reserved.

Figures 6.11, 6.14: From Thomas J. Gagliano, Gagliano Graphics, Albuquerque, NM

CHAPTER 7

Figure 7.2: From Sylvia S. Mader, *Biology*, 4th edition. Copyright © 1993 Wm. C. Brown Communications, Inc. Reprinted by permission of Times Mirror Higher Education Group, Inc., Dubuque, Iowa. All Rights Reserved.

Figures 7.3, 7.5, 7.11: From Thomas J. Gagliano, Gagliano Graphics, Albuquerque, NM

CHAPTER 8

Figures 8.1, 8.10, 7.11: From Thomas J. Gagliano, Gagliano Graphics, Albuquerque, NM

CHAPTER 9

Figures 9.3, 9.13, 9.15: From Thomas J. Gagliano, Gagliano Graphics, Albuquerque, NM

Figure 9.4: From Gerald J. Stine, *The New Human Genetics.*Copyright © 1989 Wm. C. Brown Communications, Inc. Reprinted by permission of Times Mirror Higher Education Group, Inc., Dubuque, Iowa. All Rights Reserved.

Figure 9.5: From Geoffrey Zubay, *Biochemistry,* 3rd edition. Copyright © 1993 Wm. C. Brown Communications, Inc. Reprinted by permission of Times Mirror Higher Education Group, Inc., Dubuque, Iowa. All Rights Reserved.

Figure 9.10a: From Sylvia S. Mader, *Biology,* 4th edition. Copyright © 1993 Wm. C. Brown Communications, Inc. Reprinted by permission of Times Mirror Higher Education Group, Inc., Dubuque, Iowa. All Rights Reserved.

CHAPTER 10

Figure 10.17: From Thomas J. Gagliano, Gagliano Graphics, Albuquerque, NM

CHAPTER 11

Figures 11.10, 11.11: From Thomas J. Gagliano, Gagliano Graphics, Albuquerque, NM

CHAPTER 12

Figure 12.8: From Thomas J. Gagliano, Gagliano Graphics, Albuquerque, NM

Figures 12.11, 12.12: From Robert F. Weaver and Philip W. Hedrick, *Genetics,* 2nd edition. Copyright © 1992 Wm. C. Brown Communications, Inc. Reprinted by permission of Times Mirror Higher Education Group, Inc., Dubuque, Iowa. All Rights Reserved.

Figure 12.14: From Robert H. Tamarin, *Principles of Genetics,* 4th edition. Copyright © 1993 Wm. C. Brown Communications, Inc. Reprinted by permission of Times Mirror Higher Education Group, Inc., Dubuque, Iowa.

All Rights Reserved.

PHOTOS

CHAPTER 1

Figure. 1.1 Courtesy Department of Library Services, American Museum of Natural History, Neg. # 219467, **Figure 1.4** Courtesy of Kleberg Cyto-Genetics Laboratory, Department of Molecular and Human Genetics, Baylor College of Medicine, Houston, TX., **Figure 1.5a** Genetics, 42 (1947): frontispiece, **Figure 1.5 c-1 top** © Carolina Biological Supply Company/Phototake, Inc., **Figure 1.5 c-2 bottom** © Carolina Biological Supply Company/Phototake, Inc., **Figure 1.7** © Carolina Biological Supply Company/Phototake, Inc., **Figure 1.10** Courtesy of the National Academy of Sciences, **Figure 1.11** Courtesy of Cold Spring Harbor Laboratory Archive, **Figure 1.12a** © M. Wurtz/SPLPhoto Researchers, Inc., **Figure 1.12b** © Lee D. Simon/Photo Researchers, Inc.

CHAPTER 2

Figure 2.5 © Barrington Brown/Photo Researchers, Inc., **Figure 2.5 b-1 top** © Margot Bennet/Cold Spring Harbor Laboratory Archives, **Figure 2.5 b-2 bottom** Courtesy Washington University In St. Louis, photo permission by Joe Angeles, **Figure 2.6** Courtesy of the Biophysics Department, King's College London, **Figure 2.7a** From "The Double Helix:, by J. D. Watson, Atheneum Press, NY, 1968. Courtesy of Cold Spring Harbor Laboratory Archives, **Figure 2.7b** Courtesy of Maurice Wilkins, **Figure 2.11** © Nelson Max/LLNL/Peter Arnold, Inc.

CHAPTER 3

Figure 3.1a © AP/Wide World Photos, **Figure 3.1b** © AP/Wide World Photos, **Figure 3.5 a-1** © Bill Longcore/Photo Researchers, Inc., **Figure 3.5 a-2** © Bill Longcore/Photo Researchers, Inc., **Figure 3.7** From Symington, J., Commoner, R., and Yamada, M. (1962) P.N.A.S. 48:1676, **Figure 3.10** Courtesy Dr. Elena Kiselva/The Institute of Cytology and Genetics Russia, **Figure 3.12b** Courtesy Nelson Max, Lawrence Livermore National Laboratory, University of California and the Department of Energy.

CHAPTER 4

Figure 4.1 a,b Courtesy General Electric Research and Development Center, **Figure 4.3a** Courtesy of Harold W. Fisher, University of Rhode Island and Robley C. Williams, University of California at Berkeley, **Figure 4.3b** © Dennis Kunkel/Phototake, Inc., **Figure 4.3c** © Thomas Broker/Phototake, Inc., **Figure 4.6** Courtesy Charles C. Brinton, Jr. and Judith Carnahan, **Figure 4.7** © Drs. T.F. Anderson, F. Jacob and E. Wollman/Photo Researchers, Inc., **Figure 4.8a** left © UPI/The Bettmann Archive/Corbis, **Figure 4.8a** center Courtesy Hamilton O. Smith, **Figure 4.8a** right Courtesy Daniel Nathans, **Figure 4.12a** Courtesy Paul Berg, **Figure 4.12b** Courtesy Stanley N. Cohen, **Figure 4.13** © K.G. Murti/Visuals Unlimited, **Figure 4.16** Courtesy National Academy of Science Archive **Figure 4.17a** © Coral L. Brierley/Visuals Unlimited, **Figure 4.17b,c** Courtesy Society for Industrial Microbiology, **Figure 4.17d** © Dan McCoy/Rainbow.

CHAPTER 5

Figure 5.1 © Dan McCoy/Rainbow, **Figure 5.2** © Jim Olive/Peter Arnold, Inc., **Figure 5.5** Courtesy of Richard Kolodner, **Figure 5.7** Courtesy of Dr. Roger Hendrix, **Figure 5.8** Courtesy John Harrington Athersys Inc., Cleveland OH, **Figure 5.9a** © David M. Phillips/Visuals Unlimited, **Figure 5.9b** © Dr. Tony Brain/SPL/Photo Researchers, Inc., **Figure 5.9c** © David Scharf/Peter Arnold, Inc., **Figure 5.11** © Matt Meadows/Peter Arnold, Inc.

CHAPTER 6

Figure 6.1 © SIU/Visuals Unlimited **Figure 6.3a** background, inset, **Figure 6.3b** Courtesy of Dr. Daniel C. Williams, Lilly Research Laboratories, Indianapolis, IN, SCIENCE 215:687-89, Feb. 1982., **Figure 6.10** Courtesy Massachusetts General Hospital, Offfice of Public Affairs, **Figure 6.12a** Courtesy Armed Forces Institute of Pathology, AFIP 74-5195, **Figure 6.12b** Courtesy John E. Donelson, **Figure 6.13** Courtesy Park, Davis & Company, a Division of Warner-Lambert Company,**Figure 6.16** © SPL/Photo Researchers, Inc., **Figure 6.17a** Courtesy Inhale Therapeutic Systems,**Figure 6.17b** Courtesy Edith Mathioivitz, Brown University Providence, RI,**Figure 6.18a** left Courtesy Dr. Stephen Johnston, **Figure 6.18 a** right Courtersy K. Walsh/OLR/UCSC.

CHAPTER 7

Figure 7.4 Courtesy Affymetric Corporation, **Figure 7.7** Courtesy U.S. Government Rocky Mountain Lab,

Hamilton, MT, **Figure 7.8** Courtesy Dr. Victor McKusick, Johns Hopkins Univ., **Figure 7.10** © Sam Ogden Photography, **Figure 7.11** © Custom Medical Stock, **Figure 7.12** Courtesy Mannual Graeber, Imperial College School of Medicine, London. From Neurogenetics, Vol. 1, No. 1, pg. 73-80 © 1987 Oxforfd University Press, **Figure 7.13** © The Bettmann Archive/Corbis, **Figure 7.14** © 1995 Amgen Inc., **Figure 7.15** © James King-Holmes/SPL/Photo Researchers, Inc., **Figure 7.16** Courtesy Pall Corporation.

CHAPTER 8

Figure 8.3 © Baylor College of Medicine/Peter Arnold, Inc., **Figure 8.4** Courtesy National Heart, Lung and Blood Institute, NIH, **Figure 8.5** © Ted Thai/Time Magazine, **Figure 8.6** © Abe Frajnklich/Sygma, **Figure 8.8** © M. Eddy/Don W. Fawcett/Photo Researchers, Inc., **Figure 8.9** © Biophoto Assoc./Photo Researchers, Inc., **Figure 8.11** © Bill Brinkhous.

CHAPTER 9

Figure 9.1 Courtesy Lifecodes Corporation, **Figure 9.2** Courtesy Professor Sir Alec Jeffreys, **Figure 9.7** From G. Vassart et al, " A Sequence in M13 Phage Detects Hypervariable Monisiatallites in Human and Animal DNA". SCIENCE Feb. 6, 1987, p. 684, Figure 1A. © 1987 by the AAAS, **Figure 9.8** © Corbis, **Figure 9.9** Courtesy Jeffrey Taubenberger/AFIP, **Figure 9.10** © Tom McHugh/Researchers, Inc., **Figure 9.11** Courtesy Hendrik Poinar, Max Planck Inst. For Evolutionary Anthropology, Leipzig, Germany, **Figure 9.12a** Courtesy Department of Library Services, American Museum of Natural History, Transp. #K13079, **Figure 9.12b** Courtesy Department of Library Services, American Museum of Natural History, Painting by Charles Knight, Transp. #2435, **Figure 9.14** © Dr. Keith Porter,**Figure 9.16** © Argum, **Figure 9.17** © Tim Davis/Tony Stone Images, **Figure 9.18** © The Bettmann Archive/Corbis.

CHAPTER 10

Figure 10.1a,b,c,d Courtesy Detlef Weigel, Salk Institute, Biological Sciences, LaJolla, CA, **Figure 10.2** Courtesy of the Agricultural Research Service, USDA, **Figure 10.3** Photo Courtesy of Robert Turgeon and Gillian Turgeon, **Figure 10.6** © George Godfrey/Animals, Animals/Earth Scenes, **Figure 10.7** © UPI/The Bettmann Archive/Corbis, **Figure 10.8** Courtesy of the late Chris Hannay and Phil Fitz-James, **Figure 10.9 a,b** Courtesy of Monsanto Company, **Figure 10.10a-1,a-2** Courtesy Dr. Detlef Weigel, Salk Institute, Biol. Sciences,.La Jolla, CA, **Figure 10.10b** Courtesy Dr. C.G. Tauer, **Figure 10.11** Courtesy Monsanto Agricultural Company, **Figure 10.13** © Norm Thomas/Photo Researchers, Inc., **Figure 10.14** © Dr. Jeremy Burgess/SPL/ Photo Researchers, Inc., **Figure 10.16** © Robert E. Daemmrich/Tony Stone Images.

CHAPTER 11

Figure 11.1 © Hank Morgan/Photo Researchers, Inc., **Figure 11.3** Courtesy of Ralph L. Brinster, School of Veterinary Medicine, University of Pennsylvania, **Figure 11.5** © Ira Wyman/Sygma, **Figure 11.6** © Mark Lyons/NYT Pictures, **Figure 11.7** Courtesy Pharming Group N.V. International, **Figure 11.8a** Courtesy Genzyme Transgenic Corporation, **Figure 11.9a** © Colin McPherson/Corbis/AFP, **Figure 11.9b** © AP Photo/PA, **Figure 11.12a,b** Courtesy Masatu Okabe, Osaka Research Institute for Microbial Diseases, **Figure 11.13** Courtesy of the Agricultural Research Service, USDA, **Figure 11.14a** © John D. Cunningham/ Visuals Unlimited, **Figure 11.14b** © Holt Studios/Photo Researchers, Inc.

CHAPTER 12

Figure 12.1 © Sam Ogden, **Figure 12.3** © Bill Geddes/Cold Spring Harbor Laboratory Archives, **Figure 12.4** Courtesy Norton Zinder, photo by Robert Reichert, Rockefeller University, **Figure 12.5a** © Jen and Des Bartlett/Photo Researchers, Inc., **Figure 12.5b** © David Scharf/Peter Arnold, Inc., **Figure 12.5c** © Dr. Jonathan Eisenback/Phototake, Inc., **Figure 12.5d** © David M. Phillips/ Visuals Unlimited, **Figure 12.5e** © Jack Dermid/Photo Researchers, **Figure 12.5f** © Tim Hauf/Visuals Unlimited, **Figure 12.6** Courtesy Francis Collins, Director, National Human Genome Research Institute, NIH, **Figure 12.14a** © Steve Murez/Blackstar, **Figure 12.14 b** © Sam Ogden, **Figure 12.18a** © David M. Phillips/Photo Researchers, Inc., **Figure 12.18b** Courtesy Dr. J. Craig Venter, **Figure 12.19** © C. Lenars/Photo Researchers, Inc.

Index